Lecture Notes in Computer Science   1558
Edited by G. Goos, J. Hartmanis and J. van Leeuwen

Springer
*Berlin
Heidelberg
New York
Barcelona
Hong Kong
London
Milan
Paris
Singapore
Tokyo*

H. Jaap van den Herik   Hiroyuki Iida (Eds.)

Computers and Games

First International Conference, CG'98
Tsukuba, Japan, November 11-12, 1998
Proceedings

Series Editors

Gerhard Goos, Karlsruhe University, Germany
Juris Hartmanis, Cornell University, NY, USA
Jan van Leeuwen, Utrecht University, The Netherlands

Volume Editors

H. Jaap van den Herik
Department of Computer Science, University of Maastricht
Maastricht, The Netherlands
E-mail: herik@cs.unimaas.nl

Hiroyuki Iida
Department of Computer Science, Shizuoka University
Hamamatsu, Japan
E-mail: iida@cs.inf.shizuoka.ac.jp

Cataloging-in-Publication data applied for

Die Deutsche Bibliothek - CIP-Einheitsaufnahme

Computers and games : first international conference ; proceedings / CG'98, Tsukuba, Japan, November 11 - 12, 1998. Jaap van den Herik ; Hiroyuki Iida (ed.). - Berlin ; Heidelberg ; New York ; Barcelona ; Hong Kong ; London ; Milan ; Paris ; Singapore ; Tokyo : Springer, 1999
  (Lecture notes in computer science ; 1558)
  ISBN 3-540-65766-5

CR Subject Classification (1998): G, F.2, I.2.1, I.2.6, I.2.8

ISSN 0302-9743
ISBN 3-540-65766-5 Springer-Verlag Berlin Heidelberg New York

This work is subject to copyright. All rights are reserved, whether the whole or part of the material is concerned, specifically the rights of translation, reprinting, re-use of illustrations, recitation, broadcasting, reproduction on microfilms or in any other way, and storage in data banks. Duplication of this publication or parts thereof is permitted only under the provisions of the German Copyright Law of September 9, 1965, in its current version, and permission for use must always be obtained from Springer-Verlag. Violations are liable for prosecution under the German Copyright Law.

© Springer-Verlag Berlin Heidelberg 1999
Printed in Germany

Typesetting: Camera-ready by author
SPIN: 10702793   06/3142 – 5 4 3 2 1 0   Printed on acid-free paper

Preface

This book contains the papers presented at the First International Conference on Computers and Games (CG'98) held at the Electrotechnical Laboratory (ETL), in Tsukuba, Japan, on November 11-12, 1998.

The CG'98 focuses on all aspects of research related to computers and games. Relevant topics include, but are not limited to, the current state of game-playing programs. The book contains new theoretical developments in game-related research, general scientific contributions produced by the study of games, social aspects of computer games, mathematical games, cognitive research of how humans play games, and so on. As this volume shows, CG'98 is an international conference, with participants from many different countries who have different backgrounds and hence exhibit different views on computers and games.

The Conference was the first one in a series of conferences on this topic. It was a direct follow-up of many successful computer-games-related events held in Japan, such as the series of four Game Programming Workshops (GPW'94 to GPW'97) and the IJCAI-97 Workshop on Computer Games.

The technical program consisted of a keynote lecture, titled: *Predictions* (by H.J. van den Herik), and 21 presentations of accepted papers. The conference was preceded by an informal Workshop on November 10, 1998. The Program Committee (PC) received 35 submissions. Each paper was sent to three referees, who were selected on the basis of their expert knowledge. Twelve papers were accepted immediately, 12 papers were not accepted, and 11 papers were returned to the authors with the request to improve them, and with the statement that they would be refereed again. Finally, with the help of many referees (see the end of this preface), the PC accepted 21 papers for presentation and publication.

Originally, we tried to sequence the contributions in some logical order, such as: from mathematical games via computer science to cognitive sciences, but we failed. Neither did the listing of the contents as mentioned above solve the problem of ordering the papers. In a way it is a fortunate coincidence that such an order could not be established, since it shows that the topic of computers and games has an interdisciplinary nature. Nevertheless, to structure the book to some extent we distinguish, somewhat arbitrarily, between four sections: (1) Search and Strategies, (2) Learning and Pattern Acquisition, (3) Theory, and (4) Go, Tsume Shogi, and Heian Shogi.

Search and Strategies

In the proceedings of this conference, the first set of six contributions deals with search and strategies. The editors believe that search is an important factor when trying to solve simple games or to play complex games. Although this is disputable for the game of Go, it certainly is true for chess-like games. Moreover, any strategy defined as a combination of straightforward movement, indirect approaches, and prophylaxis, is based on search and knowledge. The nature of a game determines which factor is predominant.

The first paper by Junghanns and Schaeffer, titled: *Relevance Cuts: Localizing the Search*, deals with single-agent search. The authors apply their new pruning technique on Sokoban. The idea is to use the influence of a move as a measure of relevance. Hence, they distinguish between local (relevant) moves and non-local (not relevant) moves, with respect to the sequence of moves leading to the current state. The new pruning technique uses the m previous moves to decide if a move is relevant in the current context; if not, the move must be cut off. The application of the technique on a 90-problem test set using search, limited to 20 million nodes, leads to 44 solutions. So, much more research is needed to solve all 90 problems.

The contribution by Björnsson and Marsland, titled: *Multi-Cut Pruning in Alpha-Beta Search*, examines the benefits of investing additional search effort at cut-nodes by expanding other move alternatives as well. Their results when applied to the game of chess show a strong correlation between the number of promising move alternatives at cut-nodes and an emerging new principal variation. This correlation can also be exploited otherwise. Hence, there is still a great deal of research to be done on other innovative methods based on investigating other move options.

Breuker, Van den Herik, Uiterwijk, and Allis treat the well-known graph-history-interaction (GHI) problem. Their contribution, titled: *A Solution to the GHI Problem for Best-First Search*, introduces the notion of twin nodes, which makes it possible to distinguish nodes according to their history. The implementation of this idea, called BTA (Base-Twin Algorithm), is performed to proof-number search. Experimental results in the field of computer chess confirm the claim that the GHI problem has been solved for best-first search.

Under the heading: *Optimal Play against Best Defence: Complexity and Heuristics*, Frank and Basin investigate the best defence model of an imperfect information game. They prove that finding optimal strategies for such a model is NP-complete in the size of the game tree. The introduction of two new heuristics, viz. beta-reduction and iterative biasing, appears to work well. The general idea is that there is a reduction of non-locality due to the introduction of mutual relationship between the various choices at MAX nodes. The heuristics are applied to a Bridge problem set and actually outperform the human experts who produced the model solutions.

Gao, Iida, Uiterwijk, and Van den Herik present a generalization of OM search, called (D,d)-OM search. Their paper, titled: *A Speculative Strategy*, investigates whether it is worthwhile to deviate from the objectively best path when knowing that the opponent only searches to a depth d, whereas the player (e.g., the program) searches to a depth $D > d$. It is shown that a difference in search depth can be exploited by deliberately chosing a suboptimal move in order to gain a larger advantage than when playing the optimal move. Some experiments in the domain of Othello confirm the effectiveness of the proposed strategy.

In their paper: *An Adversarial Planning Approach to Go*, Willmott, Richardson, Bundy, and Levine propose an alternative to the usual procedure of searching a tree of possible move sequences combined with an evaluation function. They model the goals of the players and their strategies for achieving these goals. It implies searching the space of possible goal expansions, which is typically much smaller than the space of move sequences. They describe how adversarial hierarchical task network planning can provide a framework for goal-directed game playing. The program GOBI has been successfully tested on two test sets of Go problems taken from Yoshinori's four-volume series. It was observed that strengthening GOBI's defensive knowledge led to an improvement in attacking plans, and vice versa. This reflects the fact that the better opponent model is more likely to find refutations for poor attacking plans.

Learning and Pattern Acquisition

The second set of five contributions deals with learning and pattern acquisition. The techniques described are applied to the following games: Shogi, Othello, Tsume Go (twice), and Checkers.

The first paper of this set, by Beal and Smith, attempts to determine whether sensible values for Shogi pieces can be obtained in the same manner as for western chess pieces. Under the heading: *First Results from using Temporal Difference Learning in Shogi*, the authors arrive at values that perform well in matches against programs with handcrafted values. They stress the fact that the Shogi piece values were learnt from self-play without any domain-specific knowledge being supplied. It is remarkable to note that Shogi experts are traditionally reluctant to assign values to the pieces. The authors claim that the method is also applicable to learning an appropriate weight for positional evaluation terms in Shogi.

Even more advanced is the topic of learning features to be used in evaluation functions. This topic is treated by Buro in his paper *From Simple Features to Sophisticated Evaluation Functions*. He discusses a practical framework for the semi-automatic construction of evaluation functions for games. Based on a structured evaluation-function representation, a procedure for exploring the feature space is presented. So, new features are discovered in a computationally feasible way. Convincing experimental results for Othello are given and several theoretical issues are discussed.

In their paper: *A Two-Step Model of Pattern Acquisition: Application to Tsume-Go*, Kojima and Yoshikawa carry out a cognitive study. The first step is the pattern acquisition step, which uses only positive examples. The second step, the pattern refinement step, uses both positive and negative examples. The combination of positive and negative examples leads to precise conditions and also to a way of conflict resolution. Three distinct algorithms are introduced for the first step, and two for the second one. The domain of application is Tsume-Go (life and death problems). The performances of six conditions are compared. The best performance is achieved by a condition which gives 31% of the answers correctly. This result equals the achievement of a one-dan human player.

Sasaki, Sawada, and Yoshimura focus on Tsume-Go problems positioned on a 9×9 board which has a unique solution. Under the heading: *A Neural Network Program of Tsume-Go*, they describe a network with 543 neurons dealing with Kurosen-Shiroshi problems. The backpropagation method is applied and the performance of the network is roughly equivalent to a one-dan human player. The authors claim that their neural network can be used as a component of the strong Tsume-Go and Go programs.

Although the Checkers program CHINOOK (by Schaeffer *et al.*) has been crowned as the champion of man-machine contests, the game has not lost any of its attractiveness as a research domain. In their paper: *Distributed Decision Making in Checkers*, Giráldez and Borrajo use the game as a testing ground for techniques for distributed decision making and learning by Multi-Agent DEcision Systems (MADES). They propose a new architecture for knowledge-based systems dedicated to Checkers playing. MADES should learn how to combine individual decisions, in such a way that it outperforms programs without a priori knowledge of the quality of each model.

Theory

Theory is an outstanding tool for the verification of ideas. We admit that good ideas in the context of computers and games must be implementable, but if the implemented ideas contain unexpected errors, they give computers a bad reputation. So, the theoretical contributions constitute an important part of this book. We arranged five papers under this heading. They deal with solution trees, heap games, impartial games, complexity, and thermography.

Pijls and De Bruin show in their contribution: *Game Tree Algorithms and Solution Trees*, that the concept of solution tree is the basic idea underlying the minimax principle. They distinguish between two types of solution trees: max trees and min trees. Subsequently, they formulate a cut-off criterion in terms of solution trees, which eliminates nodes from the search without affecting the result. Moreover, they show that any algorithm actually constructs a superposition of a max and a min solution tree. At the end of their paper they discuss solution trees in relation to alphabeta, SSS*, and MT-SSS.

Fraenkel and Zusman analyse an extension of Wythoff's game and provide a polynomial-time strategy. Their contribution titled: *A New Heap Game*, deals with k heaps of tokens ($k \geq 3$). It is a two-player game with the following rules: a move is either taking a positive number of tokens from at most $k-1$ heaps, or removing the same positive number of tokens from all the k heaps. The authors remark that the Sprague-Grundy function g of a game provides a strategy for the sum of several games. They express their interest in computing the g-function for this new heap game, but state that they are unaware of the complexity of the problem.

The contribution: *Infinite Cyclic Impartial Games*, by Fraenkel and Rahat, treats the family of locally path-bounded digraphs, which is a class of infinite digraphs. The authors show that it is relatively easy to compute an optimal strategy for a combinatorial game on this particular class of graphs. Whenever possible, they achieve a win in a finite number of moves. This is done by proving that the Generalised Sprague-Grundy function is unique and has finite values on this class of graphs.

On the Complexity of Tsume-Go is the title of Crâşmaru's contribution. With the game of Go as a starting point, the author embarks upon an analysis of the concept of alive vs. dead, for which he proposes a mathematical model. Tsume-Go problems are investigated and it is shown that this kind of problem is NP-complete.

In *Extended Thermography for Multiple Kos in Go*, Spight discusses the concept thermography. Many Go positions give rise to combinatorial games. The mean value of the game corresponds to the count, and its temperature to the value of the play. Thermography determines the mean value and the temperature of a combinatorial game. Moreover, thermography has been generalized to include positions containing a single ko. Spight extends the notion of thermography even further, namely to include positions with multiple kos. He also introduces a method for pruning redundant branches of the game tree.

Go, Tsume Shogi, and Heian Shogi

The last set of five contributions deals with Go, Tsume Shogi, and Heian Shogi. All five papers provide relevant information on the games and put them in perspective.

Although the interest in Go research has increased considerably in the last decade, the playing strength of Go programs is still mediocre. Among the Go researchers, a feeling has emerged that developments in the world of chess also may crop up in the world of Go.

As a first step, Müller contributes to this feeling in his contribution: *Computer Go: a Research Agenda*. The author suggests that the obstacles to progress are posed by the current structure of the Go community and are at least as serious as the purely technical challenges. He introduces three proposals for large-scale Go projects, viz. (1) form teams funded by a large company (such as DEEP BLUE), (2) make public-domain source code available (such as GNU CHESS and CRAFTY), and (3) initiate as many university projects as possible. His main concern is to overcome the lack of critical human resources. Having seen the enthusiasm of the Go researchers at the CG'98, the editors believe that Go research has a bright future.

In Go, the position evaluation is very important, but also very complex. So far, no good evaluation functions have been developed. One of the major factors for the evaluation of a position is the strength of a group. Tajima and Sanechika describe a new method for estimating the strength of a group, in their paper: *Estimating the Possible Omission Number for Groups in Go by the Number of n-th Dame*. The authors have developed a simple method for making a rough estimation. They define a PON (Possible Omission Number) as a precise measure for the strength of groups. Using PON, their method calculates n-th dame (liberties). Experiments support the claim of the effectiveness of the method.

The way of using Go terms while playing Go depends on the player's skill. Not every player uses the same notion to indicate a certain board characteristic. Yoshikawa, Kojima, and Saito have performed extensive cognitive research in this area. They report on their research in the paper: *Relations between Skill and the Use of Terms - An Analysis of Protocols of the Game of Go*. Three experiments are described in full detail. Starting with a profound analysis of their results, the authors developed a hypothesis, which they call the iceberg model, implying that the bulk of knowledge is not known to human players. Since it is crucial to make the knowledge of how to evaluate a Go position explicitly available for computer programs, protocol analyses and the modelling of thought processes remain an important issue for future research.

Grimbergen provides a very readable overview of Tsume-Shogi programs, titled: *A Survey of Tsume-Shogi Programs using Variable-Depth Search*. Tsume-Shogi is the name for mating problems in Japanese chess. He discusses six different Tsume-Shogi programs. Difficult Tsume-Shogi problems have solution sequences which are longer than 20 plies. Hence, all programs have a variable search depth and use hashing techniques. The combination of transposition, domination, and simulation leads to strong programs that outperform human experts. The best program is able to solve Microcosmos, a Tsume-shogi problem with a solution sequence of 1525 plies.

Finally, in the contribution: *Retrograde Analysis of the KGK Endgame in Shogi: Its Implications for Ancient Heian Shogi*, Iida, Yoshimura, Morita, and Uiterwijk examine the evolutionary changes that have occurred in the game of Shogi. They go back to the ancient game of Heian Shogi and investigate the game results of the KGK endgame (King and Gold vs. King) on $N \times N$ boards. Since Heian Shogi is only briefly described in the literature, the authors must guess which rules were applicable under which circumstances. The paper focuses on a logical interpretation of the change of rules at the time that the 8×8 board was replaced by a 9×9 board. Moreover, the authors demonstrate that the 10×10 board is the largest $N \times N$ board on which the KGK endgame is a deterministic win (of course, with the exception of trivially drawn cases in which the Gold can be captured). Future research will focus on the relation between the given analysis of the KGK endgames and the reuse rule of captured pieces in modern Shogi.

Acknowledgements

The CG'98 conference and this volume would not have been possible without the generous sponsorship of Electrotechnical Laboratory (ETL), Telecommunications Advancement Foundation (TAF), Osaka University of Commerce, Shizuoka University, and the Foundation for Fusion of Science & Technology (FOST), and without the cooperation of the following organizations: Computer Shogi Association (CSA), Computer Go Forum (CGF), International Computer Chess Association (ICCA), and IEEE, Tokyo. Moreover, CG'98 has benefited from the efforts of many people. Among them are Elwyn Berlekamp (University of California, Berkeley), Aviezri Fraenkel (The Weizmann Institute of Science), Jurg

Nievergelt (ETH Zurich), Monty Newborn (McGill University), Tony Marsland (University of Alberta).

The editors gratefully acknowledge the expert assistance of the Advisory Committee, the Program Committee, the Organizing Committee, the Local Arrangements Committee and the referees. Their names are mentioned on the next pages.

Finally, the editors would like to express their sincere gratitude to Ms. A. Oshima, who assisted us during the preparation of the conference; Ms. S. Iida and Ms. Y. Tajima, who assisted us in organizing the local arrangements; Mr. Nobusuke Sasaki and Mrs. Sabine Vanhouwe for all their efforts and support, especially in producing these proceedings.

                          H. Jaap van den Herik   Hiroyuki Iida
                          Maastricht, The Netherlands   Hamamatsu, Japan
                          December 1, 1998

Organization

CG'98 is organized by many persons and organizations. Below we have listed them by function and by support.

Organizing Committee

Conference Chairs:	Yoshiyuki Kotani (Tokyo University of Agriculture and Technology, Japan)
	Takenobu Takizawa (Waseda University, Japan)
Program Chairs:	H. Jaap van den Herik (Universiteit Maastricht, The Netherlands)
	Hiroyuki Iida (University of Shizuoka, Japan)
Workshop:	Ian Frank (ETL, Japan)

Advisory Committee

Elwyn Berlekamp (UC Berkeley, USA)
Hans Berliner (CMU, USA)
Jurg Nievergelt (ETH Zurich, CH)
Monty Newborn (McGill Univ., Can.)
Tony Marsland (University of Alberta, Can.)

Local Arrangements Committee

Chair:	Hitoshi Matsubara (ETL, Japan)
	Atsushi Yoshikawa (NTT Basic Research Laboratories, Japan)
	Reijer Grimbergen (ETL, Japan)
	Ian Frank (ETL, Japan)
	Martin Müller (ETL, Japan)
Secretary-Treasurer:	Morihiko Tajima (ETL, Japan)

Program Committee

Program Chair: H. Jaap van den Herik (Universiteit Maastricht, NL)
Program Co-chair: Hiroyuki Iida (University of Shizuoka, Japan)

Don F. Beal (Queen Mary and Westfield College, UK)
Michael Buro (NEC Institute, USA)
Susan Epstein (Hunter College, USA)
Ian Frank (ETL, Japan)
Ralph Gasser (Microsoft, USA)
Reijer Grimbergen (ETL, Japan)
Richard Korf (UC Los Angeles, USA)
Kenji Koyama (NTT Communication Science Lab, Japan)
Shaul Markovitch (Technion, Israel)
Martin Müller (ETL, Japan)
Kohei Noshita (University of Electro-Communications, Japan)
Barney Pell (NASA, USA)
Wim Pijls (Erasmus University Rotterdam, NL)
Noriaki Sanechika (ETL, Japan)
Jonathan Schaeffer (University of Alberta, Canada)
Takao Uehara (Tokyo Engineering University, Japan)
Jos Uiterwijk (Universiteit Maastricht, NL)
Janet Wiles (Queenland University, AU)

Referees (Different from PC Members)

H. Bal	E. A. Heinz	J. Nievergelt
J. Baxter	R. M. Hyatt	A. Plaat
E. Berlekamp	A. Junghanns	C. Posthoff
Y. Bjornsson	T. Kaneko	J. Romein
A. de Bruin	H. Kaindl	N. Sasaki
M. Campbell	T. Kojima	M. Tajima
T. Cazenave	R. E. Korf	T. Takizawa
K. Chen	Y. Kotani	A. N. Walker
R. Feldmann	B. Levinson	J. Weill
A. Fraenkel	U. Lorenz	T. Yaguchi
J. Fuernkranz	T. Marsland	A. Yoshikawa
M. L. Ginsberg	H. Matsubara	A. J. van Zanten
R. Grimbergen	K. Nakamura	
D. Hartmann	M. Newborn	

Sponsoring Institutions

Electrotechnical Laboratory (ETL)
Telecommunications Advancement Foundation (TAF)
Osaka University of Commerce
Shizuoka University
Foundation for Fusion of Science & Technology (FOST)

Cooperative Organizations

Computer Shogi Association (CSA)
Computer Go Forum (CGF)
International Computer Chess Association (ICCA)
IEEE, Tokyo

Table of Contents

Search and Strategies

Relevance Cuts: Localizing the Search 1
  Andreas Junghanns and Jonathan Schaeffer

Multi-cut Pruning in Alpha-Beta Search 15
  Yngvi Björnsson and Tony Marsland

A Solution to the GHI Problem for Best-First Search 25
  Dennis M. Breuker, H. Jaap van den Herik, Jos W.H.M. Uiterwijk, and L. Victor Allis

Optimal Play against Best Defence: Complexity and Heuristics 50
  Ian Frank and David Basin

A Speculative Strategy 74
  Xinbo Gao, Hiroyuki Iida, Jos W.H.M. Uiterwijk, and H. Jaap van den Herik

An Adversarial Planning Approach to Go 93
  Steven Willmott, Julian Richardson, Alan Bundy, and John Levine

Learning and Pattern Acquisition

First Results from Using Temporal Difference Learning in Shogi 113
  Donald F. Beal and Martin C. Smith

From Simple Features to Sophisticated Evaluation Functions 126
  Michael Buro

A Two-Step Model of Pattern Acquisition: Application to Tsume-Go 146
  Takuya Kojima and Atsushi Yoshikawa

A Neural Network Program of Tsume-Go 167
  Nobusuke Sasaki, Yasuji Sawada, and Jin Yoshimura

Distributed Decision Making in Checkers............................ 183
  J. Ignacio Giráldez and Daniel Borrajo

Theory

Game Tree Algorithms and Solution Trees 195
  Wim Pijls and Arie de Bruin

A New Heap Game 205
    Aviezri S. Fraenkel and Dmitri Zusman

Infinite Cyclic Impartial Games 212
    Aviezri S. Fraenkel and Ofer Rahat

On the Complexity of Tsume-Go 222
    Marcel Crâşmaru

Extended Thermography for Multiple Kos in Go 232
    William L. Spight

Go, Tsume Shogi, and Heian Shogi

Computer Go: A Research Agenda 252
    Martin Müller

Estimating the Possible Omission Number for Groups in Go by the
Number of n-th Dame 265
    Morihiko Tajima and Noriaki Sanechika

Relations between Skill and the Use of Terms - An Analysis of
Protocols of the Game of Go - 282
    Atsushi Yoshikawa, Takuya Kojima, and Yasuki Saito

A Survey of Tsume-Shogi Programs Using Variable-Depth Search 300
    Reijer Grimbergen

Retrograde Analysis of the KGK Endgame in Shogi: Its Implications
for Ancient Heian Shogi 318
    Hiroyuki Iida, Jin Yoshimura, Kazuro Morita, and
    Jos W.H.M. Uiterwijk

Author Index 337

Relevance Cuts: Localizing the Search

Andreas Junghanns and Jonathan Schaeffer

Department of Computing Science
University of Alberta
Edmonton, Alberta
CANADA T6G 2H1
{andreas, jonathan}@cs.ualberta.ca

Abstract. Humans can effectively navigate through large search spaces, enabling them to solve problems with daunting complexity. This is largely due to an ability to successfully distinguish between relevant and irrelevant actions (moves). In this paper we present a new single-agent search pruning technique that is based on a move's *influence*. The influence measure is a crude form of relevance in that it is used to differentiate between local (relevant) moves and non-local (not relevant) moves, with respect to the sequence of moves leading up to the current state. Our pruning technique uses the m previous moves to decide if a move is relevant in the current context and, if not, to cut it off. This technique results in a large reduction in the search effort required to solve Sokoban problems.

Keywords: single-agent search, heuristic search, Sokoban, local search, IDA*

1 Introduction and Motivation

It is commonly acknowledged that the human's ability to successfully navigate through large search spaces is due to their meta-level reasoning [4]. The relevance of different actions when composing a plan is an important notion in that process. Each next action is viewed as one logically following in a series of steps to accomplish a (sub-)goal. An action judged as irrelevant is not considered.

When searching small search spaces, the computer's speed in base-level reasoning can effectively overcome the lack of meta-level reasoning by simply enumerating large portions of the search space. However, it is a trivial matter to pose a problem to the computer that is easy for a human to solve (using reasoning) but is exponentially large to solve using standard search algorithms. We need to enhance computer algorithms to be able to reason at the meta-level if they are to successfully tackle these larger search tasks. In the world of computer games (two-player search), a number of meta-level reasoning algorithmic enhancements are well known, such as null-move searches [5] and futility cut-offs [11]. For single-agent search, macro moves [9] are an example.

In this paper, we introduce *relevance cuts*. The search is restricted in the way it chooses its next action. Only actions that are relevant to previous actions can be performed, with a limited number of exceptions being allowed. The exact definition of relevance is domain dependent.

Consider an artist drawing a picture of a wildlife scene. One way of drawing the picture is to draw the bear, then the lake, then the mountains, and finally the vegetation. An alternate way is to draw a small part of the bear, then draw a part of the mountains, draw a single plant, work on the bear again, another plant, maybe a bit of lake, etc. The former corresponds to how a human would draw the picture: concentrate on an identifiable component and work on it until a desired level of completeness has been achieved. The latter corresponds to a typical computer method: the order in which the lines are drawn does not matter, as long as the final result is achieved.

Unfortunately, most search algorithms do not follow the human example. At each node in the search, the algorithm will consider all legal moves regardless of their relevance to the preceding play. For example, in chess, consider a passed "a" pawn and a passed "h" pawn. The human will analyze the sequence of moves to, say, push the "a" pawn to queen. The computer will consider dubious (but legal) lines such as push the "a" pawn one square, push the "h" pawn one square, push the "a" pawn one square, etc. Clearly, considering alternatives like this is not cost-effective.

What is missing in the above examples is a notion of *relevance*. In the chess example, having pushed the "a" pawn and then decided to push the "h" pawn, it seems silly to now return to considering the "a" pawn. If it really was necessary to push the "a" pawn a second time, why weren't both "a" pawn moves considered *before* switching to the "h" pawn? Usually this switching back and forth (or "ping-ponging") does not make sense but, of course, exceptions can be constructed.

In other well-studied single-agent search domains, such as the N-puzzle and Rubik's Cube, the notion of relevance is not important. In both these problems, the geographic space of moves is limited, i.e. all legal moves in one position are "close" (or local) to each other. For two-player games, the effect of a move may be global in scope and therefore moves almost always influence each other (this is most prominent in Othello, and less so in chess). In contrast, a move in the game of Go is almost always local. In non-trivial, real-world problems, the geographic space might be large, allowing for local and non-local moves.

This paper introduces relevance cuts and demonstrates their effectiveness in the one-player game Sokoban. For Sokoban we use a new influence metric that reflects the structure of the maze. A move is considered relevant if it is influencing all the previous m moves made. The search is only allowed to make relevant moves with respect to previous moves and only a limited number of exceptions is permitted. With these restrictions in place, the search is forced to spend its effort locally, since random jumps within the search area are discouraged. In the meta-reasoning sense, forcing the program to consider local moves is making it adopt a pseudo-plan; an exception corresponds to a decision to change plans. This results in a decrease of the average branching factor of the search tree.

For our Sokoban program *Rolling Stone*, relevance cuts result in a large reduction of the search space. These reductions are on top of an already highly efficient[1] searcher. On a standard set of 90 test problems, relevance cuts allow *Rolling Stone* to increase the number of problems it can solve from 39 to 44. Given that the problems increase exponentially in difficulty, this relatively small increase in the number of problems solved represents a large increase in search efficiency.

2 Sokoban and Related Work

Single-agent search (A*) has been extensively studied in the literature. There are a plethora of enhancements to the basic algorithm, allowing the application developer to customize their implementation. The result is an impressive reduction in the search effort required to solve challenging applications (see [10] for a recent example). However, the applications used to illustrate the advances in single-agent search efficiency are "easy" in the sense that they have some (or all) of the following properties:

1. effective, inexpensive lower-bound estimators,
2. small branching factor in the search tree, and
3. moderate solution lengths.

The sliding-tile puzzles are the best known examples of these problems. Problem domains such as these also have the important property that given a solvable starting state, every move preserves the solvability (although not necessarily the optimality).

Sokoban is a popular one-player computer game. The game originated in Japan, although the original author is unknown. The game's appeal comes from the simplicity of the rules and the intellectual challenge offered by deceptively easy problems.

Figure 1 shows a sample Sokoban problem.[2] The playing area consists of rooms and passageways, laid out on a rectangular grid of size 20x20 or less. Littered throughout the playing area are *stones* (shown as circular discs) and *goals* (shaded squares). There is a *man* whose job it is to move each stone to a goal square. The man can only push one stone at a time and must push from behind the stone. A square can only be occupied by one of a wall, stone or man at any time. Getting all the stones to the goal squares can be quite challenging; doing this in the minimum number of moves is much more difficult.

To refer to squares in a Sokoban problem, we use a coordinate notation. The horizontal axis is labeled from "A" to "T", and the vertical axis from "a" to "t"

---

[1] Of course, "highly efficient" here is meant in terms of a computer program. Humans shake their heads in disbelief when they see some of the ridiculous lines of play considered in the search.
[2] This is problem 1 of the standard 90-problem suite available at http://xsokoban.lcs.mit.edu/xsokoban.html.

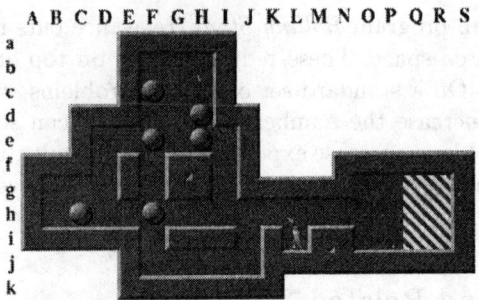

He-Ge Hd-Hc-Hd Fe-Ff-Fg Fh-Gh-Hh-Ih-Jh-Kh-Lh-Mh-Nh-
Oh-Ph-Qh-Rh-Rg   Fg-Fh-Gh-Hh-Ih-Jh-Kh-Lh-Mh-Nh-Oh-
Ph-Qh-Qi-Ri   Fc-Fd-Fe-Ff-Fg-Fh-Gh-Hh-Ih-Jh-Kh-Lh-Mh-
Nh-Oh-Ph-Qh-Qg Ge-Fe-Ff-Fg-Fh-Gh-Hh-Ih-Jh-Kh-Lh-Mh-
Nh-Oh-Ph-Qh-Rh Hd-He-Ge-Fe-Ff-Fg-Fh-Gh-Hh-Ih-Jh-Kh-
Lh-Mh-Nh-Oh-Ph-Pi-Qi Ch-Dh-Eh-Fh-Gh-Hh-Ih-Jh-Kh-Lh-
Mh-Nh-Oh-Ph-Qh

Fig. 1. Sokoban Problem 1 With One Solution

(assuming the maximum sized 20x20 problems), starting in the upper left corner. A move consists of pushing a stone from one square to another. For example, in Figure 1 the move *Fh-Eh* moves the stone on *Fh* left one square. We use *Fh-Eh-Dh* to indicate a sequence of pushes of the same stone. A move, of course, is only legal if there is a valid path by which the man can move behind the stone and push it. Thus, although we only indicate stone moves (such as *Fh-Eh*), implicit in this is the man's moves from its current position to the appropriate square to do the push (for *Fh-Eh* the man would have to move from *Li* to *Gh* via the squares *Lh*, *Kh*, *Jh*, *Ih* and *Hh*).

Unlike most single-agent search problems studied in the literature, a single Sokoban move can change a problem from being solvable to unsolvable. For example, in Figure 1, making the move *Fh-Fg* creates an unsolvable problem. It requires a non-trivial analysis to verify this deadlock. This is a simple example, since deadlock configurations can be large and span the entire board. Identifying deadlock is critical to prevent a lot of futile searching.

The standard 90 problems range from easy (such as problem 1 above) to difficult (requiring hundreds of stone pushes). A global score file is maintained showing who has solved which problems and how efficient their solution is (also at http://xsokoban.lcs.mit.edu/xsokoban.html). Thus solving a problem is only part of the satisfaction; improving on your solution is equally important.

Sokoban has been shown to be PSPACE-complete [2, 3]. Dor and Zwick show that the game is an instance of a motion planning problem, and compare the game to other motion planning problems in the literature [3]. For example, Sokoban is similar to Wilfong's work with movable obstacles, where the man is allowed to hold on to the obstacle and move with it, as if they were one object [12]. Sokoban can be compared to the problem of having a robot in a warehouse move a number of specified goods from their current location to their final destination, subject to the topology of the warehouse and any obstacles in the way. When viewed in this context, Sokoban is an excellent example of using a game as an experimental test-bed for mainstream research in artificial intelligence.

Sokoban is a difficult problem domain for computers because of the following reasons:

1. it has a complex lower-bound estimator ($O(n^3)$, given n goals),
2. the branching factor is large and variable (potentially over 100),
3. the solution may be very long (some problems require over 500 moves to solve optimally),
4. the search space complexity is $O(10^{98})$ for problems restricted to a 20x20 area only, and
5. some reachable states are unsolvable (deadlock).

For sliding-tile puzzles, there are algorithms for generating a non-optimal solution. In Sokoban, because of the presence of deadlock, often it is very difficult to find *any* solution.

Our previous attempts to solve Sokoban problems using standard single-agent search techniques are reported in [7]. There, using our program *Rolling Stone*, we compare the different techniques and their usefulness with respect to the search efficiency when solving Sokoban problems. IDA* [8] was augmented with a sophisticated lower bound estimator, transposition tables, move ordering, macro moves and deadlock tables. Even though each of the standard single-agent search enhancements we investigated resulted in significant improvements (often several orders of magnitude in search-tree size reduction), at the time we were able to solve only 20 problems of a 90-problem test suite.

In [6] we introduced a new search enhancement, *pattern searches*, a method that dynamically finds deadlocks and improved lower bounds. Since a single move can introduce a deadlock, before playing a move we perform a *pattern search* to analyze if deadlock will be introduced by that move. The pattern search attempts to identify the conditions for a deadlock and, if all the conditions are satisfied, saves a pattern of stones that is the minimal board configuration required for the deadlock. During the IDA* search, a new position can be matched with these patterns to see if it contains a deadlock. As a side benefit, these pattern searches can also identify arbitrary increases to the lower bound (e.g. a deadlock increases the lower bound to ∞).

The notion of bit (stone) patterns is similar to the Method of Analogies [1]. Pattern searches are a conflict-driven top-down proof of correctness, while the Method of Analogies is a bottom-up heuristic approximation.

Pattern searches allow us to now solve 39 of the 90 problems [6][3]. Although pattern searches can be enhanced to make them more efficient, we concluded

---

[3] Note that [6] reports slightly different numbers than this paper, caused by subsequent refinements to the pattern searches and bug fixes.

that they are inadequate to successfully solve all 90 Sokoban test positions. Even with all the enhancements, and the cumulative improvements of several orders of magnitude in search efficiency, the search trees are still too deep and the effective branching factor too high. Hence, we need to find further ways to improve the search efficiency.

3 Relevance Cuts

Analyzing the trees built by an IDA* search quickly reveals that the search algorithm considers move sequences that no human would ever consider. Even completely unrelated moves are tested in every legal combination – all in an effort to prove that there is no solution for the current threshold. How can a program mimic an "understanding" of relevance? We suggest that a reasonable approximation of relevance is influence. If two moves are not influencing each other then they are very unlikely to be relevant to each other. If a program had a good "sense" of influence, it could assume that in a given position all previous moves belong to a (unknown) plan of which a continuation can only be a move that is relevant – in our approximation, is influencing whatever was played previously.

Thus, the general idea for relevance cuts is to prevent the program from trying all possible move sequences. Moves tried have to be relevant to previously executed moves. This can be achieved in different, domain specific, ways. The following shows one implementation for the domain of Sokoban. Even though the specifics aren't necessarily applicable to other domains, the basic philosophy of the approach is.

3.1 Influence

When judging how two squares in a Sokoban maze are influencing each other, Euclidean distance is not adequate. Taking the structure of the maze into account would lead to a simple geographic distance which is still not proportional with influence. For example, consider two squares connected by a tunnel; the squares are equally influencing each other, no matter how long the tunnel is. Figure 1 shows several tunnels of which one consists of the squares Ff and Fg. Prolonging the tunnel without changing the general topology of the problem would change the geographic distance, but not the influence.

The following is a list of properties we would like the influence measure to reflect:

Alternatives: The more alternatives that exist on a path between two squares, the less they influence each other. That is, squares in the middle of a room where stones can go in all 4 directions should decrease influence more than squares in a tunnel, where no alternatives exist.
Goal-Skew: Squares on the optimal path to any goal should have stronger influence than squares off the optimal path.

Connection: Two neighboring squares connected such that a stone can move between them should influence each other more than two squares connected such that only the man can move between them.

Tunnel: In a tunnel, influence remains the same: It does not matter how long the tunnel is (one could, for example, collapse a tunnel into one square).

Our first implementation of relevance cuts used small off-line searches to statically precalculate a (20x20)x(20x20) table containing the influence values for each square of the maze to every other square in the maze. Between every pair of squares, a breadth-first search is used to find the path(s) with the largest influence. The algorithm is similar to a shortest-path finding algorithm, except that we are using influence here and not geographic distance. The smaller the influence number, the more two squares are influencing each other.

Note that influence is not necessarily symmetric ($dist(a,b) \neq dist(b,a)$). A square close to a goal influences squares further away more than it is influenced by them. Furthermore, $dist(a,a)$ is not necessarily 0. A square in the middle of a room will be less influenced by each of its many neighbors than a square in a tunnel. To reflect that, squares in the middle of a room receive a larger bias than more restricted squares.

The exact numbers used in our implementation are the following (with the name of the wish-list item following in parenthesis). Each square on the path between the start and goal squares adds 2 for each direction (off the path considered) a stone can be pushed and 1 for each direction the man can go. Thus, the maximum one square can add for alternatives is 4 (alternatives). However, every square that is part of an optimal path towards any of the goals from the start square will add only half of that amount (goal-skew). If the connection from the previous square on the path to the current squares can be taken by a stone only 1 is added, else 2 (connection). If the previous square is in a tunnel, 0 is added (tunnel), regardless of all other properties.

3.2 Relevance Cut Rules

Given the above influence measure, we can now proceed to explain how to use that information to cut down on the number of moves considered in each position. To do this, we need to define *distant moves*. Given two moves, $m1.from$-$m1.to$ and $m2.from$-$m2.to$, move $m2$ is distant with respect to move $m1$ if the from squares of the moves ($m1.from$ and $m2.from$) do not influence each other. More precisely, two moves influence each other if

$$InfluenceTable[\ m1.from\ ][\ m2.from\ ] < d$$

where *InfluenceTable* is the table of precalculated values and d is a tunable threshold.

Relevance cuts eliminate some moves that are distant from the previous moves played, and therefore are considered not relevant to the search. There are two ways that a move can be cut off:

1. If within the last m moves more than k distant moves were made. This cut will discourage arbitrary switches between non-related areas of the maze.
2. A move that is distant with respect to the previous move, but not distant to a move in the past m moves. This will not allow switches back into an area previously worked on and abandoned just briefly.

In our experiments, we set k to 1. This way, the first cut criterion will entail the second. The parameters d and m are set according to the following properties of the maze. The maximal influence distance, d, is set to half the average influence value from all squares to the squares on optimal paths to any goal, but not less than 6. The length of history used, m, is set to the average influence value of all squares to all other non-dead squares in the maze, but not less than 10.

3.3 Example

Figure 2 shows an example where humans immediately identify that solving this problem involves solving two separate sub-problems. Solving the left and right side of the problem is completely independent. An optimal solution needs 82 moves; *Rolling Stone*'s lower bound estimator returns a value of 70. Standard IDA* will need 7 iterations to find a solution (our lower-bound estimator preserves the odd/even parity of the solution length). In each of the iterations but the last, IDA* will try every possible (legal) move combination with moves from both sides of the problem. This way IDA* proves for each of the 6 iterations i that the problem cannot be solved with $70 + 2*i$ moves, regardless of the order of the considered moves. Clearly, this is unnecessary and inefficient. Solving one of the sub-problems requires only 4 iterations, since the lower bound is off by only 6. Considering this position as two separate problems will result in an enormous reduction in the search complexity.

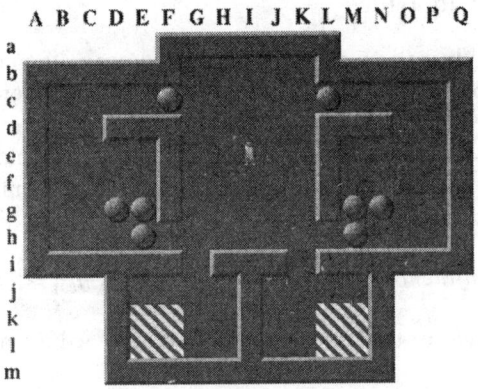

Fig. 2. Example Maze With Locality

Our implementation considers all moves on the left and on the right side as distant from each other. This way only a limited number of switches is considered during the search. Our parameter settings allow for only one non-local move per 9-move sequence. For this contrived problem, relevance cuts decrease the number of nodes searched from 32,803 nodes to 24,748 nodes while still returning an optimal solution (the pattern searches were turned off for simplicity). Although this is a significant reduction, it is only a small step towards achieving all the possible savings. For example, one of the sub-problems can be solved by itself in only 329 nodes! The difference between 329 and 32,803 illustrates why IDA* in its current form is inadequate for solving large, non-trivial real-world problems. Clearly, more sophisticated methods are needed.

3.4 Discussion

Further refinement of the parameters used are certainly possible and necessary if the full potential of relevance cuts is to be achieved. Some ideas with regards to this issue will be discussed in the future work section.

The overhead of the relevance cuts is negligible, at least for our current implementation. The influence of two moves can be established by a simple table lookup. This is in stark contrast to our pattern searches, where the overhead dominates the cost of the search for most problems.

4 Experimental Results

Rolling Stone has been tested using the 90-problem test set using searches limited to 20,000,000 nodes. Our previous best version of *Rolling Stone* was capable of solving 39 of the test problems. With the addition of relevance cuts, the number of problems solved has increased to 44[4]. Table 1 shows a comparison of *Rolling Stone* with and without relevance cuts for each of the 44 solved problems.

For each program version in Table 1, the third column gives the number of IDA* iterations that the program took to solve the problem. Note that problems #9, #21 and #51 are now solved non-optimally, taking at least one iteration longer than the program without relevance cuts. This confirms the unsafe nature of the relevance cuts. However, since none of the problems solved before is lost and 5 more are solved, the gamble paid off. Long ago we abandoned our original goal of obtaining optimal solutions to Sokoban problems. The size of the search space dictates radical pruning measures if we want to have any chance of solving some of the tougher problems.

Of the 5 new problems solved, #11 is of interest. Without relevance cuts, only 17 IDA* iterations could be completed within our pre-set limit of 20,000,000 nodes. Relevance cuts allow *Rolling Stone* to search 19 iterations and solve the

---

[4] Note that we "cheat" with problem #46, as we allow it to go 47,000 nodes beyond the 20 million node limit. A bug fix pushed it beyond the 20 million limit and we wanted it to count in the statistics. We tested all the unsolved problems without the relevance cuts to 50 million nodes and no other problem was solved.

#	without relevance cuts			with relevance cuts		
	top level nodes	total nodes	# iterations	top level nodes	total nodes	# iterations
1	32	270	2	32	270	2
2	200	3,251	2	177	2,764	2
3	392	10,486	2	301	10,395	2
4	394	10,556	1	392	10,554	1
5	5,999	121,502	3	6,079	152,082	3
6	170	1,593	3	151	1,574	3
7	12,378	156,334	5	6,821	68,202	5
8	152,919	3,066,098	6	89,838	1,806,540	6
9	11,572	234,454	5	14,963	307,006	8
10	834,147	16,678,800	4	340,935	6,815,512	4
11	> 998,299	> 20,000,000	17	337,143	6,759,590	19
17	1,160	14,891	7	1,250	14,740	7
19	890,100	17,829,863	9	88,575	1,814,505	9
21	38,371	765,392	9	47,854	970,776	10
34	> 1,651,897	> 20,000,000	9	227,525	4,495,028	9
38	333,257	844,882	42	236,351	748,024	42
40	> 998,236	> 20,000,000	7	311,618	6,239,163	8
43	112,610	2,270,703	8	60,120	1,215,022	8
45	729,333	14,646,623	9	250,157	5,059,596	9
46	> 2,022,198	> 20,000,000	12	1,004,325	20,047,197	15
49	> 17,074,823	> 20,000,000	11	2,303,495	3,047,672	13
51	725	2,611	1	2,049	18,368	2
53	182	3,737	1	185	3,740	1
54	283,609	4,381,171	9	255,827	3,884,844	9
55	1,696,996	3,194,830	3	603,190	1,349,908	3
56	4,318	34,429	6	3,817	31,388	6
57	61,900	1,084,732	5	45,339	797,766	5
60	5,929	116,103	3	1,252	24,403	3
62	2,720	71,578	5	1,984	46,534	5
63	10,195	197,922	3	21,131	390,422	3
64	194,846	3,900,639	10	32,706	652,857	10
65	364	12,971	5	364	12,971	5
67	239,515	2,177,787	13	160,936	2,103,866	13
68	128,716	2,651,559	11	16,814	355,306	11
70	841,495	15,003,603	3	94,792	1,949,842	3
72	1,908	43,260	5	3,168	72,647	5
73	11,371	247,816	3	14,791	292,799	3
78	75	809	1	75	783	1
79	362	4,017	5	200	3,512	5
80	805	15,513	1	2,081	48,220	1
81	1,251	40,806	4	1,074	29,698	4
82	8,500	181,571	5	4,643	97,406	5
83	635	15,423	1	389	13,106	1
84	272,160	521,068	4	153,220	443,508	4
	> 29,637,064	> 190,559,653		6,748,129	72,210,106	

Table 1. Experimental Data

problem. Given that the cost of an extra iteration is large (and can typically be a factor of about 5,000 *per iteration* [6]), a gain of 2 iterations represents a massive improvement.

The tree size for each program version given in Table 1 is broken into two numbers. *Top-level nodes* refers to that portion of the search tree that IDA* is applied to. *Total nodes* includes the top-level nodes and the pattern search nodes. Clearly, for some problems (such as #45) the cost of performing pattern searches overwhelms the search effort, whereas in other problems (such as #53) they are a small investment. Further details on pattern searches and when they are executed can be found in [6].

The magnitude of the top-level nodes can be misleading; superficially it looks like these problems can be "trivially" solved with few nodes. Using standard IDA* with our sophisticated lower bound estimator fails to solve *any* of the 90 test problems within our limit of 20,000,000 nodes. Consequently, we added a plethora of enhancements to the program, including transposition tables, macro moves, move ordering and deadlock tables, *each* of which is capable of reducing the search tree size by one or more orders of magnitude [7]! Thus the small top-level node counts reported in the table are the result of extensive improvements to the search algorithm.

Relevance cuts reduce the number of top-level nodes by at least a factor of 4.5. Note that since the program not using relevance cuts cannot solve 5 problems, this factor may be a gross underestimation of the actual impact.

With respect to the total search nodes, relevance cuts improve search efficiency by almost a factor of three. Again, this is a lower bound. In particular, problem #11 still requires an enormous amount of search, given that it still has 2 iterations to go before it can find the solution.

Comparing node numbers of individual searches is difficult because of many volatile factors in the search. For example, a relevance cut might eliminate a branch from the search justifiably, but a pattern search there would have uncovered valuable information that would have been useful for reducing the search in other parts of the tree. Problem #80 is one such example: despite the relevance cuts the node count goes up from 99 to 123 nodes; an important discovery was not made and the rest of the search increases. However, the overall trend is in favor of the relevance cuts. An excellent example is problem #70: the top level node count is cut down to 3,006 nodes and a solution is found. Previously 579,037 nodes were considered without finding a solution.

Figures 3 and 4 plot the amount of effort to solve a problem, using the numbers from Table 1 sorted by total nodes. An additional data point is given with a curve that shows what the program's performance was with all the standard single-agent search techniques implemented, before pattern searches where added. Figure 3 shows the impact of the relevance cuts. The exponential growth in difficulty with each additional problem solved is dampened, allowing for more problems solved with the same number of nodes. Figure 4 is a logarithmic representation of Figure 3. The figure more clearly shows that up to about the 25th problem (ordered according to number of nodes needed to solve) there is very

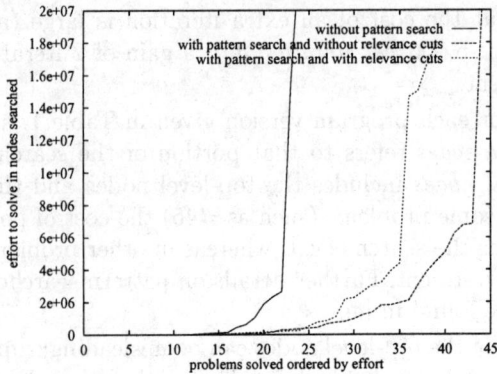

Fig. 3. The Effect of Relevance Cuts

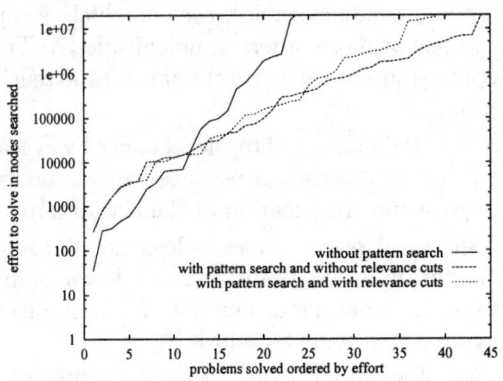

Fig. 4. The Effect of Relevance Cuts (Log Scale)

little difference in effort required; the relevance cuts do not save significant portions of the small search trees. However, with larger search trees, the success of relevance cuts gets more pronounced.

5 Conclusions and Future Work

Relevance cuts provide a crude approximation of human-like problem-solving methods by forcing the search to favor local moves over global moves. This simple idea provides large reductions in the search tree size, at the expense of possibly returning a longer solution. Given the breadth and depth of Sokoban search trees, finding optimal solutions is a secondary consideration; finding *any* solution is challenging enough.

There are several ideas on how to improve the effectiveness of relevance cuts.

- Use different distances depending on crowding. If many stones are crowding an area, it is likely that the relevant area is larger than it would be with less stones blocking each other.
- The current influence measure can most likely be improved. A thorough investigation of all the parameters used could lead to substantial improvements.
- There are several parameters used in the relevance cuts. The setting of those is already dependent of properties of the maze. These parameters are critical for the performance of the cuts and are also largely responsible for increased solution lengths. More research on those details is needed to fully exploit the possibilities relevance cuts are offering.
- So far, *Rolling Stone* is painting locally, but is not yet "object oriented". If a flower and the bear are close, painting both at the same time is very likely. Better methods are needed to further understand subgoals, rather than localizing by area.

Although relevance cuts introduce non-optimality, this is not an issue. Once humans solve a Sokoban problem, they have two choices: move on to another problem (they are satisfied with the result), or try and re-solve the same problem to get a better solution. *Rolling Stone* could try something similar. Having solved the problem once, if we want a better solution, we can reduce the probability of introducing non-optimality in the search by decreasing the aggressiveness of the relevance cuts. This will make the searches larger but, on the other hand, the last iteration does not have to be searched, since a solution for that threshold was already found.

Relevance cuts are yet another way to significantly prune Sokoban search trees. We have no shortage of promising ideas, each of which potentially offers another order of magnitude reduction in the search tree size. Although this sounds impressive, our experience suggests that each factor of 10 improvement seems to only yield another 4 or 5 problems being solved. At this rate, we will have do a lot of research if we want to successfully solve all 90 problems!

Acknowledgements

The authors would like to thank the German Academic Exchange Service, the Killam Foundation and the Natural Sciences and Engineering Research Council of Canada for their support.

References

[1] G. Adelson-Velskiy, V. Arlazarov, and M. Donskoy. Some methods of controlling the tree search in chess programs. *Artificial Intelligence*, 6(4):361–371, 1975.
[2] J. Culberson. Sokoban is PSPACE-complete. Technical Report TR97-02, Department of Computing Science, University of Alberta, Edmonton, Alberta, Canada, 1997. ftp://ftp.cs.ualberta.ca/pub/TechReports/1997/TR97-02.

[3] D. Dor and U. Zwick. SOKOBAN and other motion planning problems, 1995. At: http://www.math.tau.ac.il/~ddorit.
[4] M. Ginsberg. *Essentials in Artificial Intelligence*. Morgan Kaufman Publishers, San Francisco, 1993.
[5] G. Goetsch and M.S. Campbell. Experiments with the null-move heuristic. In T.A. Marsland and J. Schaeffer, editors, *Computers, Chess, and Cognition*, pages 159–181, New York, 1990. Springer-Verlag.
[6] A. Junghanns and J. Schaeffer. Single-agent search in the presence of deadlock. In *AAAI-98*, pages 419–424, Madison/WI, USA, July 1998.
[7] A. Junghanns and J. Schaeffer. Sokoban: Evaluating standard single-agent search techniques in the presence of deadlock. In *Advances in Artificial Intelligence*, pages 1–15. Springer Verlag, 1998.
[8] R.E. Korf. Depth-first iterative-deepening: An optimal admissible tree search. *Artificial Intelligence*, 27(1):97–109, 1985.
[9] R.E. Korf. Macro-operators: A weak method for learning. *Artificial Intelligence*, 26(1):35–77, 1985.
[10] R.E. Korf. Finding optimal solutions to Rubik's Cube using pattern databases. In *AAAI-97*, pages 700–705, 1997.
[11] J. Schaeffer. *Experiments in Search and Knowledge*. PhD thesis, Univ. of Waterloo, Canada, 1986.
[12] G. Wilfong. Motion planning in the presence of movable obstacles. In *4th ACM Symposium on Computational Geometry*, pages 279–288, 1988.

Multi-cut Pruning in Alpha-Beta Search

Yngvi Björnsson and Tony Marsland

University of Alberta, Department of Computing Science,
Edmonton AB, Canada T6G 2H1
{yngvi,tony}@cs.ualberta.ca

Abstract. The efficiency of the $\alpha\beta$-algorithm as a minimax search procedure can be attributed to its effective pruning at so called cut-nodes; ideally only one move is examined there to establish the minimax value. This paper explores the benefits of investing additional search effort at cut-nodes by expanding other move alternatives as well. Our results show a strong correlation between the number of promising move alternatives at cut-nodes and a new principal variation emerging. Furthermore, a new forward pruning method is introduced that uses this additional information to ignore potentially futile subtrees. We also provide experimental results with the new pruning method in the domain of chess.

1 Introduction

The $\alpha\beta$-algorithm is the most popular method for searching game-trees in such adversary board games as chess, checkers and Othello. It is much more efficient than a plain brute-force minimax search because it allows a large portion of the game-tree to be pruned off, while still backing up the correct game-tree value. However, the number of nodes visited by the algorithm still increases exponentially with increasing search depth. This obviously limits the scope of the search, since game-playing programs must meet external time-constraints: often having only a few minutes to make a decision. In general, the quality of play improves the further the program looks ahead[1].

Over the years the $\alpha\beta$-algorithm has been enhanced in various ways and more efficient variants have been introduced. For example, the basic algorithm explores all continuations to some fixed depth, but in practice it is not used that way. Instead various heuristics allow variations in the distance to the search horizon (often called the search depth or search tree height), so that some move sequences can be explored more deeply than others. "Interesting" continuations are expanded beyond the nominal depth, while others are terminated prematurely. The latter case is referred to as *forward pruning*, and involves some additional risk of overlooking a good continuation. The rationale behind the approach is

---

[1] Some artificial games have been constructed where the opposite is true; when backing up a minimax value the decision quality actually decreases as we search deeper. This phenomenon has been studied thoroughly and is referred to as *pathology* in game-tree search [6]. However, such pathology is not seen in chess or the other games we are investigating.

that the time saved by pruning non-promising lines is better spent searching other lines more deeply, in an attempt to increase the overall decision quality.

To effectively apply forward-pruning, good criteria are needed to determine which subtrees to ignore. In here we show that the number of good move alternatives a player has at cut-nodes can be used to identify potentially futile subtrees. Furthermore, we introduce *multi-cut $\alpha\beta$-pruning*, a new forward pruning method that makes its pruning decisions based on the number of promising moves at cut-nodes. In the minimax sense it is enough to find one refutation to an inferior line of play. However, instead of finding one such refutation, our method uses shallow searches to identify moves that "look" good. If there are several moves available that seem good enough to refute the current line of play, multicut-pruning prevents that particular line from being expanded more deeply.

In the following section we introduce the basic idea behind the new pruning method, and then show how the idea is implemented in an actual game-playing program. Experimental results follow; first the promise of the new pruning criterion is established, and second the new pruning method is tested in the domain of chess. Finally, before drawing our conclusions we explain how related works use complementary ideas.

2 Multi-cut Idea

In a traditional $\alpha\beta$-search, if a move returns a value greater or equal to β there is no reason to examine that position further, and the search can return. This is often referred to as a β-cutoff (we are using here Knuth's [3] nega-max formulation of the $\alpha\beta$-algorithm, where there is no distinction between α- and β-cutoffs). Intuitively, this means that the player to move has found a way to refute the current line of play, so there is no need to find a better refutation. By way of explanation, and to introduce our terminology, we are seeking the principal variation (pv): the best sequence of moves from the root node (current position in the game) to the best of the accessible nodes on the search horizon. We expect β-cutoffs to occur at so called cut-nodes (that is, nodes that are refuted). The root node of a game-tree is a pv-node, the first child of a pv-node is also a pv-node, while the other children are cut-nodes. All children of a cut-node are all-nodes (where every successor must be explored) and vice versa. In a perfectly ordered tree only one child of a cut-node is expanded. If a new best move is found at a pv-node, the node it leads to also becomes a pv-node. At pv- and all-nodes every successor is examined. Most often it is the first child that causes the cutoff, but if it fails to do so the sibling nodes are expanded in turn, until either one returns a value greater or equal to β or all the children have been searched. If none of the moves causes a cutoff, a cut-node becomes an all-node.

For a new principal variation to emerge, every expected cut-node on the path from a leaf-node to the root must become an all-node. In practice, however, it is common that if the first move does not cause a cutoff, one of the alternative moves will. Therefore, *expected cut-nodes, where many moves have a good potential of causing a β-cutoff, are less likely to become all-nodes, and conse-*

quently such lines are unlikely to become part of a new principal-variation. This observation forms the basis for the new forward pruning scheme we introduce here, *multi-cut $\alpha\beta$-pruning*. Before explaining how it works, let us first define an mc-prune (multi-cut prune).

Definition 1 (mc-prune). *When searching node v to depth $d + 1$ using $\alpha\beta$-search, and if at least c of the first m children of v return a value greater or equal to β when searched to depth $d - r$, an mc-prune is said to occur and the search can return.*

In multi-cut $\alpha\beta$-search, we try for an mc-prune only at expected cut-nodes (we would not expect it to be successful elsewhere). Figure 1 shows the basic idea. At node v, before searching v_1 to a full depth d like a normal $\alpha\beta$-search does, the first m successors of v are expanded to a reduced depth of $d - r$. If c of them return a value greater or equal to β an mc-prune occurs and the search returns the value of β, otherwise the search continues as usual exploring v_1 to a full depth d. The subtrees below $v_2, ..., v_m$, to depth $(d - r)$, represent extra search overhead introduced by mc-prune, and would not be expanded by normal $\alpha\beta$-search, but the dotted area of the subtree below node v_1 shows the savings that are possible if the mc-prune is successful. However, if the pruning condition is not satisfied, we are left with the overhead but no savings. By searching the subtree of v_1 to less depth, there is of course some risk of overlooking a tactic that would make v_1 become a new principal variation. We are willing to take that risk, because we expect at least one of the c moves that returns a value greater or equal to β when searched to a reduced depth, will cause a genuine β-cutoff if searched to a full depth.

3 Multi-cut Implementation

Figure 2 is a C-code version of a null-window search (NWS) routine using multi-cut. For clarity we have omitted details about search extensions, transposition table lookups, null-move searches, and history heuristic updates that are irrelevant to our discussion. The *NWS* routine [4] is an integral part of the *Principal Variation Search* algorithm. Multi-cut could equally well be implemented in other enhanced $\alpha\beta$-variants like *NegaScout* [7]. The parameter *depth* is the remaining length of search for the position, and β is an upper-bound on the value we can achieve. There is no need to pass α as a parameter, because it is always equal to β - 1. On the other hand, the new parameter, *node_type*, shows the expected type of the node we are currently looking at. In a null-window search we are dealing only with either cut-nodes (CUT) or all-nodes (ALL).

As is normal, the routine starts by checking whether the horizon has been reached, and if so uses a quiescence search (QS) to return the value of the position. Otherwise, we look for useful information about the position in the transposition table. This is followed by a null-move search (most chess programs use this powerful technique). Normally a standard null-window $\alpha\beta$ search would follow, if the null-move does not cause a cut-off. Instead we insert here a multicut

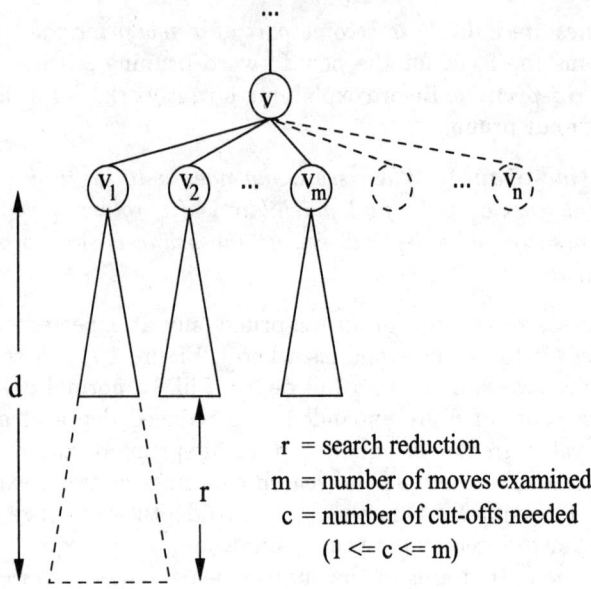

Fig. 1. Applying the mc-prune method at node v

search to see if the mc-prune condition applies. The parameters mc_M, mc_R, and mc_C stand for m (number of moves to look at), r (search reduction), and c (number of cutoffs needed), respectively. Although they are shown here as constants, they could be determined more dynamically and be allowed to vary during the search.

We do not check for the mc-prune conditions at every node in the tree. First, we only test for them at expected cut-nodes. Second, they are not applied at the levels of the search tree close to the horizon thus reducing the time overhead involved in this method. Finally, there are some game-dependent restrictions that apply. In Figure 2 these latter restrictions are encapsulated in the function $TryMultiCut()$. In our experiments in the domain of chess (see later) the pruning is disabled when the endgame is reached, since there are usually few viable move options there and the mc-searches are therefore not likely to be successful. Also, the positional understanding of chess programs in the endgame is generally poorer than in the earlier phases of the game. Therefore the programs rely more heavily on the search to guide them, and any forward pruning scheme is therefore more likely to be harmful. Furthermore, the pruning is not done if the side to move is in check, or if search extensions have been applied for any of the three previous moves leading to the current position.

```
#define mc_M    10   //Multi-Cut: # of moves to look at
#define mc_C    3    //            # of cuts to cause an mc-prune
#define mc_R    2    //            depth reduction

#define CUT     2
#define ALL     3
#define CHILDTYPE(t) (((t)==CUT) ? ALL : CUT)

VALUE NWS(int depth, VALUE beta, NODETYPE node_type)
{
  VALUE score;
  MOVE  move;

      ... Search extension code omitted ...

  if ( depth <= 0 ) return QS(beta-1, beta);

      ... Transposition table lookup and nullmove-search code omitted ...

  // Multi-Cut pruning
  if ( (node_type == CUT) && (depth > mc_R) && TryMultiCut() ) {
    int m = 0, c = 0;
    move = MoveFirst();
    while ( m < mc_M && move ) {
      MakeMove( move );
      score = -NWS(depth-mc_R-1, -beta+1, ALL);
      RetractMove( move );
      if ( score >= beta ) {
        c++;
        if ( c == mc_C ) return beta;
      }
      m++;
      move = MoveNext();
    }
  }

  // Standard null-window search
  move = MoveFirst();
  while ( move ) {
    MakeMove( move );
    score = -NWS(depth-1, -beta+1, CHILDTYPE(node_type));
    RetractMove( move );
    if ( score >= beta ) break;
    move = MoveNext();
  }
      ... Update transposition/history table code omitted ...

  return score;
}
```

Fig. 2. Multi-Cut component within Principal Variation Search

4 Criteria Selection Process

The multi-cut idea stands or falls with the hypothesis that nodes having many promising move alternatives are more likely to cause a β-cutoff than those with fewer promising move alternatives. In turn, this implies that we have a method for determining which moves are promising. In mc-pruning, shallow searches are used to identify promising moves. This scheme was chosen based on the experiments presented in this section.

4.1 Identifying False Cut-Nodes

We will refer to any node where a β-cutoff is anticipated as an expected cut-node. Only after searching the node do we know if it actually causes a cutoff; if it does we call it a True cut-node, otherwise a False cut-node. What we seek is a scheme that accurately predicts which expected cut-nodes are False. We experimented with the following four different ways of anticipating cut nodes:

1. Number of legal moves (NM):
   The most straight forward approach is simply to assume that every move has the same potential of causing a β-cutoff. Therefore, the more children an expected cut-node has, the more likely it is to be a true cut-code. Although this assumption is not realistic, it can serve as a baseline for comparison.
2. History heuristic ($HH > \Delta$):
   A more sensible approach is to distinguish between good and bad moves. For example, by using information from the *history-heuristic* table [8]. Moves with a positive history-heuristic value are known to be useful elsewhere in the search-tree. This method defines moves with a history-heuristic value greater than a constant Δ as potentially good. One advantage of this scheme is that no additional search is required.
3. Quiescence search ($QS() \geq \beta - \delta$):
   Here quiescence search is used to determine which children of a cut-node have a potential for causing a cutoff. If the quiescence search returns a value greater or equal to $\beta - \delta$ the child is considered promising. The constant δ, called the β-cutoff margin, can be either positive or negative. Although, this scheme may require additional search, it will hopefully give a better estimate than the previous schemes.
4. Null-window search ($NS(d-r) \geq \beta - \delta$):
   This scheme is much like the one above, except instead of using quiescence search to estimate the merit of the children, a null-window search to a closer horizon at distance $d - r$ is used.

Next, we established how well the number of promising moves, as judged by each of the above schemes, correlates to an expected cut-node being a True cut-node or not. To do this, the four test functions were implemented in a chess program, *TheTurk*[2].

---

[2] TheTurk is a chess program developed at University of Alberta by Yngvi Björnsson and Andreas Junghanns.

When the program visits an expected cut-node it calculates the number of promising move alternatives in the position according to each scheme. Then, after searching the node to a full depth to determine if it really is a cut-node, the number of promising moves information is logged to a file along with a flag indicating whether the node is a True cut-node.

4.2 Experimental Results

Fifty middle-game positions, taken from the Botvinnik vs. Tal world championship match in 1960, were used in the experiment. The positions were sampled every 5th move from move 10 to move 30. Furthermore, for sake of consistency, we only used positions from games when Tal was playing the White side and it was his turn to move. The program performed a search for each position logging relevant information at expected cut-nodes, as described above.

The resulting data was classified into two groups, one with the True cut-nodes, and the other with the False cut-nodes. We gathered statistics about 100,000 expected cut-nodes, and of these only 2.5% were classified incorrectly (i.e. were False cut-nodes). The average number of promising moves, as judged by each scheme, is presented in Table 1. The second column shows the average for the True cut-node group and the third column the average for the False cut-node group. By comparing the averages and the standard deviations (also shown in the table) of the two groups we can determine the scheme that can best predict False cut-nodes. That is, we are looking for the scheme that has the greatest difference between the averages for the two groups, and the lowest standard deviation.

Table 1. Comparison of different schemes for identifying False cut-nodes

Method	True cut-nodes		False cut-nodes	
	$\bar{x}$	σ	$\bar{x}$	σ
NM	35.60	11.74	24.83	14.46
$HH > 0$	22.27	8.87	16.35	9.77
$HH > 100$	9.15	5.72	7.13	5.33
$QS() > \beta$	20.48	15.03	0.32	1.44
$QS() > \beta$-25	23.70	14.08	1.66	4.20
$NS(\text{d-2}) > \beta$	20.62	14.88	0.17	0.55
$NS(\text{d-2}) > \beta$-25	23.75	14.00	1.46	3.75

In Table 1, it is interesting to note that even a simplistic scheme like looking at the number of legal moves shows a difference in the averages. However, the difference is relatively small and the standard deviation is high. The history heuristic schemes have lower standard deviation, but unfortunately the averages are too similar. This renders them useless. The methods that rely on search, $QS()$ and $NS()$, do much better, especially those where δ (the β-cutoff margin) is set to zero[3]. Not only are the averages for the two groups far apart, but the standard deviation is also very low. From the data in Table 1 the two schemes look almost equally effective. Therefore, to discriminate between them further, we filtered the data for the False cut-nodes looking only at non-zero data-points (that is, we only consider data-points where at least one promising move alternative is found by either scheme). The result using the filtered data is given in Table 2. Now we can see more clearly that the null-window (NS) scheme is a better predictor of False cut-nodes. Not only does it show on average fewer false promises, but the standard deviation is also much lower. This means that it only very infrequently shows False cut-nodes as having more than several promising move alternatives. Even in the worst case there never were more than 6 moves listed as promising, while for the $QS()$ scheme at least one position had 32 false indicators.

Table 2. Comparison of selected schemes using filtered data

Method	False cut-nodes	
	$\bar{x}$	σ
$QS() > \beta$	2.31	3.20
$NS(d-2) > \beta$	1.45	0.86

The above experiments clearly support the hypothesis that there is a way to discriminate between nodes that are likely to become true cut-nodes and those that are not. As a result we selected the shallow null-window searches as the scheme for finding promising moves in the multi-cut $\alpha\beta$-pruning.

5 Multi-cut in Practice

Ultimately, we want to show that game-playing programs using the new pruning method can achieve increased playing strength. To test the idea in practice, multi-cut $\alpha\beta$-pruning was implemented in *The Turk*. Two versions of the program were matched against each other, one with multi-cut pruning and the other without. Three matches, with 80 games each, were played using different time controls. To prevent the programs from playing the same game over and over, forty well-known opening positions were used as a starting point. The programs played each opening once from the white side and once as black. Table 3 shows the match results. T stands for the unmodified version of the program and $T_{mc(c,m,r)}$ for the version with multi-cut implemented. We experimented with the case $m = 10, r = 2$, and $c = 3$ (i.e. 10 moves searched with a depth reduction of 2 ply and with 3 β-cutoffs required to achieve the mc-prune condition). These parameter values are somewhat arbitrary, based on experience and a few test trials.

---

[3] In *The Turk*, a δ value of 25 is equivalent to a quarter of a pawn.

Table 3. Summary of 80-game match results

$T_{mc(3,10,2)}$ versus T		
Time control	Score	Winning %
40 moves in 5 minutes	41 - 39	51.3
40 moves in 15 minutes	40 - 40	50.0
40 moves in 25 minutes	44 - 36	55.0

The multi-cut version shows a slight improvement over the unmodified version. In tournament play this winning percentage would result in about 15 points difference in the players' rating. Since more than 1,000 games are typically needed to obtain a standard error of less than 10 rating points [1], we cannot claim that the multi-cut version is the stronger, based only on this single set of experiments.

One final insight, the programs gathered statistics about the behavior of the multi-cut pruning. The search spends about 25%-30% if its time (in terms of nodes visited) in shallow multi-cut searches, and an mc-prune occurs in about 45%-50% of its attempts. Obviously, the tree expanded using multi-cut pruning differs significantly from the tree expanded when it is not used.

6 Related Work

The idea of exploring additional moves at cut-nodes is not entirely new. There exist at least two other variants of the $\alpha\beta$-algorithm that consider more than one move alternative at cut-nodes: one is *Singular Extensions* [2] and the other McAllester's *Alpha-Beta-Conspiracy Search* [5].

The singular extensions algorithm extends "singular" moves more deeply than others. A move is defined as singular if its evaluation is higher than all its alternative moves by some specified margin, called the singular margin. Moves that fail-high, i.e. cause a β-cutoff, automatically become candidates for being singular (the algorithm also checks for singular moves at pv-nodes). To determine if a candidate move that fails-high really is singular, all its sibling moves are explored to a reduced depth. The move is declared singular only if the value of all the alternatives is significantly (as defined by the singular margin) lower than the value of the principal variation. Singular moves are "remembered" and extended one additional ply on subsequent iterations. This method improved the playing strength of Deep Thought (predecessor of Deep Blue) by about 30 USCF rating points [1]. One might think of multi-cut as the complement of singular-extensions: instead of extending lines where there is seemingly only one good move, it prunes lines where there are many promising (refutation) moves available.

The Alpha-Beta-Conspiracy algorithm is essentially an $\alpha\beta$-search that uses conspiracy depth, instead of classical ply depth, to decide how deep to search. When determining whether to terminate the search, static evaluations of all siblings of nodes that lie on the current search path are used. However, empirical results using this algorithm were not favorable.

In addition, there are some best-first search methods that use the notion of having options. They are not discussed here.

7 Conclusions

We feel that our experimental results give a rise to optimism. Although the self-play matches do not prove in a statistically significant way that the new method is better, they clearly show that a search method expanding a radically different tree than the $\alpha\beta$-algorithm seemingly has at least equal playing strength.

The multi-cut method is still in its infancy. We are experimenting here with a preliminary implementation of the idea. There is still much scope for improvement through further tuning and enhancement, by experimenting with different settings for the parameters c, m, and r, for example. Also, the current implementation defines the parameters as constants. Instead, we might need to determine them more dynamically, and have the possibility to adjust the values as the game/search progresses.

Our experiments show the feasibility of the idea and indicate a strong correlation between the number of promising move alternatives available at an expected cut-node, and the node becoming a True cut-node. The multi-cut idea as described and implemented here is not the only way of exploiting this correlation, and is by no means necessarily the best. There is still much room for innovative methods to be developed based on looking at other move options.

References

[1] T. Anantharaman. *A Statistical Study of Selective Min-Max Search in Computer Chess*. PhD thesis, Carnegie-Mellon University, Pittsburgh, PA, May 1990.

[2] T. Anantharaman, M. S. Campbell, and F. Hsu. Singular extensions: Adding selectivity to brute-force searching. *Artificial Intelligence*, 43(1):99–109, 1990.

[3] D. E. Knuth and R. W. Moore. An analysis of alpha-beta pruning. *Artificial Intelligence*, 6(4):293–326, 1975.

[4] T. A. Marsland. Single-Agent and Game-Tree Search. In A. Kent and J. G. Williams, editors, *Encyclopedia of Computer Science and Technology*, volume 27, pages 317–336, New York, 1993. Marcel Dekker, Inc.

[5] D. A. McAllester and D. Yuret. Alpha-beta-conspiracy search, 1993. URL: http://www.research.att.com/~dmac/abc.ps.

[6] D. S. Nau. Pathology on game trees: A summary of results. In *Proceedings of the ACM National Conference on Artificial Intelligence*, pages 102–104, 1980.

[7] A. Reinefeld. An improvement to the Scout tree search algorithm. *ICCA Journal*, 6(4):4–14, 1983.

[8] J. Schaeffer. The history heuristic and alpha-beta search enhancements in practice. *IEEE Transactions on Pattern Analysis and Machine Intelligence*, 11(1):1203–1212, 1989.

A Solution to the GHI Problem for Best-First Search

Dennis M. Breuker[1], H. Jaap van den Herik[1], Jos W.H.M. Uiterwijk[1], and L. Victor Allis[2]

[1] Department of Computer Science
Universiteit Maastricht
P.O. Box 616
6200 MD Maastricht, The Netherlands
{breuker,herik,uiterwijk}@cs.unimaas.nl
[2] Quintiq B.V.
Van Grobbendoncklaan 83
5213 AV 's Hertogenbosch
The Netherlands
victor@quintiq.nl

Abstract In a search graph a node's value may be dependent on the path leading to it. Different paths may lead to different values. Hence, it is difficult to determine the value of any node unambiguously. The problem is known as the *graph-history-interaction* (GHI) problem. This paper provides a solution for best-first search. First, we give a precise formulation of the problem. Then, for best-first search and for other searches, we review earlier proposals to overcome the problem. Next, our solution is given in detail. Here we introduce the notion of *twin nodes*, enabling a distinction of nodes according to their history. The implementation, called BTA (*Base-Twin Algorithm*), is performed for pn search, a best-first search algorithm. It is generally applicable to other best-first search algorithms. Experimental results in the field of computer chess confirm the claim that the GHI problem has been solved for best-first search.
Keywords: graph-history interaction (GHI) problem, best-first search, base-twin algorithm (BTA).

1 The GHI Problem

Search algorithms are used in many domains, ranging from theorem proving to computer games. The algorithms are searching in a state space containing problem states (positions), often represented as nodes. A move which transforms a position into a new position is represented as an edge connecting the two nodes.

In a search tree, it may happen that identical nodes are encountered at different places. If these so-called *transpositions* are not recognized, the search algorithm unnecessarily expands identical subtrees. Therefore, it is profitable to recognize transpositions and to ensure that for each set of identical nodes, only one subtree is expanded.

In computer-chess programs using a *depth-first* search algorithm, this idea is realized by storing the result of a node's investigation in a transposition table (e.g., [9], [12]). If an identical node is encountered in the search process, the result is retrieved from the transposition table and used without further investigation.

If a (selective) *best-first* search algorithm (which stores the whole search tree in memory) is used, the search tree is converted into a search graph, by joining identical nodes into one node, thereby merging the subtrees.

These common ways of dealing with transpositions contain an important flaw: determining whether *nodes* are identical is not the same as determining whether the *search states* represented by the nodes are identical. For two reasons, the path leading to a node cannot be ignored. First, the history of a node may partly determine the *legitimacy of a move*. For instance, in chess, castling rights are not only determined by the position of the pieces on the board, but also by the knowledge that in the position under investigation the King and Rook have not moved previously. Second, the history of a node may play a role in determining the *value of a node*. For instance, a position may be declared a draw by its three-fold repetition or by the so-called k-move rule [10].

We refer to the first problem as the *move-generation problem*, and to the second problem as the *evaluation problem*. The combination of these two problems is referred to as the *graph-history-interaction* (GHI) problem (cf. [13] and [8]).

The GHI problem is a noteworthy problem not only in chess but in the field of game playing in general. Its applicability extends though to all domains where the history of states is important. To mention just one example: in job-shop scheduling problems the costs of a task may be dependent on the tasks done so far, e.g., the cost of preparing a machine for performing some process depends on the state left after the previous process.

A possible solution to the GHI problem is to include in all nodes the status of the relevant properties of the history of the node, i.e., the properties which may influence either the move generation or the evaluation of the node. A disadvantage of such a solution is that too many properties may be relevant, resulting in the need of storing large amounts of extra information in each node. For chess, we can distinguish four relevant properties of the history of a position (the first two being relevant for the move-generation problem, and the last two for the evaluation problem):

1. the castling rights (Kingside and Queenside for both players),
2. the *en-passant* capturing rights,
3. the number of moves played without a capture or a pawn move, and
4. the set of all positions played on the path leading to this node.

The first two properties can be included in each node, without much overhead. The third property can be included in each node, but will reduce the frequency of transpositions drastically. The inclusion of the fourth property, necessary to determine whether a draw by three-fold repetition has been encountered, would require too much overhead. As a result, in most chess programs, the first two properties are included in a node, while the last two are not.

Depending on which properties are included in a node, the probability of two nodes being identical will be reduced. If not all relevant properties are included and transpositions are used, it is possible that incorrect conclusions are drawn from the transpositions. Campbell mentioned that, contrary to best-first search (which he calls selective search), in depth-first search the GHI problem occurs relatively infrequently [8].

In this paper we give a solution to the GHI problem for best-first search with only a few relevant properties included in a node. In Section 2 an example of the GHI problem is given. Previous work on the GHI problem is discussed in Section 3. In Section 4 the general solution to the GHI problem for best-first search is described. A formalized description and the pseudo-code for the implementation in proof-number (pn) search is given in Section 5. Section 6 lists experiments with the new algorithm. It is compared to three other pn-search variants. The results are presented in Section 7. Finally, Section 8 provides conclusions.

2 An Example of the GHI Problem

Figure 1 shows a pawn endgame position, taken from [8], where the GHI problem can occur. White (to move) has achieved a winning position. However, we show that it is possible to evaluate this position incorrectly as a draw. In this paper we assume that a single repetition of positions evaluates as a draw, in contrast with the FIDE ruling which stipulates that the same position must occur three times.

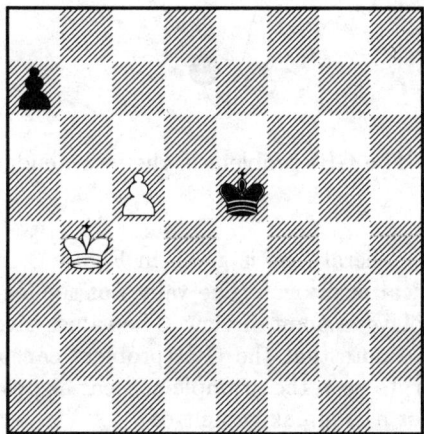

Fig.1. A pawn endgame (WTM).

In Figure 2 the relevant part of the search tree is pictured. In this article we follow the notation of [2], i.e., for all AND/OR trees (or graphs) white squares represent OR nodes (positions with the first player to move), and black circles represent AND nodes (positions with the second player to move).

After the move sequence **1. ♔b5? ♚e6? 2. ♔a6? ♚d5 3. ♔b5 ♚e6** the position after move 1 is repeated (node E), and evaluated as a draw. Since White does not have any better alternative on the third move, the position after **2. ♔a6** (node H) should be evaluated as a draw. Backing up this draw leads to the incorrect conclusion that node A evaluates as a draw. However, after the winning move sequence **1. ♔a5! ♚e6 2. ♔a6!** the same position (node H) is reached, which is (now) evaluated as a win after **2. ..., ♚d5 3. ♔b5 ♚e6 4. ♔c6!** (node G). Backing up this win leads to the correct conclusion that node A evaluates as a win.

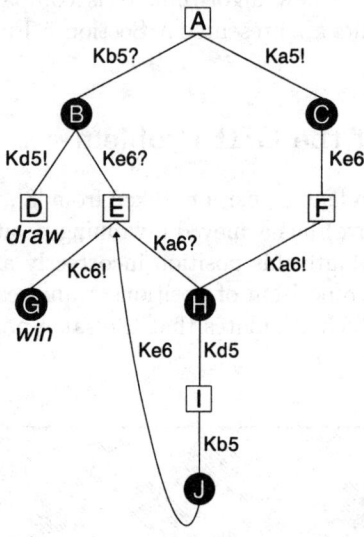

Fig.2. The GHI problem in the pawn endgame.

An example of the general case is given in Figure 3. It shows an AND/OR search tree with identical positions[1]. The values of the leaves (given in italics) are seen from the OR player's point of view. The values given next to the nodes are back-up values. We note that the GHI problem can occur in any type of AND/OR tree. However, to keep the example as clear as possible we have chosen to show the example for a minimax game tree.

The terminal nodes E and G are a win for the OR player, and the terminal nodes C and F are evaluated as a draw because of repetition of positions. Propagating the evaluation values of the terminal nodes through the search tree

---

[1] In games such as chess, a repetition of positions is impossible after only two ply (node C in the left subtree of node B and node F in the subtree of node D). Our example disregards this characteristic for simplicity's sake.

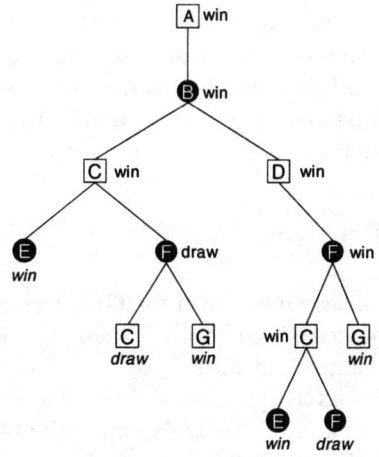

Fig.3. A search tree with repetitions.

results in a win at the root. When making use of transpositions, every node should occur only once in the tree. Assume that a parent generates its children and that one of its children already exists in the tree. Then a connecting edge from the parent to the existing node is made. This transforms the search tree into a Directed Cyclic Graph (DCG) (Figure 4).

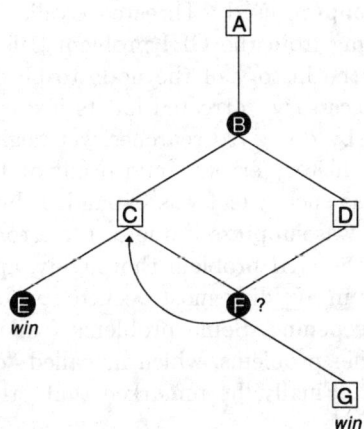

Fig.4. The DCG corresponding with the tree of Figure 3.

In this DCG it is difficult to determine unambiguously the value of node F due to the GHI problem. The value of this node is dependent on the path leading to it. Following the path A–B–C–F, child C of node F is a repetition and hence

F is evaluated as a draw, but following the path A–B–D–F, child C is not a repetition and is not evaluated as a draw. Thus, in the DCG, node F has two different values. Hence, in this example it is not possible to determine the value of root A, since in the first case it is a draw, and in the second case it is a win, due to the values of E and G.

3 A Review of Previous Work

Although several authors have mentioned the GHI problem, so far no solution to this problem has been described. Only provisional ideas have been given. Below, we review the five most important ideas[2].

Palay first identified the GHI problem [13]. He suggested two "solutions": (1) refrain from using graphs, and (2) recognize when the GHI problem occurs and handle accordingly. The first "solution" (apart from not being a real solution, it merely ignores the problem) had as a drawback that large portions of the graph would be duplicated every time a duplicate node occurred, wasting a large amount of time and memory. The second solution worked as follows. When the positions suffering from the GHI problem were recognized, the path from the repetition node upwards to the ancestor with multiple parents was split into separate paths. He did not implement this strategy, since he conjectured that such positions only occurred occasionally (the GHI problem occurred in three out of 300 test positions). A disadvantage of this solution is that the recognition of positions suffering from the GHI problem is not straightforward.

Another idea for a solution originates from Thompson [8]. While building a tactical analyzer, Thompson used a Directed Cyclic Graph (DCG) representation. He saw it suffering from the GHI problem [19]. He cured the problem by taking into account the history of the node to be expanded. The value of this node was then, if necessary, corrected for its history. The newly-generated children were evaluated by doing $\alpha\beta$ searches, yet neglecting their history. As a consequence, the only history errors could occur at the leaves. These errors were corrected as soon as such a leaf was expanded, but it could happen that the expansion of a node was suppressed due to the error.

Campbell discussed the GHI problem thoroughly, applying it to depth-first search only [8]. The key in avoiding most occurrences of the GHI problem appears to be iterative deepening. Some problems (called "draw-first") can be overcome[3]. However, other problems, which he called "draw-last" could not be solved by his approach[4]. Finally, he remarked that "the GHI problems occur

---

[2] Berliner and McConnell suggested the use of conditional values as an idea to solve the GHI problem [4]. They promised details in a forthcoming paper.

[3] In the draw-first case node F in Figure 4 is first reached through path A–B–C–F (and the value of node F is based on child C being a repetition) and later in the search node F is reached through path A–B–D–F and the previous value of node F is used.

[4] In the draw-last case node F in Figure 4 is first reached through path A–B–D–F (and the value of node F is based on child C being *no* repetition) and later in the

much more frequently in selective search programs, and require some solution in order to achieve reasonably general performance. Both Palay's and Thompson's approaches seem to be acceptable." We conclude that Campbell gave a partial solution for depth-first search, and no solution for best-first search.

Baum and Smith stumbled on the GHI problem, when implementing their best-first search algorithm BPIP (Best Play for Imperfect Players) [3]. Baum and Smith completely store the DCG in memory and grow it by using "gulps". In each gulp a fraction of the most interesting leaves is expanded. For each parent-child edge e a subset $S(e)$ was defined as the intersection of *all* ancestor nodes and *all* descendant nodes of edge e. A DCG was claimed to be legitimate (i.e., no nodes have to be split) if and only if, for all children C with more than one parent P, $S(e_{PC})$ is independent of P. Their solution was as follows. Each time a new leaf was created three possibilities were distinguished: (1) if the leaf was a repetition it was evaluated as a draw, else (2) if a duplicate node existed in the graph, these two nodes were merged on the condition that the resultant DCG was legitimate, else (3) the node was evaluated normally. After leaf expansion it was exhaustively investigated whether every node C with multiple parents passed the $S(e)$ test. If not, such a node C was split into several nodes C', C'', ..., with distinct subsets $S(e_{PC})$. Then, the subtrees of the newly-created nodes had to be rebuilt and re-evaluated. Baum and Smith gave this idea as a solution to the GHI problem without the support of an implementation. Moreover they remarked that "Implementation in a low storage algorithm would probably be too costly". We believe that the overhead introduced by our idea, described in the next section, is much less than the overhead introduced by the idea of Baum and Smith.

Schijf *et al.* investigated the problem [17] in the context of proof-number search (pn search) [1]. They examined the problem in Directed Acyclic Graphs (DAGs) and DCGs separately. They noted that, when the pn-search algorithm for trees is used in DAGs, the proof and disproof numbers are not necessarily correctly computed, and the most-proving node is not always found. Schijf proved that the most-proving node always exists in a DAG [16]. Furthermore, he formulated an algorithm for DAGs that correctly determines the most-proving node. However, this algorithm is only of theoretical importance, since it has an unfavourable time-and-memory complexity. Therefore, a practical algorithm was developed. Surprisingly, only two minor modifications to the pn-search algorithm for trees are needed for a practical algorithm for DAGs. The first modification is that instead of updating only *one* parent, *all* parents of a node have to be updated. The second modification is that when a child is generated, it has to be checked whether this node is a transposition (i.e., if it was generated earlier). If this is the case, the parent has to be connected to this node that has already been generated. Schijf *et al.* note that this algorithm contains two flaws [17]. First, the proof and disproof numbers do not represent the cardinality in the smallest proof and disproof set, but these numbers are upper bounds to the real

---

search node F is reached through path A–B–C–F and the previous value of node F is used.

proof and disproof numbers. Second, the node selected by the function Select-MostProvingNode is not always equal to a most-proving node. However, it still holds that if the node chosen is proved, the proof number of the root decreases, whereas if this node is disproved, the disproof number of the root decreases. In either case the proof or disproof number may decrease by more than unity, as a result of the transpositions present. This algorithm has been tested on tic-tac-toe [16]. For the problem of applying pn search to a DCG Schijf et al. give a time-and-memory-efficient algorithm, which, however, sometimes inaccurately evaluates nodes as a draw by repetition [17]. They remark that, as a consequence, their algorithm is sometimes unable to find the goal, even though it should have found it.

4 BTA: An Enhanced DCG Algorithm

In this section we describe a new and correct algorithm (denoted BTA: Base-Twin Algorithm) for solving the GHI problem for best-first search. The BTA algorithm is based on the distinction of two types of node, termed *base nodes* and *twin nodes*. The purpose of these types is to distinguish between equal positions with different history. Although it was known in the DCG algorithm described by Schijf et al. [17] that nodes sometimes *may* be incorrectly evaluated as a draw, their algorithm was unable to note *when* this occurs. We have devised an alternative in which a sufficient set of relevant properties for correct evaluation is recorded. We have chosen to include in a node only a small number of relevant properties. The reasons for not including *all* relevant properties are:

- some properties are only relevant for a *small* number of nodes,
- the more properties are included, the lower the frequency of transpositions, and
- some properties require too much overhead and/or take up too much space when included in a node.

The move-generation problem (cf. Section 1) can easily be solved by including the relevant properties (in chess these are the castling rights and the *en-passant*-capturing rights) into each node. Hence, only the evaluation problem (cf. Section 1) needs to be solved. We have chosen to describe the solution of repetition of positions, since repetition of positions occurs in many search problems, and the k-move rule is a special rule which seldomly shows up in practice. As mentioned before, we assume that a single repetition of positions results in a draw.

We further distinguish between terminal nodes and leaves. A *terminal node* represents a terminal position, i.e., a position where the rules of the game determine whether the result is a win, a draw, or a loss. *Leaves* are nodes which do not have children (yet). Leaves include terminal nodes and nodes which are not yet expanded.

4.1 Our Representation of a DCG

Basically the GHI problem occurs because the search tree is transformed into a DCG by merging nodes representing the same position, but having a different history. To avoid such an undesired coalescence, we propose an enhanced representation of a DCG. In the graph we distinguish two types of node: *base* nodes and *twin* nodes. After a node is generated, it is looked up in the graph by using a pointer-based table. If it does not exist, it is marked as a *base node*. If it exists, it is marked as a *twin node*, and a pointer to its base node is created. Thus, any twin node points to its base node, but a base node does not point to any of its twin nodes. Only base nodes can be expanded. The difference with the "standard implementation" of a DCG is that if two or more nodes are representing the same position (ignoring history) they are not merged into one node. However, their subtree is generated only once. In general, a twin node may have a value different from its base node, although they represent the same position.

Figure 5 exhibits our implementation of the DCG given in Figure 4 (assuming that the position corresponding with node F is first generated as child of node C and only later as child of node D). Nodes in upper-case are base nodes, nodes in lower-case are twin nodes. The dashed arrows are pointers from twin nodes to base nodes. The problem mentioned in Figure 4 can now be handled by assigning separate values to nodes F and f, and to C and c, depending on the paths leading to the corresponding positions.

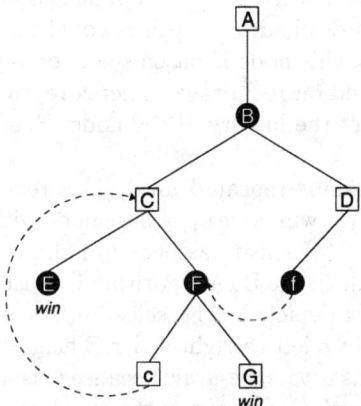

Fig. 5. Our DCG with base nodes and twin nodes corresponding with the DCG of Figure 4.

4.2 The BTA Algorithm as Solution

As stated before, encountering a repetition of positions in node p does not mean that the repetition signals a *real* draw (defined as the inevitability of a repetition of positions under optimal play). To handle the distinction, we introduce the new concept of *possible-draw*. Node p is marked as a *possible-draw* if the node is a repetition of a node P in the search path. (Whether a possible draw also is a real draw depends on the history.) Then the depth of node P in the search path (termed the *possible-draw depth*) is stored in node p.

The BTA algorithm for best-first search consists of three phases. Phase 1 deals with the selection of a node. Phase 2 evaluates the selected node. Phase 3 backs up the new information through the search path. The three phases are repeatedly executed until the search process is terminated.

Phase 1: Select the Best Node In phase 1 a node is selected for evaluation.[5] or First, the root is selected (for further selection, see below). Then, for each selected node, two cases exist:

1. if a child of the selected node is marked as a *possible-draw*, and the remaining children are either real draws, or marked as *possible-draws*, then the selected node is marked as a *possible-draw* and the corresponding possible-draw depth is set to the minimum of the possible-draw depths of the children. Subsequently, all possible-draw markings from the children are removed and the parent of the selected node is re-selected for investigation;
2. otherwise, the best child is selected for investigation, ignoring the children which are either real draws, or marked as *possible-draws*.

Assume that a node at depth d in the search path is marked as a *possible-draw* and the corresponding possible-draw depth is equal to d. This implies that the possible-draw marking of this node is based solely on repetitions of positions *in the subtree of the node* and on real draws. Therefore, the node is a real draw by repetition, independent of the history of the node. Hence the node is evaluated accordingly.

The selection of a node is repeated until (1) a real draw by repetition has been encountered, or (2) (a twin node of) a base node with known game-theoretic value has been found[6], or (3) a leaf has been found.

The selection of a node in the BTA algorithm is illustrated below. In Figure 6 part of a search graph is depicted. The selection starts at the root (node A). Assume the traversal is in a left-to-right order. Then, at a certain point, node c is selected, and marked as a *possible-draw* because it is a repetition of node C at depth two in the search path (see Figure 6; the equal sign represents the possible-draw marking and the subscript two represents the possible-draw depth).

After marking node c as a *possible-draw*, the parent of this node (node D) is re-selected and marked as a *possible-draw*, with the same possible-draw depth as node c. Further, the possible-draw marking of node c is removed. After marking node D as a *possible-draw*, its parent C is re-selected. The next best child (not

---

[5] We assume that the selection of a node proceeds in a top-down fashion.
[6] This is possible, because a base node does not point to its twin nodes. If the game-theoretic value of a twin node becomes known, its corresponding base node is evaluated accordingly, but other twin nodes remain unchanged.

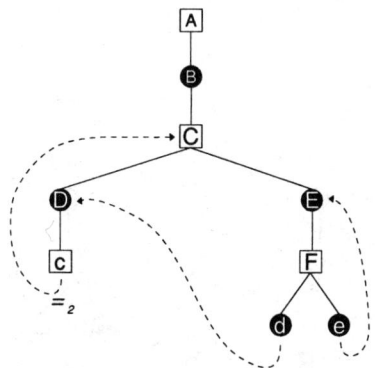

Fig.6. Encountering the first repetition c.

marked as a *possible-draw*) E is selected. Continuing this procedure, at a certain point child d of node F is selected. The child c of twin node d is found by directing the search to the base node D of node d. Node c is (again) marked as a *possible-draw* because it is a repetition of node C at depth two in the search path. See Figure 7.

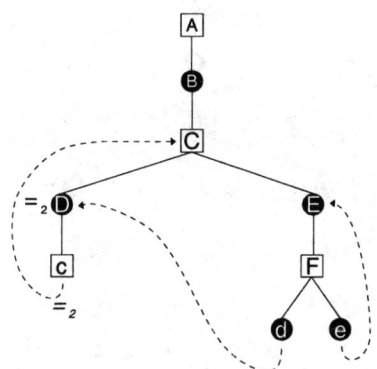

Fig.7. Encountering the second repetition c.

After the re-marking of node c as a *possible-draw*, the parent of this node (twin node d) is re-selected and marked as a *possible-draw*, with the same possible-draw depth as node c. Thereafter, the possible-draw marking of node c is removed (for the second time). After marking node d as a *possible-draw*, its parent F is re-selected. The next best child (not marked as a *possible-draw*) e is selected. This node is a repetition of node E at depth three in the search path, and is now marked as a *possible-draw*. See Figure 8.

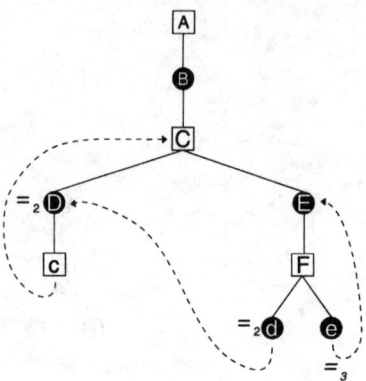

Fig.8. Encountering the repetition e.

After marking node e as a *possible-draw*, the parent of this node (node F) is re-selected. All its children are marked as a *possible-draw*. Therefore, node F is also marked as a *possible-draw*, with a possible-draw depth of two (the minimum of the possible-draw depths of the children). Further, the possible-draw markings of all children are removed. See Figure 9.

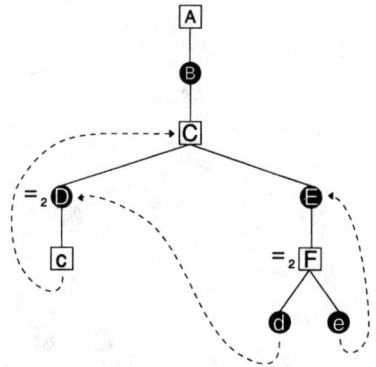

Fig.9. Marking node F as a *possible-draw*.

After marking node F as a *possible-draw*, the parent of this node (node E) is re-selected and marked as a *possible-draw*, with the same possible-draw depth as node F. Subsequently, the possible-draw marking of node F is removed. After marking node E as a *possible-draw*, its parent (node C) is re-selected. However, all its children are marked as a *possible-draw*. Therefore, node C is also marked as a *possible-draw*, with a possible-draw depth of two (the minimum of the possible-draw depths of the children). Again, the possible-draw markings of all children are removed. See Figure 10.

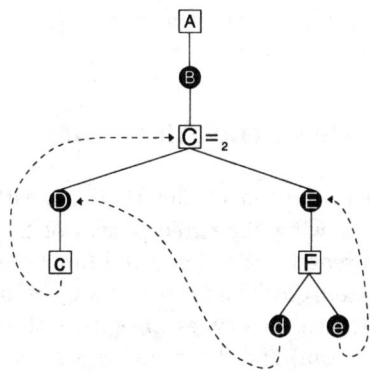

Fig.10. Marking node C as a *possible-draw*.

Now the selection process finishes, since node C is marked as a *possible-draw* and its corresponding possible-draw depth is equal to the depth of the node in the search path. This means that *all* continuations from C lead, in one or another way, to repetitions occurring
in the subtree of node C. Therefore, node C is evaluated as a real draw by repetition, independent of the history of the node, but on the basis of its potential continuations.

Phase 2: Evaluate the Best Node In phase 2 the selected node (say P) is evaluated. values Three cases are distinguished.

1. If P is a real draw by repetition, it is evaluated as a draw. The corresponding base node (if existing) is also evaluated as a draw.
2. If P is a twin node and its corresponding base node is a terminal node, P becomes a terminal node as well and is evaluated as such.
3. If P is a leaf, it is expanded, the children are evaluated and P is evaluated using the evaluation values of the children.

Phase 3: Back up the New Information In phase 3 the value of the selected node is updated to the root[7] and all possible-draw markings are removed. more lead the is In contrast to the tree algorithm, in the BTA updating process nodes marked as a *possible-draw* may occur. The back-up value of a node is determined by using only the evaluation values of children not marked as a *possible-draw*. in Thus, the children marked as a *possible-draw* are ignored, because in the next iteration the search could be mistakenly directed to one of these children, whereas this child was a repetition in the current path, not giving any new information.

---

[7] In a DCG there can exist more than one path from a node to the root. However, only the path along which the node was selected is taken into account. Other paths, if any, may be updated after other selection processes.

After establishing the back-up value of a node, the possible-draw markings of the children are removed.

5 The Pseudo-Code of the Algorithm

In this section an implementation of the BTA algorithm in pn search [1] is given. An explanation following the three phases of Section 4 provides details on the seven relevant pn-search procedures and functions. For chess, the goal of pn search is finding a mate. A loss and a real draw are in this respect equivalent (i.e., they are no win). Hence, two types of node with a known game-theoretic value exist: proved nodes (*win*) and disproved nodes (*no win possible*). A proved or disproved node is called a solved node.

5.1 Phase 1: Select the Most-Proving Node

Phase 1 of the algorithm deals with the selection of a (best) node for evaluation. This node is termed the most-proving node. In Figure 11 the main BTA pn-search algorithm is shown. The only parameter of the procedure is root, being the root of the search tree. The BTA algorithm resembles the tree algorithm described in [1], a difference being that procedure UpdateAncestors is called with the *parent* of the most-proving node as parameter instead of the most-proving node itself, since the most-proving node already has been evaluated in procedure ExpandNode.

```
procedure BTAProofNumberSearch( root )
  Evaluate( root )
  SetProofAndDisproofNumbers( root )
  root.expanded := false
  root.depth := 0

  while root.proof ≠ 0 and root.disproof ≠ 0 and
    ResourcesAvailable() do begin
    mostProvingNode := SelectMostProvingNode( root )
    ExpandNode( mostProvingNode )
    UpdateAncestors( mostProvingNode.parent, root )
  end

  if root.proof=0 then root.value := true
  elseif root.disproof=0 then root.value := false
  else root.value := unknown /* resources exhausted */
end /* BTAProofNumberSearch */
```

Fig.11. The BTA pn-search algorithm for DCGs.

The procedures Evaluate and SetProofAndDisproofNumbers and the function ResourcesAvailable are identical to the same procedures and function in the standard tree algorithm (see [7]), and not detailed here. The function SelectMostProvingNode finds a most-proving node according to certain conditions. The function is given in Figure 12. The only parameter of the function is node, being the root of the (sub)tree where the most-proving node is located.

```
function SelectMostProvingNode( node )
  if NodeHasBaseNode( node )
  then baseNode := BaseNode( node )
  else baseNode := node
  /* 1: Base node has been solved */
  if baseNode.proof=0 or baseNode.disproof=0
  then return node
  elseif Repetition( node )
  then begin /* 2: Repetition of position */
    MarkAsPossibleDraw( node )
    ancestorNode := FindEqualAncestorNode( node )
    node.pdDepth := ancestorNode.depth
    return SelectMostProvingNode( node.parent )
  end elseif not baseNode.expanded then /* 3: Leaf */
    return node
  else begin /* 4: Internal node; look for child */
    bestChild := SelectBestChild( node, baseNode, pdPresent )
    if bestChild=NULL then begin
      if pdPresent then begin
        MarkAsPossibleDraw( node )
        node.pdDepth := ∞
        for i:=1 to baseNode.numberOfChildren do begin
          if PossibleDrawSet( baseNode.children[ i ] ) then
            if baseNode.children[ i ].pdDepth < node.pdDepth then
              node.pdDepth := baseNode.children[ i ].pdDepth
            UnMarkAsPossibleDraw( baseNode.children[ i ] )
        end
        if node.depth = node.pdDepth then return node
        else return SelectMostProvingNode( node.parent )
      end else begin
        /* All children are solved, so choose any one */
        baseNode.proof := baseNode.child[ 1 ].proof
        baseNode.disproof := baseNode.child[ 1 ].disproof
        return node
      end
    end else begin
      bestChild.depth := node.depth+1
      return SelectMostProvingNode( bestChild )
    end
  end
end /* SelectMostProvingNode */
```

Fig. 12. The function SelectMostProvingNode.

The function starts to examine whether the node under investigation (say P) is a twin node. If so, then the investigation proceeds with the associated base node.

If P has been solved (case 1), P is returned, because the graph has to be backed up using this new information.

If P has not been solved, it is examined whether P is a repetition in the current path (case 2). If so, it is marked as a *possible-draw*. Its ancestor transposition node in the current path is looked up, and the pdDepth (possible-draw depth) of the node becomes equal to the depth in the search path of the ancestor node[8]. Since it is not useful to examine a repetition node further, the selection of the most-proving node is directed to the parent of P.

If P has not been solved and is not a repetition in the current path, it is examined whether P is a leaf (case 3). If so, P is the most-proving node which has to be expanded, and P is returned.

Otherwise (case 4), the best child is selected by the function SelectBestChild, to be discussed later. If no best child was found, it means that every child is either solved (proved in case of an AND node, and disproved in case of an OR node) or is marked as a *possible-draw*. If any of the children is marked as a *possible-draw*, P is marked as a *possible-draw* as well. The pdDepth of the node is set to the minimum of the children's pdDepths and the markings of *all* children are removed, *etc.* See Section 4. is

In Figure 13 the function SelectBestChild is listed. The function has three parameters. The first parameter (node) is the parent from which the best child will be selected. The second parameter (baseNode) is the base node of that parent[9]. Finally, the third parameter (pdPresent, meaning possible draw present) indicates whether one of the children is marked as a *possible-draw*. The parameter pdPresent is initialized by the function SelectBestChild. If the node is an OR node, a child marked as a *possible-draw* will not be selected as best child, since it gains nothing and the goal (win) cannot be reached. A best child (of an OR node) is a child with the lowest proof number. If the node is an AND node, a child marked as a *possible-draw* is a best child, since the player to move in the AND node is satisfied with a repetition (thereby making it impossible for the opponent to reach the goal). Otherwise, a best child (of an AND node) is a child with the lowest disproof number. This best child is returned. If the best child is either solved or marked as a *possible-draw*, NULL is returned.

5.2 Phase 2: Evaluate the Most-Proving Node

After the most-proving node has been found, it has to be expanded and evaluated. Phase 2 of the algorithm performs this task. Figure 14 provides the procedure ExpandNode. The only parameter is node, being the node to be expanded.

---

[8] The variable pdDepth will act as an indicator of the lowest level in the tree at which there are nodes having repetition nodes in their subtrees.

[9] We note that if the parent is a base node itself, then the base node is equal to the parent.

```
function SelectBestChild( node, baseNode, pdPresent )
  bestChild := NULL
  bestValue := ∞
  pdPresent := false
  if node.type=OR then begin /* OR node */
    for i := 1 to baseNode.numberOfChildren do begin
      if PossibleDrawSet( baseNode.children[ i ] ) then
        pdPresent := true
      elseif baseNode.children[ i ].proof < bestValue
      then begin
        bestChild := baseNode.children[ i ]
        bestValue := bestChild.proof
      end
    end
  end else begin /* AND node */
    for i := 1 to baseNode.numberOfChildren do begin
      if PossibleDrawSet( baseNode.children[ i ] ) then begin
        pdPresent := true
        break
      end
      if baseNode.children[ i ].disproof < bestValue
      then begin
        bestChild := baseNode.children[ i ]
        bestValue := bestChild.disproof
      end
    end
  end

  return bestChild
end /* SelectBestChild */
```

Fig.13. The function SelectBestChild.

The procedure starts establishing the base node of the node.[10] If the base node is solved (case 1), the node is evaluated accordingly.

Otherwise, if the node is marked as a *possible-draw* (case 2) (and since it was chosen by function SelectMostProvingNode), it is evaluated as a real draw.

In case 3 the node has to be expanded. All children are generated, and evaluated. If a generated child has no corresponding base node, the attribute expanded is initialized to false; if it has a corresponding base node, the attribute expanded has been initialized before. Then the node itself is initialized by procedure SetProofAndDisproofNumbers.

---

[10] We note that if the node is a base node itself, then the base node is equal to the node.

```
procedure ExpandNode( node )
  if NodeHasBaseNode( node )
  then baseNode := BaseNode( node )
  else baseNode := node

  if baseNode.proof=0 or baseNode.disproof=0 then begin
    /* 1: base node already solved */
    node.proof := baseNode.proof
    node.disproof := baseNode.disproof
  end elseif PossibleDrawSet( node ) then begin
    /* 2: node has become a real draw */
    node.proof := ∞
    node.disproof := 0
    baseNode.proof := ∞
    baseNode.disproof := 0
  end else begin
    /* 3: node has to be expanded */
    GenerateAllChildren( baseNode )
    for i:=1 to baseNode.numberOfChildren do begin
      Evaluate( baseNode.children[ i ] )
      SetProofAndDisproofNumbers( baseNode.children[ i ] )
      if not NodeHasBaseNode( baseNode.children[ i ] ) then
        baseNode.children[ i ].expanded := false
    end
    SetProofAndDisproofNumbers( baseNode )
    baseNode.expanded := true
    node.proof := baseNode.proof
    node.disproof := baseNode.disproof
  end
end /* ExpandNode */
```

Fig.14. The procedure ExpandNode.

5.3 Phase 3: Back up the New Information

Phase 3 of the algorithm has as task to back up the evaluation value of the most-proving node. The procedure to update the values of the nodes in the path is listed in Figure 15. The procedure has two parameters. The first parameter (node) is the node to be updated, while the second parameter (root) is the root of the search tree. Depending on the node type, UpdateOrNode (Figure 16) or UpdateAndNode (Figure 17) is performed.

The parameters of UpdateOrNode are node and baseNode. The algorithm basically is the same as the OR part of procedure SetProofAndDisproofNumbers. It only differs when a child is marked as a *possible-draw*. In that case, the child is discarded so its value is not used when calculating the back-up value of the node. Then, the *possible-draw* marking of the child is removed. If the node appears to be disproved (since all children are either disproved or marked as a *possible-draw*)

```
procedure UpdateAncestors( node, root )
  while node ≠ nil do begin
    if NodeHasBaseNode( node )
    then baseNode := BaseNode( node )
    else baseNode := node

    if node.type=OR
    then UpdateOrNode( node, baseNode )
    else UpdateAndNode( node, baseNode )

    node := node.parent /* parent in current path */
  end
  if PossibleDrawSet( root ) then
    UnMarkAsPossibleDraw( root )
end /* UpdateAncestors */
```

Fig.15. The procedure UpdateAncestors.

```
procedure UpdateOrNode( node, baseNode )
  min := ∞
  sum := 0
  pdPresent := false
  for i:=1 to baseNode.numberOfChildren do begin
    if PossibleDrawSet( baseNode.child[ i ] ) then begin
      pdPresent := true
      proof := ∞
      disproof := 0
      UnMarkAsPossibleDraw( baseNode.child[ i ] )
    end else begin
      proof := baseNode.child[ i ].proof
      disproof := baseNode.child[ i ].disproof
    end
    if proof < min then min := proof
    sum := sum + disproof
  end
  if min=∞ and pdPresent then
    SetProofAndDisproofNumbers( node )
  else begin
    node.proof := min
    node.disproof := sum
  end
  if node.proof=0 or node.disproof=0
  then begin /* node solved */
    baseNode.proof := node.proof
    baseNode.disproof := node.disproof
  end
end /* UpdateOrNode */
```

Fig.16. The procedure UpdateOrNode.

and a repetition child exists, the value of the node is calculated by procedure SetProofAndDisproofNumbers. Otherwise, the value has been calculated correctly. If the node has been solved, its base node is evaluated accordingly.

```
procedure UpdateAndNode( node, baseNode )
  min := ∞
  sum := 0
  for i:=1 to baseNode.numberOfChildren do begin
    proof := baseNode.child[ i ].proof
    disproof := baseNode.child[ i ].disproof
    sum := sum + proof
    if disproof < min then min := disproof
  end

  node.proof := min
  node.disproof := sum
  if node.proof=0 or node.disproof=0
  then begin /* node solved */
    baseNode.proof := node.proof
    baseNode.disproof := node.disproof
  end
end /* UpdateAndNode */
```

Fig.17. The procedure UpdateAndNode.

The two parameters of UpdateAndNode are equal to the parameters of procedure UpdateOrNode. The procedure differs from the AND part of the procedure SetProofAndDisproofNumbers when the node is solved, and hence the value of its base node is evaluated accordingly[11].

6 Experimental

6.1 The Proof-Number Search Engine

The proof-number search engine has been implemented in a straightforward chess program. The only goal of the pn-search algorithm is searching for mate. We distinguish between the attacker and the defender. A position is proved if the attacker can mate, while draws (stalemate, repetition of positions, and the 50-move rule) and mates by the defender are defined to be disproved positions for the attacker. the tree information thus

---

[11] We note that it is impossible for a child of an AND node to be marked as a *possible-draw*, since in that case the search for a most-proving node would have been terminated in an earlier phase, and the parent already would have been marked as a *possible-draw*.

6.2 The Test Set

Since proof-number search operates best when searching for mates in chess [5], we used a set of mating problems [11,15]. Krabbé's 35 positions are indicated by kx, in which x refers to the diagram number in the source. The diagrams are 8, 35, 37, 38, 40, 44, 60, 61, 78, 192, 194, 195, 196, 197, 198, 199, 206, 207, 208, 209, 210, 211, 212, 214, 215, 216, 217, 218, 219, 220, 261, 284, 317, 333 and 334. Reinfeld's 82 positions are indicated by rx, x again referring to the problem number in the source, this time running over 1, 4, 5, 6, 9, 12, 14, 27, 35, 49, 50, 51, 54, 55, 57, 60, 61, 64, 79, 84, 88, 96, 97, 99, 102, 103, 104, 105, 132, 134, 136, 138, 139, 143, 154, 156, 158, 159, 160, 161, 167, 168, 172, 173, 177, 179, 182, 184, 186, 188, 191, 197, 201, 203, 211, 212, 215, 217, 218, 219, 222, 225, 241, 244, 246, 250, 251, 252, 253, 260, 263, 266, 267, 278, 281, 282, 283, 285, 290, 293, 295 and 298. This results in a test set of 117 positions.

6.3 The Setting

Our BTA algorithm, denoted by *BTA*, is compared with the following three pn-search variants:

1. the standard tree algorithm, denoted by *Tree*,
2. a DAG algorithm, developed by Schijf [16], denoted by *DAG*, and
3. an (incorrect) DCG algorithm, developed by Schijf et al. [17], denoted by *DCG*.

The results for the DAG and DCG algorithm will be taken from the literature [16,17]. In all implementations, the move ordering is identical. All four algorithms searched for a maximum of 500,000 nodes per test position. After 500,000 nodes the search was terminated and the problem was marked as not solved. Under these conditions 10 positions (k8, k40, k78, k195, k209, k210, k220, r96, r105, r201) turned out to be not solvable by any of the four algorithms. Therefore they are not taken into account in the next section.

7 Results

To verify our solution we have first tested the position given in Figure 1[12]. *Tree* finds a solution within 482,306 nodes. *DCG*, ignoring the history of a position, incorrectly states that White cannot win (due to the GHI problem). Our *BTA* does find a solution within 10,694 nodes. This provides evidence that this occurrence of the GHI problem has been correctly handled. *BTA* shows the benefit of being a DCG algorithm, as evidenced by the decrease in number of nodes investigated by a factor of roughly 40 as compared to *Tree*.

---

[12] We note that for this problem the goal for White was set to promotion to Queen (without Black being able to capture it on the next ply) instead of mate. Further, the search was restricted to the 5×5 a4–e8 board. This helps to find the solution faster, but does not influence the occurrence of the GHI problem.

Thereafter, we have performed the experiments with the test set described in Section 6.2. The outcomes are summarized in Table 1. The first column shows the four pn-search variants. The number of positions solved by each algorithm is given in the second column. Exactly 96 positions were solved by all four algorithms. In the third column the total number of nodes evaluated for the 96 positions are listed. The additional positions solved per algorithm are as follows.

- For *Tree*: k208, k215, r281;
- for *DAG*: k208, k215, k216, r168, r182, r281;
- for *DCG*: k44, k60, k217, k284, r168, r182, r252;
- for *BTA*: k44, k60, k208, k215, k216, k217, k284, r168, r182, r252, r281.

	# of pos. solved (out of 117)	Total nodes (96 positions)
Tree	99	4,903,374
DAG	102	3,222,234
DCG	103	2,482,829
BTA	107	2,844,024

Table1. Comparing four pn-search variants.

Obviously, *Tree* investigates the largest number of nodes, the easy explanation being that this algorithm does not recognize transpositions. Further, *DCG* examines the smallest number of nodes: this algorithm sometimes prematurely disproves positions; hence, on the average less nodes have to be examined. However, if such a prematurely disproved position does lead to a win and the node is important to the principal variation of the tree, the win can be missed, as happens in the positions k208, k215, k216 and r281. This is already mentioned by Schijf *et al.* [17].

From Table 1 it further follows that *BTA* performs best. It solves each position which was solved by at least one of the other three algorithms. Furthermore, the four positions which were incorrectly disproved by *DCG* were proved by *BTA*. Compared to the tree algorithm, *BTA* solves eight additional positions and uses only 58% of the number of nodes: a clear improvement. The reduction in nodes compared to *DAG* is still 11.7%. The increase in nodes searched relative to *DCG* (12.7%) is already explained by the unreliability of the latter. We feel that the advantage of the larger number of solutions found heavily outweighs the disadvantage of the increase in nodes searched. We note that the selection of the most-proving node in *BTA* can be costly in positions with many possible transpositions. However, in these types of position the reduction in the number of nodes searched is even larger than in "normal" positions.

As a case in point we present Figure 18 corresponding with Diagram 216 in Krabbé [11]. It is solved by our BTA algorithm (in 247,686 nodes) and by the DAG algorithm (in 366,336 nodes) and not by the two other algorithms (within

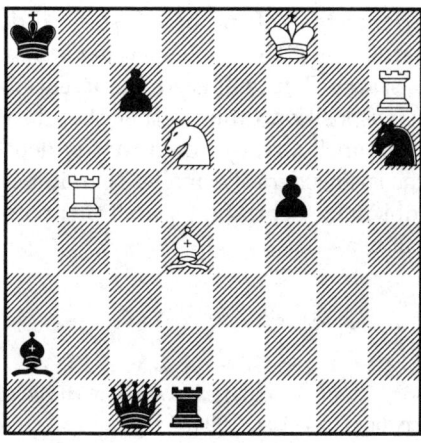

Fig.18. Mate in 14 (WTM); (J. Kriheli).

500,000 nodes). Many transpositions (and many repetitions of positions) exist, since after **1. ♖a5+ ♔b8** White has a so-called *zwickmühle* and can position the Bishop anywhere along the a7-g1 diagonal for free. For instance, after **2. ♗a7+ ♔a8 3. ♗b6+ ♔b8** almost the same position with the same player to move is reached: the Bishop has moved from d4 to b6. At any time White can choose such a manoeuvre. For the chess-playing reader, the solution is **1. ♖a5+! ♔b8 2. ♗a7+ ♔a8 3. ♗c5+! ♔b8 4. ♖b5+ ♔a8 5. ♖e7! ♗f7 6. ♖a5+ ♔b8 7. ♗a7+ ♔a8 8. ♗d4+! ♔b8 9. ♖b5+ ♔a8 10. ♖d7! ♕g5 11. ♖a5+ ♔b8 12. ♗a7+ ♔a8 13. ♗b6+ ♔b8 14. ♗xc7 mate.**

8 Conclusions

In this article we have given a solution to the GHI problem, resulting in an improved DCG algorithm for pn search, denoted BTA (Base-Twin Algorithm). It is shown that in a well-known position, in which the GHI problem occurs when a naïve DCG algorithm is used, our BTA algorithm finds the correct solution. The results on a test set of 117 selected positions support our claim. Despite the additional overhead to recognize positions suffering from the GHI problem, our BTA algorithm is hardly less efficient than other, non-reliable, DCG algorithms, and finds more solutions.

We note that, though our algorithms are confined to pn search, the strategy used is generally applicable to any best-first search algorithm. The only important criterion for application is that a DCG is being built according to the best-first principle (choose some leaf node, expand that node, evaluate the children, and back up the result). We consider the GHI problem in best-first search to be solved. The importance of this statement is that with increasing availability of computer memory a growing tendency exists to use best-first search algorithms, like SSS* [18] and variants thereof, or best-first fixed-depth algorithms [14], which no longer suffer from the GHI problem. What remains is solving the GHI problem for depth-first search. This will need a different approach, storing additional information in transposition tables rather than in the search tree/graph in memory. However, Campbell already noted that in depth-first search the frequency of GHI problems is considerably smaller than in best-first search [8]. The solution of the GHI problem for depth-first search therefore seems to be of minor importance for practical use.

References

1. L.V. Allis, M. van der Meulen, and H.J. van den Herik. Proof-Number Search. *Artificial Intelligence*, 66(1):91–124, 1994.
2. L.V. Allis. *Searching for Solutions in Games and Artificial Intelligence*. Ph.D. thesis, University of Limburg, Maastricht, The Netherlands, 1994.
3. E.B. Baum and W.D. Smith. Best Play for Imperfect Players and Game Tree Search. Submitted to *Artificial Intelligence*.
4. H.J. Berliner and C. McConnell. B* Probability Based Search. *Artificial Intelligence*, 86(1):97–156, 1996.
5. D.M. Breuker, L.V. Allis, and H.J. van den Herik. How to Mate: Applying Proof-Number Search. In H.J. van den Herik, I.S. Herschberg, and J.W.H.M. Uiterwijk, editors, *Advances in Computer Chess 7*, pages 251–272, Maastricht, The Netherlands, 1994. University of Limburg.
6. D.M. Breuker, H.J. van den Herik, L.V. Allis, and J.W.H.M. Uiterwijk. A Solution to the GHI Problem for Best-First Search. *Technical Report* #CS 97-02, Computer Science Department, Universiteit Maastricht, Maastricht, The Netherlands, 1997.
7. D.M. Breuker. *Memory versus Search in Games*. Ph.D. thesis, Universiteit Maastricht, The Netherlands, 1998.
8. M. Campbell. The Graph-History Interaction: On Ignoring Position History. In *1985 Association for Computing Machinery Annual Conference*, pages 278–280, 1985.
9. R.M. Hyatt, A.E. Gower, and H.L. Nelson. Cray Blitz. In D.F. Beal, editor, *Advances in Computer Chess 4*, pages 8–18, Oxford, United Kingdom, 1984. Pergamon Press.
10. B. Kažić, R. Keene, and K.A. Lim. *The Official Laws of Chess and Other FIDE Regulations*. B.T. Batsford Ltd., London, United Kingdom, 1985.
11. T. Krabbé. *Chess Curiosities*. George Allen and Unwin Ltd., London, United Kingdom, 1985.
12. T.A. Marsland. A Review of Game-Tree Pruning. *ICCA Journal*, 9(1):3–19, 1986.
13. A.J. Palay. *Searching with Probabilities*. Ph.D. thesis, Boston University, Boston MA, USA, 1985.
14. A. Plaat, J. Schaeffer, W. Pijls, and A. de Bruin. Best-First Fixed-Depth Minimax Algorithms. *Artificial Intelligence*, 87(2):255–293, 1996.
15. F. Reinfeld. *Win at Chess*. Dover Publications Inc., New York NY, USA, 1958. Originally published (1945) as *Chess Quiz* by David McKay Company, New York NY, USA.
16. M. Schijf. *Proof-Number Search and Transpositions*. M.Sc. thesis, University of Leiden, Leiden, The Netherlands, 1993.

17. M. Schijf, L.V. Allis, and J.W.H.M. Uiterwijk. Proof-Number Search and Transpositions. *ICCA Journal*, 17(2):63–74, 1994.
18. G. Stockman. A Minimax Algorithm Better than Alpha-beta? *Artificial Intelligence*, 12:179–196, 1979.
19. K. Thompson. *Personal communication*, 1995.

Optimal Play against Best Defence: Complexity and Heuristics

Ian Frank[1] and David Basin[2]

[1] Complex Games Lab, ETL, Umezono 1-1-4, Tsukuba, Ibaraki, Japan 305
ianf@etl.go.jp
[2] Institut für Informatik, Universität Freiburg, Am Flughafen 17, Freiburg, Germany
basin@informatik.uni-freiburg.de

Abstract. We investigate the best defence model of an imperfect information game. In particular, we prove that finding optimal strategies for this model is NP-complete in the size of the game tree. We then introduce two new heuristics for this problem and show that they outperform previous algorithms. We demonstrate the practical use and effectiveness of these heuristics by testing them on random game trees and on a hard set of problems from the game of Bridge. For the Bridge problem set, our heuristics actually outperform the human experts who produced the model solutions.

Keywords: imperfect information, heuristics, complexity, Bridge

1 Introduction

We analyse the problem of finding optimal strategies for two player games of the form illustrated in Figure 1. In this game, played between MAX (square nodes) and MIN (circular nodes), the terminal values indicate the payoff to MAX of reaching each leaf node under any of five possible *worlds* $w_1, \cdots, w_5$. The state of the world that actually holds depends on information that MAX does not know, but to which he can attach a probability distribution (for example the toss of a coin or the deal of a deck of cards). For a more general game with n possible worlds, every leaf node of the tree has n payoffs, each corresponding to the utility for MAX of reaching that node in one of the n worlds.

An example of a real-life game that can be fit to the model of Figure 1 is Bridge (for a concrete illustration, see §2). Recently, the assumptions used by human experts to analyse Bridge problems have been formalised in a *best defence model* [6], which we summarise here:

A–I MIN has perfect information.
A–II MIN chooses his strategy after MAX.
A–III The strategy adopted by MAX is a pure strategy.

This model is described as 'best defence' because it represents the strongest possible assumptions about the opponent — that MIN knows the actual state of the world (A-I) and chooses his strategy in the knowledge of MAX's choices

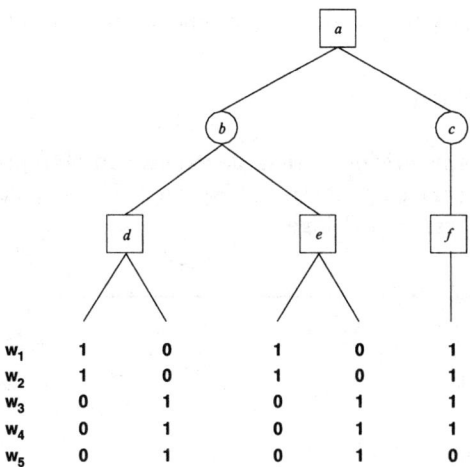

Fig. 1. A game tree with five possible worlds

(A-II). It is used by human players because modelling the strongest possible opponents provides a lower bound on the payoff that can be expected when the opponents are less informed. The assumption that MAX chooses a pure strategy (A-III) also restricts the set of possible solutions to a finite (though possibly very large) set.

In this paper we review existing algorithms for analysing trees such as that of Figure 1. These algorithms are all heuristic in nature. To show that such heuristics are the best that can realistically be hoped for, we examine the complexity of the problem and demonstrate that finding optimal strategies for the best defence model is NP-complete in the size of the game tree. This means that arbitrary (imperfect information) *trees* are hard to analyse. This is in contrast to perfect information games, where arbitrary trees are easy to analyse (the minimax algorithm returns results in time linear in the size of the tree), but arbitrary *games* are hard to analyse, at least when their definitions generate trees of exponential size.

To combat such intractability, we introduce two new heuristics, which we call *beta-reduction* and *iterative biasing*. Both of these tackle the *non-local* nature [6] of the types of game shown in Figure 1 by introducing dependencies between the choices made at MAX nodes. We demonstrate the practical importance of our new heuristics theoretically and experimentally, using both simple game trees and a large database of problems from the game of Bridge. For the Bridge database, we find that a combination of our heuristics actually outperforms the human experts that produced the model solutions. In the past, special-purpose Bridge programs for identifying complex positions such as *squeezes* have been developed [16], but our heuristics represent the first general tree search algorithm capable of consistently performing at and above expert level in actual card play.

2 A Bridge Example

To show how problems arising in real-life games can be represented in the form of Figure 1, consider the example tree of Figure 2. This depicts a simple situation in a single suit of a game of Bridge.

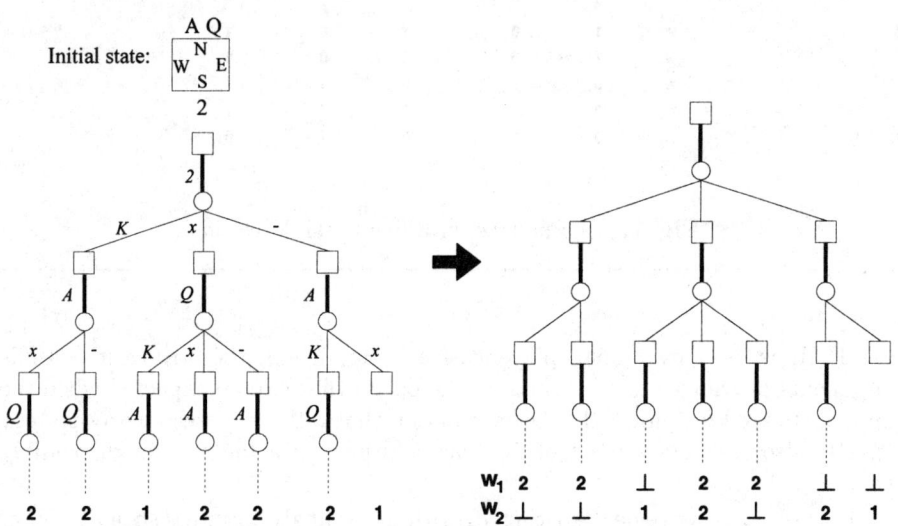

Fig. 2. The optimal MAX strategy for a single-suit Bridge problem represented (left) using constraints on the MIN branches, and (right) using possible worlds at leaf nodes

Such single-suit problems are common in the Bridge literature; the task is to find the optimal way to play just the cards in one suit, ignoring possible influences from other suits (such as *ruffing* or *entry* requirements). It is also assumed that the opponents do not initiate play in the suit. In our example, a single MAX player controls both the North and the South cards (Ace, Queen, and 2) against the two defenders East and West (who are assumed to hold the King and the remaining nine low cards).

To produce the tree of possible moves in Figure 2, we have made the natural simplification (which does not affect the analysis of the problem) that East and West have at most three distinct options at any point in the game: to play the King ('K'), a low card ('x'), or a card from a different suit ('-'). Even then, however, the complete tree for this problem has 76 leaf nodes, so we have also made the further simplification of displaying only the MAX branches that form part of the optimal strategy. This optimal strategy starts by playing the two from the South hand, and then playing the Ace from North hand if West plays a King or a card from a different suit. If West plays a low card, however, the Queen is played from the North hand (this is an example of a manœuvre called a *finesse*). After a play by the second opponent (East), the highest of the four cards played to date is said to win a *trick* for either North-South or East-West, and the next trick begins. In our example, North-South will then either be able to gain a trick by cashing the Ace or the Queen, or will have no further options. The total number of North-South tricks is indicated at the leaf nodes of the tree.

The game tree formed in this way is somewhat nonstandard in that all of the branches at MIN nodes can only be followed if certain conditions are true. For instance, East can only play the King if East actually holds the King. Applying the standard minimax algorithm to the tree without respecting these conditions results in a value of 1. However, we have some knowledge about the constraints on MIN's available moves, namely that they are the result of a chance move (the deal of the cards). If we do not distinguish between the nine low cards, this chance move can result in 20 distinct possibilities (for either East or West, they may hold between 0 and 9 low cards, and either hold the King or not). Rather than list the payoffs for each of these twenty worlds in the figure, we have again simplified the presentation by instead considering the two mutually exclusive possibilities of "West holds the King" (w_1) and "East holds the King" (w_2). On the right of our figure, we have included a second game tree with payoffs at the leaf nodes for just these two worlds, using the symbol $\perp$ to represent branches that cannot be followed. This shows that two tricks can always be made when West holds the King, but only one trick when East holds the King.

The second game tree of Figure 2 is now in a format similar to that of Figure 1. Indeed, the only substantial difference is that some of the payoffs take the undefined value $\perp$. In terms of basic game theory [18,14], both this tree and the tree of Figure 1 are just compact ways of representing the extensive form of a particular kind of two-person, zero-sum game with imperfect information. For the example of Figure 1, the extensive form tree would have a single chance move at the root, and $n = 5$ identically shaped subtrees (the payoff n-tuples can be assigned at each leaf node so that the ith component is the payoff for that leaf in the ith subtree). In the Bridge example, there is also a single chance move (the deal of the cards), but the moves that are possible in each world are different, giving rise to some $\perp$ values as payoffs.

So, game trees like those given in Figures 1 and 2 can be used to represent interesting and nontrivial games. The rest of this paper will therefore investigate the complexity of playing such games optimally, and introduce new algorithms for finding optimal strategies. Throughout our analysis of such games, we will assume that the only move with an unknown outcome is the chance move that starts the game. Thus, a single node in one of our trees represents between 1 and n *information sets*: one if the player whose turn it is to move knows the exact outcome of the chance move, and n if the player has no knowledge. Since our analysis will make repeated reference to algorithms such as minimaxing, we will make one further notational convention. Throughout the paper, the normal min and max functions will be extended so that $\min(\bot,\bot) = \max(\bot,\bot) = \bot$, and $\min(x,\bot) = \min(\bot,x) = \max(x,\bot) = \max(\bot,x) = x$ for all $x \neq \bot$.

3 Review of Existing Algorithms

The play of imperfect information games has been analysed by a number of researchers [4,17,12]. Here, we review the well-established technique of Monte-carlo sampling, and then describe two algorithms that we have recently demonstrated [8] are better at finding optimal strategies. This section contains no new results, but since the developments in this field are relatively recent, we summarise them in some detail.

3.1 Monte-Carlo Sampling

One technique for handling imperfect information is Monte-carlo sampling [3]. This approach has been used in games such as Scrabble (see [4]) and Bridge, where it was proposed by Levy [13] and recently implemented by Ginsberg [10]. In the context of game trees like that of Figure 1, Monte-carlo sampling consists of guessing a possible world and then finding a solution to the game tree for this complete information sub-problem. This is much easier than solving the original game, since if attention is restricted to just one world, the minimax algorithm can be used to find the best strategy. By guessing different worlds and repeating this process, it is hoped that an action that works well in a large number of worlds can be identified.

To make this description more concrete, let us consider a general MAX node with branches $M_1, M_2, \cdots$ in a game with n worlds. Now, let us say that we can find the minimax value, e_{ij}, of the node under branch M_i in world w_j with the minimax algorithm. We can then construct a scoring function, f, such as:

$$f(M_i) = \sum_{\substack{j=1 \\ e_{ij} \neq \bot}}^{n} e_{ij} \Pr(w_j), \qquad (1)$$

where $\Pr(w_j)$ represents MAX's assessment of the probability of the actual world being w_j. Monte-carlo sampling can then be viewed as selecting a move by using the minimax algorithm to generate values of the e_{ij}s, and determining the M_i for which the value of $f(M_i)$ is greatest. If there is sufficient time, all the e_{ij} can be generated, but in practice only some 'representative' sample of worlds is examined.

As an example, consider how the tree of Figure 1 is analysed by the above characterisation of Monte-carlo sampling. If we examine world w_1, the minimax values below node a are as shown in Figure 3 (these correspond to e_{11} and e_{21}

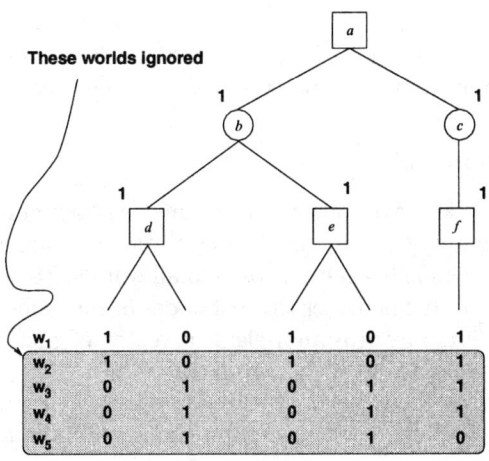

Fig. 3. Finding the minimax value of world w_1

for this tree). It is easy to check that the minimax value at node b is again 1 if we examine any of the remaining worlds, and that the value at node c is 1 in worlds w_2, w_3, and w_4, but 0 in world w_5. Thus, if we assume equally likely worlds (that is, the function Pr returns 1/5 for each world), Monte-carlo sampling using (1) to make its branch selection will choose the left-hand branch at node a whenever world w_5 is included in its sample.

Unfortunately, this is not the best strategy for this tree. The best return that MAX can hope for when making this choice of the left-hand branch at node a is a payoff of 1 in just three worlds (under the best defence model, and indeed for any reasonable assumptions about a rational opponent). Choosing the right-hand branch at node a, however, offers a payoff of 1 in four worlds.

The reason for this error can be understood by examining the implicit assumptions made by a Monte-carlo sampling approach. At first sight, the algorithm appears to model very closely the best defence assumptions given in the Introduction. For instance, it identifies pure MAX strategies that make no use of probabilities (A-III). Moreover, by repeatedly applying minimaxing it is assumed that MIN can respond optimally to MAX's moves in each individual world, for which perfect knowledge of the world state is a prerequisite (A-I).

However, as well as perfect knowledge for MIN, the repeated application of minimaxing also makes the assumption that MAX will be able to play optimally in each individual world. It is this false assumption that leads to incorrect analyses such as that in Figure 3. At nodes d and e in this example, MAX makes *different choices* in *different worlds*. In reality, however, MAX can only attach probabilities to the state of the world and must make a *single choice* for all worlds at node d and another single choice for all worlds at node e. Combining the minimax values of separate choices results in an over-optimistic analysis of node b. This is an example of the problem of *strategy fusion* formalised recently in [6]: the false assumption that MAX can play optimally in each world allows the results of different moves — or strategies — to be 'fused' together.

3.2 Vector Minimaxing

In previous work [8], we have shown how to remove strategy fusion from Montecarlo sampling with a *vector minimaxing* algorithm. This algorithm ensures that at any MAX node a single branch is chosen in all worlds. To do this, it requires a function *payoff-vector*, defined over the leaf nodes of the game tree. For any leaf-node, ν, *payoff-vector*(ν) returns an n-element vector $\boldsymbol{K}$ such that $\boldsymbol{K}[j]$ (the jth element of $\boldsymbol{K}$) takes the value of the payoff at ν in world w_j ($1 \leq j \leq n$). Vector minimaxing then uses these payoff vectors as shown in Figure 4 to identify a strategy for a tree t, where $sub(t)$ computes the set of t's immediate subtrees.

Algorithm *vector-mm*(t):
Take the following actions, depending on t.

Condition	Result
t is leaf node	*payoff-vector*(t)
root of t is a MIN node	$\min\limits_{t_i \in sub(t)}$ *vector-mm*(t_i)
root of t is a MAX node	$\max\limits_{t_i \in sub(t)}$ *vector-mm*(t_i)

Fig. 4. The vector minimaxing algorithm

In this algorithm, the normal min and max functions are extended so that they are defined over a set of payoff vectors. The max function returns the single vector $\boldsymbol{K}$, for which

$$\sum_{\substack{j=1 \\ \boldsymbol{K}[j] \neq \perp}}^{n} \Pr(\mathrm{w}_j)\boldsymbol{K}[j] \qquad (2)$$

is maximum, resolving equal choices randomly. In this way, vector minimaxing commits to just *one* choice of branch at each MAX node, avoiding strategy fusion (the actual strategy selected by the algorithm is just the set of the choices made at the MAX nodes). As for the min function, for a node with m branches and therefore m payoff vectors $\boldsymbol{K}_1, \cdots, \boldsymbol{K}_m$ to choose between, a player with perfect knowledge of the state of the world is modelled as follows:

$$\min_i \boldsymbol{K}_i = (\min_i \boldsymbol{K}_i[1], \min_i \boldsymbol{K}_i[2], \cdots, \min_i \boldsymbol{K}_i[n]). \qquad (3)$$

That is, the min function returns a vector in which the payoff for each possible world is the lowest possible.

As an example of vector minimaxing in practice, Figure 5 shows how the algorithm would analyse the tree of Figure 1, using ovals to represent the vectors produced at each node. The branches selected by (2) (assuming equally likely worlds) have been highlighted in bold. At node d, for example, the right-hand branch is chosen because its evaluation of 3/5 is higher than the 2/5 of the left-hand branch. Eventually, at the root of the tree, the right-hand branch is correctly chosen.

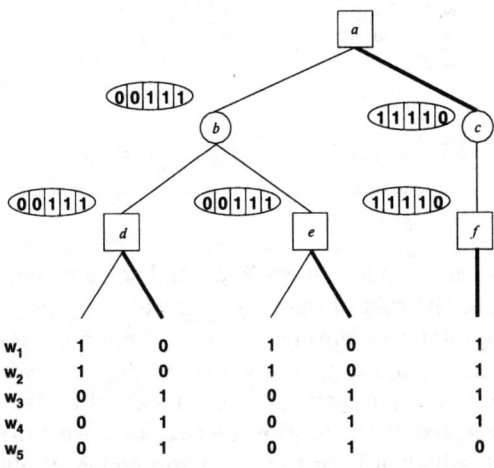

Fig. 5. Vector minimaxing applied to example tree

3.3 Payoff-Reduction Minimaxing

Vector minimaxing offers slight accuracy improvements over Monte-carlo sampling (again, see [8]). However, both Monte-carlo sampling and vector minimaxing suffer from the problem of *non-locality* [6]. This problem is illustrated in Figure 6, which depicts a game tree with just 3 worlds.

Against best defence, the optimal strategy for MAX is to choose the left-hand branch at both node a and node e. This guarantees a payoff of 1 in world w_1. In the figure, however, we have annotated the tree to show how it is analysed by vector minimaxing. The branches in bold show that the algorithm would choose the right-hand branch at node e. The vector produced at node b correctly indicates that when MAX makes this selection, a MIN player who knows the world state will always be able to restrict MAX to a payoff of 0 (by choosing

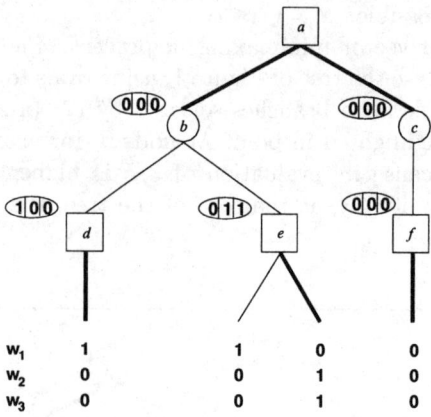

Fig. 6. Example tree with three worlds

the right-hand branch at node b in world w_1 and the left-hand branch in worlds w_2 and w_3). Thus, at the root of the tree, both subtrees have the same analysis, and vector minimaxing never wins on this tree. Applying Monte-carlo sampling to the same tree, in the limiting case where all possible worlds are examined, we see that node b has a minimax value of 1 in world w_1, so that the left-hand branch would be selected at the root of the tree. However, the same selection as vector minimaxing will then be made when subsequently playing at node d or node e. Thus, neither Monte-carlo sampling nor vector minimaxing choose the best strategy against best defence on this tree. The choice made at node e is incorrect because the situation in a different (*non-local*) subtree rooted on node d makes it impossible to actually achieve some of the payoffs under node e.

To tackle the problem of non-locality, we have previously described [8] an algorithm called *payoff-reduction minimaxing*, or *prm*. This algorithm is shown in its simplest form in Figure 7. The reduction in the second step of this algorithm

---

Algorithm $prm(t)$:

1. Use the standard minimax algorithm to conduct minimaxing of t in every world w_k. For every MIN node in t, record its minimax value in each world, m_k.
2. Examine the payoff vectors at each leaf node of t. Reduce the (non-$\perp$) payoffs p_k in each world w_k to the minimum of p_k and all the m_k of the node's MIN ancestors.
3. Apply the *vector-mm* algorithm to the resulting tree.

Fig. 7. Simple form of the *prm* algorithm

addresses the problem of non-locality by, in effect, parameterising the payoffs at each leaf node with information on the results obtainable in other portions of the tree. By using minimax values for this reduction, the game-theoretic value of the tree in each individual world is left unaltered, since no payoff is reduced to the extent that it would offer MIN a better branch selection at any node in any world.

As an example, consider how the *prm* algorithm would behave on the tree of Figure 6. The minimax value of node c is zero in every world, but all the payoffs at node f are also zero, so no reduction is possible. At node b, however, the minimax values in the three possible worlds are 1, 0, and 0, respectively. Thus, all the payoffs in each world at the leaf nodes under d and e are reduced to at most these values. This leaves only the two payoffs of 1 in world w_1 as shown in Figure 8, where the strategy selection subsequently made by vector-minimaxing has also been highlighted in bold. In this tree, then, the *prm* algorithm results in the correct strategy being chosen.

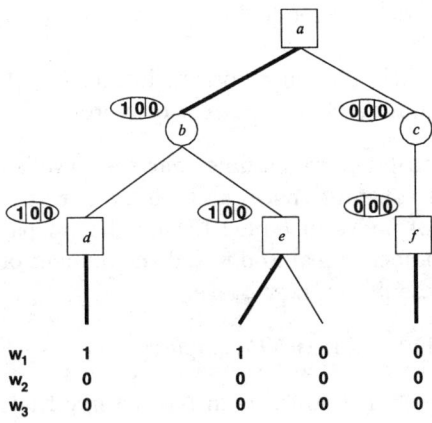

Fig. 8. Applying *vector-mm* after payoff reduction

4 Complexity Analysis of Optimal Play

Here we show that, given a game tree, finding optimal strategies under the best defence model is an NP-complete problem. Hence, unless P = NP, heuristics are required to tackle this problem in practice. As explained in the Introduction, this is fundamentally different from perfect information games (such as n-by-n checkers) which, although PSPACE-hard in the size of the initial game configuration, can be solved in linear time in the size of the game tree [9]. In contrast, our results show that for the problems we are considering, even if the complete game tree of a game is small enough to be searched by computer, finding optimal strategies may be infeasible.

Note that there have been previous analyses of the complexity of finding optimal strategies in imperfect information games [11,2], but these proofs are not applicable when considering the best defence model. For example, [2] show that it is NP-complete to determine whether or not an n-player game has a pure strategy equilibrium point, but the proof uses a reduction (from the 3-Partition problem) that cannot be reproduced with trees such as that of Figure 1. Moreover, the notion of a pure strategy equilibrium point is also not helpful in the best defence model, since A-II introduces an asymmetry: MIN's strategy is chosen after MAX has made a decision, and thus a 'stable' strategy pair is always found by simply finding the optimal response to any MAX selection.

For our proof, we first formalise the relevant problems in the format of [9].

BEST DEFENCE:
Instance: A game tree t over n worlds and a positive integer $k \leq n$.
Question: Is there is a MAX strategy that returns a payoff of 1 in at least k worlds under the best defence model?
CLIQUE:
Instance: A graph $G = (V, E)$ and a positive integer $k \leq |V|$.
Question: Does G contain a clique of size k or more?

Note that here and later, we assume that the payoffs are bounded so that storage is possible in constant space, and we measure the size of a game tree t to be the number of nodes in t plus the number of payoffs listed (which is the product of the number of leaf-nodes and the number of worlds). Given that CLIQUE is NP-complete [9], we now prove:

Theorem 1. BEST DEFENCE *is NP-complete.*

Proof: To see that BEST DEFENCE is in NP observe that given a game tree t we can guess a MAX strategy, s, (*e.g.*, by specifying the branch to be chosen at each MAX node in the tree) and correctly determine the optimal payoff in time linear in the size of t. This can be done with an algorithm that is very similar to *vector-mm*. The only modification required is at MAX nodes, where, rather than using the max operator of (2), the payoff on the branch specified by s is returned. [6] have shown that this algorithm correctly computes the payoff of s, and the time taken is clearly linear in the size of the tree.

To show NP-hardness we reduce CLIQUE to BEST DEFENCE. Let $G = (V, E)$ and k be given. We translate G to a tree t, with $n = |V|$ worlds, $w_1, \cdots, w_n$, constructed to have a payoff of 1 in at least k worlds iff G has a k clique. The root of t is a MIN node. The next layer has n MAX nodes, which we label v_1, ..., v_n, for $v_i \in V$. At each MAX node v_i there is a left and right branch, called l_i and r_i respectively. The payoff at the leaf node under each l_i is 1 in the jth world iff $i = j$ or $(v_i, v_j) \in E$. The payoff at the leaf node under each r_i is 1 in the jth world iff $i \neq j$. An example of a graph and its translation are given in

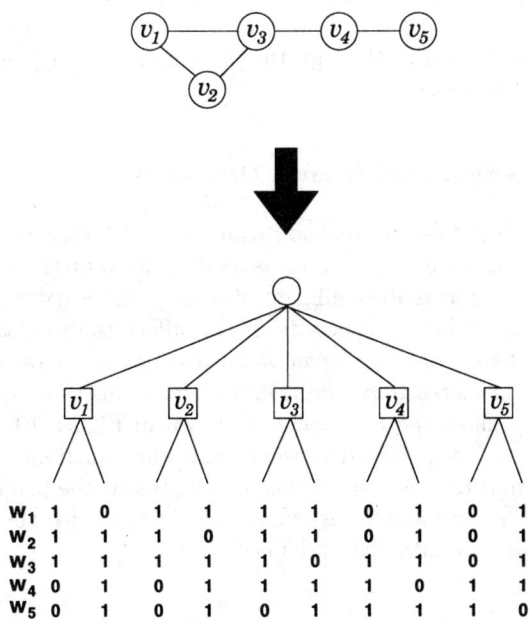

Fig. 9. Graph (top) and corresponding game tree (bottom)

Figure 9. Note that the reduction is trivially computable in time polynomial in the size of G.

Let us call a vertex v_i *selected* if MAX chooses the left branch at v_i in his strategy. Suppose G has a k clique. The clique defines a subset $V' \subseteq V$ where $k = |V'|$ and for each $v_i, v_j \in V'$, where $i \neq j$, $(v_i, v_j) \in E$. It is easy to see that the MAX strategy that selects the vertices in V' has a payoff of 1 in each world w_i, where $v_i \in V'$. Hence this strategy has a payoff of at least k.

Conversely, suppose there is a strategy for MAX with a payoff of at least k. Let W be the set of worlds in which MAX's strategy yields a payoff of 1 and let $V = \{v_i \mid w_i \in W\}$. Observe that V comprises a clique in G of size at least k since for each world $w_i \in W$, every selected vertex $v_j \in V$, $i \neq j$, must have a payoff of 1 in world w_i, which implies that $(v_i, v_j) \in E$. QED

In our example, we can select v_1, v_2, and v_3 and MIN's best strategy yields a 1 for MAX in worlds w_1, w_2 and w_3, and a 0 in the remaining two.

5 New Heuristics

We have shown that the problem of finding optimal strategies for the best defence model is NP-complete in the size of the game tree. To date, the most consistently accurate heuristic solution has been the *prm* algorithm, reviewed in §3.3. Here, we introduce two new heuristics, both with similar motivations to that of *prm*: the reduction of non-locality through the introduction of dependencies between the choices at MAX nodes.

5.1 Beta-Reduction (and Branch Ordering)

Our first new heuristic takes as its inspiration the well-known procedure of alpha-beta pruning. The alpha-beta technique is used to speed up the search of a perfect information game tree by maintaining cutoff values that are used to decide, based on the search so far, whether a new node can affect the root value of the tree. There are always two values: an *alpha* value that can never decrease, and a *beta* value that can never increase. A simple extension of this technique to game trees with multiple payoffs at the leaf nodes is shown in Figure 10. Here, the alpha and beta values are n-tuples and the max and min functions are as defined in (2) and (3). The min function is also used to represent the pruning criterion, as $\min(\boldsymbol{\alpha}, \boldsymbol{\beta}) = \boldsymbol{\beta}$. This is a simple expedient for dealing with the possibility that some payoff values may take the undefined value $\bot$.

Algorithm $vm\text{-}\alpha\beta(t, \boldsymbol{\alpha}, \boldsymbol{\beta})$:
Take the following actions, depending on t.

Condition	Result
t is a leaf node	*payoff-vector*(t)
root of t is a MIN node	for each $t_i \in sub(t)$ do $\quad \boldsymbol{\beta} \leftarrow \min(\boldsymbol{\beta}, vm\text{-}\alpha\beta(t_i, \boldsymbol{\alpha}, \boldsymbol{\beta}))$ $\quad$ if $\min(\boldsymbol{\alpha}, \boldsymbol{\beta}) = \boldsymbol{\beta}$ then return $\boldsymbol{\alpha}$ end return $\boldsymbol{\beta}$
root of t is a MAX node	for each $t_i \in sub(t)$ do $\quad \boldsymbol{\alpha} \leftarrow \max(\boldsymbol{\alpha}, vm\text{-}\alpha\beta(t_i, \boldsymbol{\alpha}, \boldsymbol{\beta}))$ $\quad$ if $\min(\boldsymbol{\alpha}, \boldsymbol{\beta}) = \boldsymbol{\beta}$ then return $\boldsymbol{\beta}$ end return $\boldsymbol{\alpha}$

Fig. 10. Vector minimaxing with alpha-beta pruning

For perfect information games, the alpha-beta algorithm represents a more efficient technique for computing the same value as standard minimax. With $vm\text{-}\alpha\beta$, however, it is not only efficiency that may be improved, but also *accuracy*. That is, $vm\text{-}\alpha\beta$ will not, in general, return the same value as *vector-mm*. For an illustration of this, let us look again at the tree of Figure 6, introduced in §3.3.

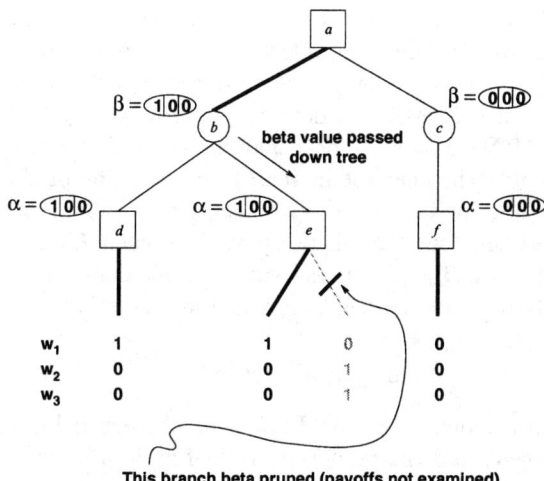

Fig. 11. A beta-pruning carried out by the vm-$\alpha\beta$ algorithm on a simple tree

Figure 11 shows how this tree is analysed by vm-$\alpha\beta$. When node d is examined, it produces an alpha value of $(1,0,0)$, which then becomes the beta of node b. This beta value is then passed down to node b's next daughter. At node e, the first daughter is a leaf node and the alpha value of node e is therefore set to the leaf node values $(1,0,0)$. Now, this alpha value is at least as good as the beta value of b (that is, $\min(\alpha, \beta)$ is now equal to β), so the remaining branches at node e can be pruned (beta pruning). Thus, vm-$\alpha\beta$ selects the correct strategy on this tree, whereas we saw in §3.3 that $vector$-mm is sub-optimal.

The explanation for vm-$\alpha\beta$'s superiority here is that the beta-pruning at node e in effect tackles the non-local nature of this game tree by preventing the second (sub-optimal) branch at node e from being examined. This is a simple example of a general effect. Non-locality occurs when choices are made at internal MAX nodes without reference to other subtrees. Since pruning decreases the number of MAX nodes that are actually examined, it also decreases the chance that non-local effects will lead to errors.

Although vm-$\alpha\beta$ represents an improvement over $vector$-mm on this particular example, it is not hard to produce a modified tree for which both algorithms find the same, incorrect solution. For instance, the small change of increasing by one the payoff under node d in world w_1 leaves the optimal strategy and its payoff unchanged. However, vm-$\alpha\beta$ (and also $vector$-mm) will not be able to find this strategy. Even for this modified tree, though, there is an adaptation of the alpha-beta technique that does improve accuracy. To see this, we simply need to realise that, when using vm-$\alpha\beta$, the branch selections made during a search are constantly reflected in the alpha and beta values passed to any node. These values therefore offer a natural way of tackling non-locality; by ensuring that payoffs rendered unachievable by branch selections in the analysed portion of the tree do not adversely affect the selections in the remainder of the search.

In particular, any beta value, β, generated at a MIN node, ν, can be used to reduce non-locality at MAX nodes in any subtree of ν that has yet to be examined. Since MIN chooses the best play in each individual world, each value $\beta[j]$ imposes a limit (that cannot increase) on the value of the optimal payoff that can now be obtained in world w_j. Thus, when making a new selection at a MAX node in a subtree of ν, all the payoffs of each K_i in world w_j should be *reduced* to at most $\beta[j]$. A simple way to implement this observation is to modify the result returned at leaf nodes in the algorithm of Figure 10 to the following:

$$\min(payoff\text{-}vector(t), \beta), \qquad (4)$$

where min is again as defined in (3). Let us call the algorithm produced by this modification *vm-beta*, and the reductions of leaf node payoffs made by (4) *beta-reductions*. This new algorithm can correctly solve the tree of Figure 11 even when the payoff in world w_1 at node d is greater than one, since the payoffs of 1 in worlds w_2 and w_3 are beta-reduced to zero.

Since beta-reduction only utilises information about branches *already selected*, it is sensitive to branch ordering. For example, consider the effect of swapping the order of the branches at node b in Figure 11. The choice between the vectors $(1,0,0)$ and $(0,1,1)$ at node e would then have to be made *before* realising that the payoffs of 1 in w_2 and w_3 could not be achieved. Thus no beta-reductions (or, of course, beta prunings) would be possible, and the optimal strategy for this reordered tree would not be found. However, note that *vm-beta* is still correct (unlike *vm-αβ*) if we simply swap the two branches at node e.

Of course, standard (perfect information) alpha beta pruning is also affected by branch ordering, at least in terms of efficiency. It is well known that the optimal branch ordering for the algorithm is for MAX's better moves to come first at MAX nodes, and for MIN's better moves to come first at MIN nodes. In fact, the same holds for *vm-beta*, but in terms of accuracy it is the ordering at MIN nodes that is most important; the earlier in the search that the payoff vectors with relatively small values are encountered, the more likely that beta-reductions will become possible.

Note that the *prm* algorithm reviewed in §3.3 can find optimal solutions for the tree of Figure 11 irrespective of branch ordering. To show that other trees exist for which *vm-beta* actually out-performs *prm*, then, consider the example of Figure 12. Here, the single MIN node has a minimax value of one in every world. Thus, *prm* cannot reduce any leaf node payoffs, and will therefore produce the same strategy (shown in the figure) as vector minimaxing. It is easy to see that this strategy has a payoff of zero in every world. Employing beta-reduction, on the other hand, will guarantee a payoff of 1 in either world w_2 or w_3 (depending on the random choice made at the second MAX node). That *prm* and *vm-beta* can find optimal strategies on different trees suggests the creation of a hybrid *prm-beta* algorithm. This is easily done by simply replacing *vector-mm* with *vm-*

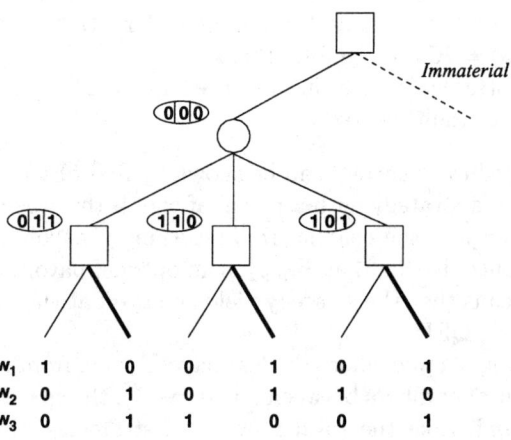

Fig. 12. An incorrect strategy selected by *vector-mm* (and by *prm*) on a tree correctly analysed by *vm-beta*

beta in step three of the *prm* algorithm of Figure 7. The insensitivity of the payoff reduction technique to branch ordering allows this hybrid algorithm to benefit fully from an ordering that favours beta-reduction.

5.2 Iterative Biasing

In our complexity proof of §4, we showed that when a MAX strategy is fixed, its payoff can be found in time linear in the size of the game tree. This suggests a heuristic for finding good strategies: simply *guess* a strategy (or even a partial strategy) and then check the strategy's payoff. This guessing can be repeated until an answer is demanded, at which point the guess with the best evaluation can be returned. However, given our extensive form game tree, we know (see, e.g., [6]) that when the nodes have at least binary branching the number of strategies for MAX is exponential in the size of tree (doubly exponential in its depth). Finding good guesses among such a large number of possibilities is unlikely to be a practical proposition in general.

However, there is something else besides strategies that can be guessed: payoffs. In fact, given an optimal payoff vector, K_{max}, we can efficiently find an optimal strategy for MAX. To see this, consider a game tree with payoff vectors K at the leaf nodes. Assume it is known that the optimal payoff vector for this game is K_{max}. We then compute an optimal strategy (which may not be unique, as there could be more than one optimal payoff vector and each such vector could also result from more than one strategy) with the following steps, which run in time linear in the size of the game tree:

1. Compare the payoff vector K at each leaf with K_{max}. Replace the vector with the integer 1 if K is at least as good for MAX as K_{max} (that is, if $\min(K, K_{max}) = K_{max}$), and 0 otherwise.
2. The optimal strategy is the one returned for MAX by applying standard minimax to the resulting tree.

That this procedure is correct can be shown by first observing that the minimax step must find a strategy with a payoff of one. If this was not the case there would be no strategy in the original tree returning a payoff that is at least as good as K_{max}, contradicting that K_{max} is an optimal payoff. Now observe that a payoff of one means that the strategy yields a payoff at least as good as K_{max} on the original tree. QED.

If we are playing a game where the leaf payoffs come from a finite m-element domain (*e.g.*, natural numbers between 0 and $m-1$), the space of possible payoff vectors has size m^n. Like the total number of strategies, this is exponential, but now the exponent is different: it is the number of worlds n. Thus, whereas guessing strategies may not be practical, guessing payoff vectors may be more feasible. In single-suit Bridge problems, for example, redundancies in the domain often reduce the number of significant worlds to a manageable number (such as the twenty worlds of the problem in Figure 2, produced by treating the low cards as indistinguishable).

We suggest the basic approach of guessing a *single* element of the optimal payoff vector to be some value v (*i.e.*, guessing that $K_{max}[k] = v$ for a particular world w_k). This guess can then be passed to a modified version of vector minimaxing that uses it to *bias* the search. This biasing is achieved by defining a new function, $\max_{v,k}$, to replace the definition of (2) in the *vector-mm* algorithm. The $\max_{v,k}$ function returns from amongst a set of vectors the one that is best according to the relation $\underset{v,k}{>}$ defined below.

Definition 1 (Biasing relation). *For any two payoff vectors, K_1 and K_2, we say that $K_1 \underset{v,k}{>} K_2$ if and only if either of the following hold:*

- *the vector K_1 offers a payoff of at least v in world w_k, but the vector K_2 does not, or*
- *if neither of K_1 or K_2 offers a payoff of at least v in w_k, or if both K_1 and K_2 offer a payoff of at least v in w_k, then K_1 must be superior to K_2 based on an expected value computation on the remaining worlds. That is,*

$$\sum_{\substack{i=1 \\ (i \neq k, K_1[i] \neq \perp)}}^{n} K_1[i] \Pr(w_i) > \sum_{\substack{i=1 \\ (i \neq k, K_2[i] \neq \perp)}}^{n} K_2[i] \Pr(w_i).$$

This definition is designed to bias a search so that, wherever possible, a branch with a payoff greater than or equal to v in world w_k is selected. Given some finite set, $\mathcal{S}$, of guesses for the pair of values $\{v, k\}$, we can then repeat the search with different biases — a technique we call *iterative biasing*. Specifically, we can create the *iterative vector minimaxing* (or *ivm*) algorithm of Figure 13.

Algorithm $ivm(t, \mathcal{S})$
Given $\mathcal{S} = \{\{v_1, k_1\}, \{v_2, k_2\}, \cdots\}$,
for each $\{v_j, k_j\} \in \mathcal{S}$
      compute $s_j = \text{biased-vm}(t, v_j, k_j)$
end
return the s_j that represents the best expected payoff

Here, $\text{biased-vm}(t, v, k)$ takes the following actions, depending on t.

Condition	Result
t is leaf node	$\text{payoff-vector}(t)$
root of t is a MIN node	$\min_{t_i \in sub(t)} \text{biased-vm}(t_i, v, k)$
root of t is a MAX node	$\max_{v,k} \text{biased-vm}(t_i, v, k)$ $t_i \in sub(t)$

Fig. 13. Iterative biasing, as carried out by the iterative vector minimaxing algorithm

Iterative biasing enables the ivm algorithm to tackle non-locality by, on each iteration, introducing a dependency between *all* MAX selections in a tree. To see that ivm can correctly analyse problems that $vector\text{-}mm$ cannot, simply consider again the tree of Figure 12. For any of the guesses $\{1, 1\}$, $\{1, 2\}$, or $\{1, 3\}$, ivm will return an optimal strategy. Each guess results in a different optimal strategy, however, that wins in just one world (w_1, w_2, or w_3, respectively).

For trees with binary payoffs, it is always possible to construct the simple set $\mathcal{S} = \{\{1, 1\}, \{1, 2\}, \cdots, \{1, n\}\}$, which guesses the value $v = 1$ for each of the n possible worlds. For games where the payoffs can take more than two values, however, we suggest the more general $\mathcal{S} = \{\{v_{max}, 1\}, \{v_{max}, 2\}, \cdots, \{v_{max}, n\}\}$. Here, v_{max} is the largest of the (perfect information) minimax values of the root of the game tree in each individual world (such values can be efficiently calculated, as in the first step of the prm algorithm, for example). The value of v_{max} is also an upper bound on the value of any entry in the optimal payoff vector, $\boldsymbol{K}_{max}$. Thus, such payoff guesses are appropriate for Bridge, where a common task is to identify the strategy with the best chance of producing the maximum possible number of tricks. In fact, a simple efficiency improvement can be made by omitting any guess $\{v_{max}, k\}$ for which the (perfect information) minimax value of the game tree in world w_k is less than v_{max}. This is justified by noting that the value of $\boldsymbol{K}_{max}[k]$ will never be v_{max} if even the best possible play in w_k itself cannot produce v_{max} tricks.

5.3 Summary of Heuristics

We have introduced the heuristics of beta-reduction and iterative biasing, demonstrating how they address the problem of non-locality by introducing dependencies between choices at MAX nodes. We described how each heuristic represents an improvement over the basic *vector-mm* algorithm, and also noted that beta-reduction could be combined with payoff-reduction to produce the *prm-beta* algorithm. In fact, there are a total of eight possible algorithms that can be produced by combinations of payoff-reduction, beta-reduction, and iterative biasing, as shown in Figure 14.

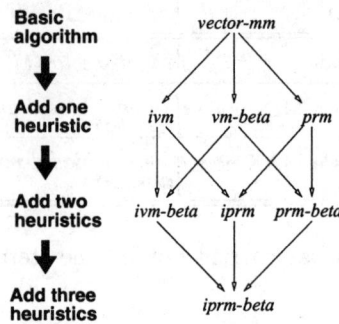

Fig. 14. Possible combinations of heuristics

In the following section we present test results that demonstrate the practical use of these algorithms. First, however, we give a further intuition on their characteristics by examining how they perform on the tree used for our complexity proof in Figure 9. The summary in Figure 15 details the node selections made on this tree by each of the eight algorithms. The original *vector-mm* algorithm can at best find a 1-clique, if it makes a fortunate guess at node v_3. Payoff-reduction cannot improve on this, as the minimax value of each individual world is one. However, both beta-reduction and iterative biasing improve the result, and when they are used together the optimal solution is found.

6 Test Results

We tested the algorithms in Figure 14 on random game trees and on a hard set of problems from the game of Bridge. The results of these tests are presented below.

6.1 Experiments on Random Trees

We follow the approach of [8] in conducting tests using complete binary trees, with $n = 10$ worlds and payoffs of just one or zero. These payoffs are assigned

Algorithm	Nodes selected	Equivalent clique
vector-mm	v_3 selected with 50% probability.	1-clique or 0-clique
prm	Same as vector minimaxing, since minimax value is one in every world.	1-clique or 0-clique
vm-beta, prm-beta	If v_3 is selected (again, a 50% chance) v_4 is also selected. If v_3 is not selected, v_4 and v_5 are selected. (If tree is analysed right to left, v_1, v_2 and v_3 are selected.)	2-clique (3-clique)
ivm, iprm	For any payoff guess of 1 in w_k, the corresponding v_k will be selected. For the guesses $k=1$, $k=2$, and $k=4$ there is a 50% chance that v_3 may also be selected.	2-clique or 1-clique
ivm-beta, iprm-beta	v_1, v_2 and v_3 only selected if the payoff guess is a 1 in world w_1. All other guesses lead to the selection of just two nodes.	3-clique

Fig. 15. Performance comparison on tree generated by complexity proof (see Figure 9)

by an application of the Last Player Theorem [15], so that the probability of a forced win for MAX in the complete information game tree in any individual world is the same for trees of all depths. The game trees are further modified by a probability, q, that determines how similar the possible worlds are. To generate a tree with n worlds and a given value of q:

- first randomly generate payoffs for n worlds, then
- generate a set of payoffs for a dummy world w_{n+1},
- and finally, for each of the original n worlds, overwrite the complete set of payoffs with the payoffs from the dummy world, with probability q.

Trees with a higher value of q tend to be easier to solve, because an optimal strategy in one world is also more likely to be an optimal strategy in another. In Figure 16 we show the results obtained when $q = 0.75$ — a value we chose because it produces similar results to the game of Bridge. This graph was produced by carrying out the following steps 1000 times for each data point of tree depth and opponent knowledge:

1. Generate a random test tree of the required depth.
2. Use each algorithm to identify a strategy. We assume that, for each algorithm, the payoffs in *all* worlds can be examined.
3. Compute the payoff of the selected strategies under the best defence model.
4. Use an inefficient, but correct, algorithm (based on examining every strategy) to find an optimal strategy and payoff.

5. For each algorithm, check whether they are in error (*i.e.*, if any of the values of the strategies found in Step 3 are inferior to the value of the strategy found in Step 4, assuming equally likely worlds).

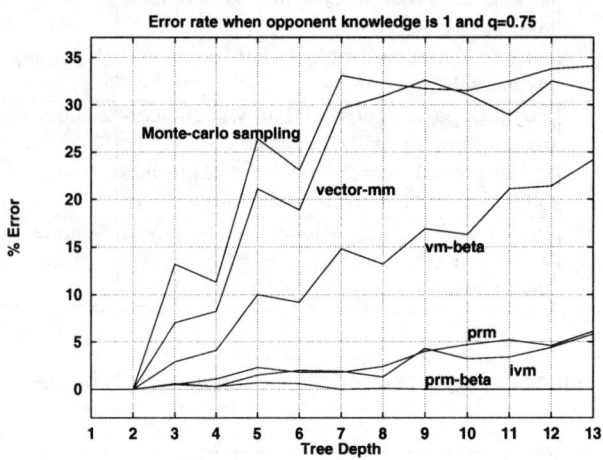

Fig. 16. Algorithm performance on random game trees where the optimal strategy in one world is more likely to be optimal in another

The graph for *vm-beta* shows that whilst it improves on simple vector minimaxing, the improvement is not as large as that produced by *ivm* or *prm*. However, a combination of heuristics performs better, with *prm-beta* performing at 100%.

6.2 Experiments on the Game of Bridge

Bridge has been heavily analysed by human experts, who have produced texts that describe the optimal play in large numbers of situations. The availability of such references provides a natural way of assessing the performance of automated algorithms. In fact, it was careful scrutiny of such expert analyses that led to the formalisation of the best defence model; thus, the game provides an excellent test of the algorithms in this paper.

To construct a Bridge test set, we used as an expert reference the Official Encyclopedia of Bridge, published by the American Contract Bridge League [1]. This book contains a 55-page section presenting optimal lines of play for a selection of 665 single-suit problems. Of these, we collected the 650 examples that gave *pure* strategies for obtaining the maximum possible payoff against best defence.[1] Using the FINESSE Bridge-playing system [7,5], we then tested Monte-carlo sampling, and the eight algorithms summarised in the previous section, against the solutions from the Encyclopedia. (For the iterative algorithms, the set of payoff guesses was produced by finding the worlds for which the perfect information minimax value of the game tree was equal to the maximum possible number of tricks, as described at the end of §5.2.)

We compared the expected payoff of the strategies produced by each algorithm (for the maximum possible number of tricks) with the expected payoff of the solution given in the Encyclopedia. The results summarised in Figure 17 show how often each algorithm was optimal. As in our tests on random trees, vector minimaxing is again slightly more accurate than Monte-carlo sampling, and correctness further improves as heuristics are added. The most effective individual heuristic is payoff-reduction (*prm* outperforms both *ivm* and *vm-beta*). When payoff-reduction, beta-reduction and iterative biasing are all combined in the *iprm-beta* algorithm, sub-optimal strategies are only generated for two problems. Given that our algorithms also revealed nine errors in the test set itself, however, this performance (and also that of *ivm-beta* and *iprm*) is actually better than the human experts that produced the model solutions. In fact, we traced the cause of *iprm-beta*'s two errors to a problem with FINESSE itself that resulted in the optimal strategies not actually being present in the search space. We intend to correct this design error in the near future.

Algorithm	Optimal	Sub-optimal	Expected Loss	Time (s)
Monte-carlo	430 (66.2%)	220 (33.8%)	17.00	8.1
vector-mm	460 (71.8%)	190 (28.2%)	12.81	3.8
vm-beta	555 (85.4%)	95 (14.6%)	6.24	4.3
ivm	613 (94.3%)	37 (5.7%)	1.61	25.5
prm	622 (95.7%)	28 (4.3%)	0.86	15.5
prm-beta	638 (98.2%)	11 (1.8%)	0.34	19.6
ivm-beta	645 (99.2%)	5 (0.8%)	0.23	96.3
iprm	645 (99.2%)	5 (0.8%)	0.13	104
iprm-beta	648 (99.7%)	2 (0.3%)	0.06	101

Fig. 17. Performance on the 650 single-suit problems from the Encyclopedia of Bridge

---

[1] The remaining fifteen examples split into four categories: six problems that give no line of play for the maximum number of tricks, four problems involving the assumption of a *mixed* strategy defence, four for which the solution relies on assumptions about the defenders playing sub-optimally by not *false-carding*, and one where there are constraints on the cards that can be played.

Our results table also includes an 'Expected Loss' column, which gives the number of times that the sub-optimal strategies produced by each algorithm can be expected to result in inferior performance. This figure measures the expected number of times that the Encyclopedia's strategies would out-perform each algorithm when playing the entire set of 650 problems once (against best defence and with a random choice among the possible holdings for the defence). The value is produced by simply summing, over every problem in the test set, the chance of success of the Encyclopedia's strategy minus the chance of success of the strategy produced by the algorithm in question. When measured by expected loss, the superiority of *iprm-beta* over Monte-carlo sampling or *vector-mm* is less marked. However, note that there is at least one task for which optimality is the crucial factor, namely the creation of tutoring systems where a computer must generate (and perhaps even explain) the best way to play a game. One natural application of *iprm-beta*, therefore, is as the basis for such a system.

The 'Time' column gives the average number of seconds required for a single problem (on a Sun SPARCstation Ultra2 running at 200MHz). We have not paid particular attention to the efficiency of our implementations (for example, none of the beta-reduction algorithms actually incorporate pruning to speed up the search). Nevertheless the speeds are acceptable, with *prm-beta*, in particular, offering a good trade-off of accuracy against speed. The iterative algorithms may appear particularly slow, but note that they can all be used in 'any-time' fashion, returning the best result encountered so far when available time is exhausted.

7 Conclusions

We have investigated the problem of finding optimal strategies under the best defence model of an imperfect information game. We demonstrated that this problem is NP-complete in the size of the game tree, and introduced the new heuristics of *beta-reduction* and *iterative biasing*. We presented test results that demonstrated the effectiveness of these heuristics, particularly when combined with payoff-reduction minimaxing to produce the *iprm-beta* algorithm. On our database of problems from the game of Bridge, *iprm-beta* actually makes less errors than the human experts that produced the model solutions. It thus represents the first general search algorithm capable of consistently performing at and above expert level on a significant aspect of Bridge card play.

References

1. ACBL. *The Official Encyclopedia of Bridge*. American Contract Bridge League, Inc., 2990 Airways Boulevard, Memphis, Tennessee 38116-3875, fifth edition, 1994.
2. J.S. Blair, D. Mutchler, and M. van Lent. Perfect recall and pruning in games with imperfect information. *Computational Intelligence*, February 1996.
3. R.A. Corlett and S.J. Todd. A Monte-carlo approach to uncertain inference. In P. Ross, editor, *Proceedings of the Conference on Artificial Intelligence and Simulation of Behaviour*, pages 28–34, 1985.

4. A. Frank. Brute force search in games of imperfect information. In D.N.L. Levy and D.F. Beal, editors, *Heuristic Programming in Artificial Intelligence 2 – The Second Computer Olympiad*, pages 204–209. Ellis Horwood, 1989.
5. I. Frank. *Search and Planning under Incomplete Information: A Study using Bridge Card Play*. PhD thesis, University of Edinburgh, 1996. Also published by Springer in the Distinguished Dissertations Series, ISBN 3-540-76257-4, 1998.
6. I. Frank and D. Basin. Search in games with incomplete information: A case study using bridge card play. *Artificial Intelligence*, 100(1–2):87–123, 1998.
7. I. Frank, D. Basin, and A. Bundy. An adaptation of proof-planning to declarer play in bridge. In *Proceedings of ECAI-92*, pages 72–76, Vienna, Austria, 1992.
8. I. Frank, D. Basin, and H. Matsubara. Finding optimal strategies for imperfect information games. In *Proceedings of AAAI-98*, pages 500–507, 1998.
9. M. R. Garey and D. S. Johnson. *Computers and intractability: a guide to the theory of NP-completeness*. W H Freeman, 1979.
10. M. Ginsberg. Partition search. In *Proceedings of AAAI-96*, pages 228–233, 1996.
11. D. Koller, N. Meggido, and B. von Stengel. Efficient computation of equilibria for extensive two-person games. *Games and Economic Behaviour*, 14(2):247–259, June 1996.
12. D. Koller and A. Pfeffer. Representations and solutions for game-theoretic problems. *Artificial Intelligence*, 94(1):167–215, July 1997.
13. D.N.L. Levy. The million pound bridge program. In D.N.L. Levy and D.F. Beal, editors, *Heuristic Programming in Artificial Intelligence – The First Computer Olympiad*, pages 95–103. Ellis Horwood, 1989.
14. R. D. Luce and H. Raiffa. *Games and Decisions—Introduction and Critical Survey*. Wiley, New York, 1957.
15. D. S. Nau. The last player theorem. *Artificial Intelligence*, 18:53–65, 1982.
16. Y. Nygate and L. Sterling. Python: An expert squeezer. *The Journal of Logic Programming*, 8(1 and 2), January/March 1990.
17. S.J.J. Smith and D.S. Nau. Strategic planning for imperfect information games. In *Games: Planning and Learning, Papers from the 1993 Fall Symposium*, pages 84–91, AAAI Press, 1993.
18. J. von Neumann and O. Morgenstern. *Theory of Games and Economic Behavior*. Princeton University Press, 1944.

A Speculative Strategy

Xinbo Gao[1], Hiroyuki Iida[1], Jos W.H.M. Uiterwijk[2], and
H. Jaap van den Herik[2]

[1] Department of Computer Science
Shizuoka University
3-5-1 Juhoku
Hamamatsu, 432 Japan
{gao,iida}@cs.inf.shizuoka.ac.jp

[2] Department of Computer Science
Universiteit Maastricht
P.O. Box 616
6200 MD Maastricht, The Netherlands
{uiterwijk,herik}@cs.unimaas.nl

Abstract. In this contribution we propose a strategy which focuses on the game as well as on the opponent. Preference is given to the thoughts of the opponent, so that the strategy might be speculative. We describe a generalization of OM search, called (D,d)-OM search, where D stands for the depth of search by the player and d for the opponent's depth of search. A difference in search depth can be exploited by deliberately chosing a suboptimal move in order to gain a larger advantage than when playing the optimal best move. The idea is that the opponent does not see the variant in sufficiently deep detail. Simulations using a game-tree model including an opponent model as well as experiments in the domain of Othello confirm the effectiveness of the proposed strategy.
Keywords: opponent modelling, speculative play, α-β^2 pruning, Othello.

1 Introduction

In minimax and its variants there is an implicit assumption that the player and the opponent use the same search strategy, i.e., (1) the leaves are evaluated by an evaluation function and (2) the values are backed up via a minimax-like procedure. The evaluation function may contain all kind of sophisticated features but it evaluates the position according to preset criteria (including also the use of quiescence search). It never changes the value of a Knight in the evaluation function, although it "knows" that the opponent has a strong reputation for playing with two Knights in the endgame. So, the evaluation function shows stability and is not speculative. The minimax back-up procedure is well-established and is as logical as one can think. So far no other idea emerged, except for one final decision of the back-up procedure. If the result is a draw (e.g., by repetition of positions) and the opponent is assumed to be weak, a contempt factor may indicate that playing the second-best move is preferred. This is the most elementary step of opponent modelling. It shows a clear deviation of the minimax-like strategy.

An extension of the idea of anticipating the opponent's weakness has been developed in opponent-model search. According to this framework a grandmaster often attempts to understand the intention behind the opponent's previous moves and then employs some form of speculative play, anticipating the opponent's weak reply [6]. Iida *et al.* modelled such grandmaster's thinking processes based on possible opponent's mistakes, and proposed OM search (short for Opponent-Model search) [4,5] as a generalized game-tree search. In OM search perfect knowledge of the opponent's evaluation function is assumed. This knowledge may lead to the conclusion that the opponent is expected to make an error in a given position. As a consequence the error may be exploited to the advantage of the player possessing the knowledge. In such an OM-search model, it is implicitly assumed that both players search to the same depth.

In actual game-playing, e.g., in Shogi tournaments, we have observed [5] that the two players may not only use different evaluation functions, but also reach different search depths. Therefore, we propose a generalization of OM search, called (D,d)-OM search, in which the difference of depth is incorporated, with D standing for the depth of search of the first player, and d for the opponent. We will show that exploiting this difference leads to a speculative strategy.

In section 2 we introduce (D,d)-OM search by some definitions and assumptions and describe a (D,d)-OM-search algorithm. Then the characteristics of (D,d)-OM search are considered in section 3, and the relationship between (D,d)-OM search, OM search, and minimax search is discussed. In section 4, an improved version in which branches are pruned is introduced. It is denoted by α-β^2 pruning. Section 5 illustrates the performance of the implicitly proposed speculative strategy with random-tree simulations as well as with experiments in the domain of Othello. How to apply this strategy efficiently to actual game positions is discussed in section 6. Finally, conclusions and limitations of this speculative strategy are given in section 7.

2 (D,d)-OM Search

In this section, (D,d)-OM search is outlined by definitions and assumptions. In addition, an example is supplied showing how a value at any position in a search tree is computed using (D,d)-OM search. By convention and for clarity, the two players are distinguished as the *max player* and the *min player*. Below, we discuss (D,d)-OM search from the viewpoint of the max player.

2.1 Definitions and Assumptions

For the description of (D,d)-OM search, we use the following definitions and assumptions.

Definition 1. [playing strategy]
A playing strategy is a three-tuple $\langle D, EV, SS \rangle$, where D is the player's search depth, EV the static evaluation function used, and SS the search strategy, i.e., the method of backing up the values from the leaves to the root in a search tree.

Definition 2. [player model]
A player model is the assumed playing strategy of a player. For any player X with search depth D_X, static evaluation function EV_X, and search strategy SS_X, we define a player model as $M_X = \langle D_X, EV_X, SS_X \rangle$.

Below we provide three assumptions for (D,d)-OM search. In the following OM stands for OM search, MM for minimax search, and P is a given position in which the max player is to move.

Assumption 1
The min player's playing strategy M_{min} is defined as $\langle d, EV_{min}, MM \rangle$, which means that the min player performs some minimax strategy at any successor of P and evaluates the leaf positions at depth $(d+1)$ in the max player's game-tree using the static evaluation function EV_{min}.

Assumption 2
The max player knows the strategy of the min player, $M_{min} = \langle d, EV_{min}, MM \rangle$, i.e., his min player's model coincides with the min player's strategy.

Assumption 3
The max player employs $\langle D, EV_{max}, (D,d)\text{-}OM \rangle$ as playing strategy, which means that the max player evaluates the leaf positions at depth D using the static evaluation function EV_{max} and backs up the values by (D,d)-OM search.

(D,d)-OM search mimics grandmaster play in that it uses speculations on what the opponent "sees". The player acquires and uses the model of the opponent to find a potential mistake, and then obtains an advantage by anticipating this error.

2.2 The Algorithm of (D,d)-OM Search

In (D,d)-OM search, a *pair* of values is computed for all positions above depth $(d+1)$. One value comes from the opponent model and one from the max player's model. Below depth $(d+1)$, the max player no longer uses any opponent model. There only one value is computed for each position; it is backed up by minimax search.

Let i, j from now on range over all immediate successor positions of a node in question. Let a node be termed a *max node* if the max player is to move, a *min node* otherwise. According to the assumptions, D is the search depth of the max player and d is the search depth of the min player as predicted by the max player. Then the function $V(P, OM(D, d))$ is defined for relevant nodes as the value considered by the max player, and $V(P, MM(d))$ as the value for the min player, predicted by the max player.

$$V(P, OM(D,d)) = \begin{cases} \max_i V(P_i, OM(D-1, d-1)) \\ \quad \text{if } P \text{ is an interior max node} \\[1em] V(P_j, OM(D-1, d-1)) \text{ with } j \text{ such that} \\ \quad V(P_j, MM(d-1)) = \min_i V(P_i, MM(d-1)) \\ \quad \text{if } P \text{ is an interior min node} \\ \quad \text{and } d \geq 0 \\[1em] \min_i V(P_i, OM(D-1, d-1)) \\ \quad \text{if } P \text{ is an interior min node} \\ \quad \text{and } d < 0 \\[1em] EV_{max}(P) \\ \quad \text{if } D = 0 \ (P \text{ is a leaf node}) \end{cases} \quad (1)$$

$$V(P, MM(d)) = \begin{cases} \max_i V(P_i, MM(d-1)) \\ \quad \text{if } P \text{ is an interior max node} \\[1em] \min_i V(P_i, MM(d-1)) \\ \quad \text{if } P \text{ is an interior min node} \\[1em] EV_{min}(P) \\ \quad \text{if } d = -1 \ (P \text{ is a "leaf" node}) \end{cases} \quad (2)$$

The pseudocode for the (D, d)-OM search algorithm is given in Figure 1.

An example of (D, d)-OM search is shown in Figure 2. The search tree shows two different root values due to the use of two different models of the players. Using $(3, 1)$-OM search yields a value of 11 and using plain minimax yields a value of 9. In this example, the max player may thus achieve a better result than by minimax. It does so by selecting the left branch. For clarity, we note that d denotes the search depth for the opponent, which is reached at depth $d+1$ in the search tree of the first player. In the example, the nodes at depth 2 thus will be evaluated for both players, while those at depth 3 will only be evaluated for the first player.

```
procedure (D, d)-OM(P,depth):
/* Iterative deepening at root P */
/* Two values are returned, according to equations (2) and (1) */

    if depth = d + 1 then begin
    /* Evaluate the min-player's leaf nodes */
        V_MM[P] ← Evaluate(P,min)
        V_OM[P] ← Minimax(P,depth)
        return (V_MM[P],V_OM[P])
    end

    {P_i|i = 1,···,n} ← Generate(P)
    /* Expand P to generate all its successors P_i */

    for each P_i do begin
        (V_MM[P_i],V_OM[P_i]) ← (D, d)-OM(P_i,depth+1)
    end

    /* Back up the evaluated values */
    if P is a max node then begin
    /* At a max node both the max player and the min player back up the maximum */
        V_MM[P] ←  max   V_MM[P_i]
                  1≤i≤n
        V_OM[P] ←  max   V_OM[P_i]
                  1≤i≤n
    end
    else begin /* P is a min node */
    /* At a min node, the min player backs up the minimum and the max player backs up the value of the
    node selected by the min player */
        V_MM[P] ← V_MM[P_j] =  min   V_MM[P_i]
                              1≤i≤n
        V_OM[P] ← V_OM[P_j]
    end
    return (V_MM[P],V_OM[P])

procedure Minimax(P,depth):
/* Iterative deepening below depth d + 1 */
/* Returns the minimax value according to the max player */

    if depth = D then begin
    /* Evaluate the max player's leaf nodes */
        V_MM[P] ← Evaluate(P,max)
        return (V_MM[P])
    end

    {P_i|i = 1,···,n} ← Generate(P)
    /* Expand P to generate all its successors P_i */

    for each P_i do begin
        V_MM[P_i] ← Minimax(P_i,depth+1)
    end

    /* Back up the evaluated values */
    if P is a max node then begin
        V_MM[P] ←  max   V_MM[P_i]
                  1≤i≤n
    end
    else begin /* P is a min node */
        V_MM[P] ←  min   V_MM[P_i]
                  1≤i≤n
    end
    return (V_MM[P])
```

Fig. 1. Pseudocode for the (D, d)-OM search algorithm.

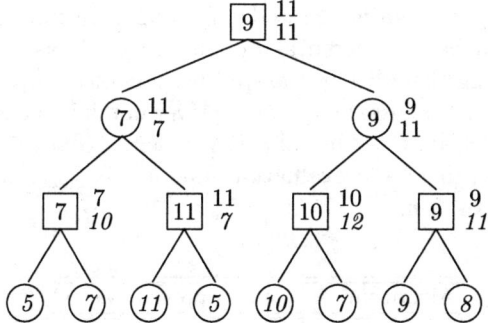

The numbers *inside* the circles/boxes represent the back-up values by minimax search from the max player's point of view. The upper numbers *beside* the circles/boxes represent the back-up values by $(3,1)$-OM search and the lower numbers the back-up values of the minimax search from the min player's point of view. The depths 3 and 2 contain the leaf positions for the max player and the min player, respectively, i.e., these values (in italics) are evaluated statically using the max player's or the min player's evaluation function.

Fig. 2. (D,d)-OM search and minimax compared, with $D=3$ and $d=1$.

We remark that the player using (D,d)-OM search always searches deeper than the opponent, i.e., $D>d$. Cases in which the opponent is modelled by a deep search using a very fast but simplistic evaluation function, and the first player is modelled as relying on a shallower search with a sophisticated evaluation function, are not treated in the above formulation.

3 Characteristics of (D,d)-OM Search

In this section, some characteristics of (D,d)-OM search are described and compared with those of minimax search. At first the relations among (D,d)-OM search, OM search, and minimax search are discussed, and two remarks are made. Then a theorem relating root values by (D,d)-OM search and minimax search is stated.

3.1 Relations among (D,d)-OM Search, OM Search, and Minimax

The (D,d)-OM search algorithm indicates that the max player performs minimax search to back up the static-evaluation-function values from depths $(d+1)$ to D, while from depths 1 to $(d+1)$ the max player performs pure OM search. So from the viewpoint of search algorithms, (D,d)-OM search can be considered as the combination of pure OM search and minimax search.

Viewed differently, all the moves determined by minimax search, OM search, and (D,d)-OM search take some opponent model into account, i.e., each choice is based on the player's own model and some opponent model. Accordingly, all three strategies can be considered as opponent-model-based search strategies. The difference among them lies in the specification of the opponent model.

The opponent models used by the max player in minimax search, OM search, and (D,d)-OM search are listed in Table 1. We assume that the max player moves first with search depth D and evaluation function EV_{max}, i.e., in a game tree the root is a max position.

Algorithm	the opponent model
minimax search	$\langle D-1, EV_{max}, \text{MM} \rangle$
OM search	$\langle D-1, EV_{min}, \text{MM} \rangle$
(D,d)-OM search	$\langle d, EV_{min}, \text{MM} \rangle$

Table 1. The opponent models used in minimax search, OM search, and (D,d)-OM search.

Table 1 shows that OM search is a generalization of minimax search (in which the opponent does not necessarily use the same evaluation function as the max player), and (D,d)-OM search is a generalization of OM search (in which the opponent does not necessarily search to the same depth as the max player). This is more precisely formulated by the following two remarks.

Remark 1
(D,d)-OM search is identical to OM search when $d = D - 1$.

Remark 2
(D,d)-OM search is identical to minimax when $d = D-1$ and $EV_{min} = EV_{max}$.

Therefore, of the opponent models used in the three search algorithms, the one in (D,d)-OM search has the highest flexibility due to the smallest limitation of the opponent's choice about search depth and evaluation function. So, (D,d)-OM search is the most universal mechanism of the three, and has in principle the largest ability for practical use.

3.2 A Theorem on Root Values

Based on the different back-up procedures of the evaluation-function values, the following theorem can be proven.

Theorem 1.
For the root position R in a game tree we have the following relation:

$$V(R, OM(D,d)) \geq V(R, MM(D)), \qquad (3)$$

where $V(R, OM(D,d))$ denotes the value at root R by (D,d)-OM search and $V(R, MM(D))$ that by minimax search with search depth D. The theorem is proven by induction on the level in the game tree.

The above theorem implies that if the max player has a perfect opponent model, (D,d)-OM search based on such a model can enable the max player to reach a position that may be better, but should never be worse, than the one yielded by the minimax search. In other words, we face the common assumption: the deeper the search, the higher the playing strength.

4 α-β^2 Pruning (D,d)-OM Search

In this section, we introduce an efficient variant of (D,d)-OM search, called α-β^2 pruning (D,d)-OM search.

As is well known, the number of nodes visited by a search algorithm increases exponentially with the search depth. This obviously limits the scope of the search, especially because game-playing programs have to meet external time constraints. Since the minimax search was introduced to game-playing, many techniques have been proposed to speed up the search process, such as the general α-β pruning [11], the null-move method for chess [1] and ProbCut for Othello [2]. On the basis of α-β pruning, Iida *et al.* proposed β-pruning as an enhancement for OM search [4].

(D,d)-OM search backs up the static-evaluation-function values from depths D to $(d+1)$ with minimax, and from depths $(d+1)$ to the root with OM search. Hence it is possible to split up (D,d)-OM search into two parts, and then achieve a search speed-up in both parts separately. To guarantee the generality, we choose α-β pruning to speed up the minimax part, and β-pruning for the OM-search part. The whole algorithm is named α-β^2 pruning.

For details about α-β [11] and β pruning [4], we refer to the literature. Pseudocode for the α-β^2 algorithm is given in Figures 3 and 4.

We note that in the M^* algorithm, the multi-model-based search strategy developed by Carmel and Markovitch [3], a similar pruning mechanism was described as our α-β^2-pruning. However, due to their recursive application of opponent modelling their pruning is not guaranteed to yield always the same result as the non-pruning analogue. Only when the evaluation functions for both players obey certain conditions, in particular when they do not differ too much, the correctness of their $\alpha\beta^*$ algorithm is proven.

```
procedure α-β²(P,α,β,depth):
/* Iterative deepening at root P */
/* Two values are returned, according to equations (2) and (1) */

    if depth = d + 1 then begin
    /* Evaluate the min-player's leaf nodes */
        V_MM[P] ← Evaluate(P,min)
        V_OM[P] ← α-β(P,α,β,d+1)
        return (V_MM[P],V_OM[P])
    end

    {P_i|i = 1,···,n} ← Generate(P)
    /* Expand P to generate all its successors P_i */

    for each P_i do begin
        (V_MM[P_i],V_OM[P_i]) ← α-β²(P_i,α,V_MM[P_i],depth+1)
        if P is a max node then begin
            /* β-pruning at the max node */
            if V_MM[P_i] >= β then begin
                return (V_MM[P],V_OM[P])
            end
        end
    end

    /* Back up the evaluated values */
    if P is a max node then begin
    /* At a max node both the max player and the min player back up the maximum */
        V_MM[P] ←   max    V_MM[P_i]
                  1≤i≤n
        V_OM[P] ←   max    V_OM[P_i]
                  1≤i≤n
    end
    else begin /* P is a min node */
    /* At a min node, the min player backs up the minimum and the max player backs
    up the value of the node selected by the min player */
        V_MM[P] ← V_MM[P_j] =  min   V_MM[P_i]
                              1≤i≤n
        V_OM[P] ← V_OM[P_j]
    end
    /* Update the the value of β */
    β ← V_MM[P]
    return (V_MM[P],V_OM[P])
```

Fig. 3. Pseudocode for the β-pruning part of the α-β^2 algorithm.

5 Experimental Results of (D, d)-OM Search

In this section, we describe two experiments on the performance of (D, d)-OM search, one with a game-tree model including an opponent model and the other in the domain of Othello. The main purpose of these experiments is to confirm the effectiveness of the proposed speculative strategy when a player has perfect knowledge of the opponent model.

```
procedure α-β(P,α,β,depth):
/* Iterative deepening below depth d + 1 */
/* Returns the minimax value according to the max player */

    if depth = D then begin
    /* Evaluate the max player's leaf nodes */
        V_MM[P] ← Evaluate(P,max)
        return (V_MM[P])
    end

    {P_i | i = 1, ···, n} ← Generate(P)
    /* Expand P to generate all its successors P_i */

    for each P_i do begin
        V_MM[P_i] ← α-β(P_i,α,β,depth+1)
        if P is a max node then begin
            if V_MM[P_i] > α then begin
                α ← V_MM[P_i]
            end
            if α >= β then begin
                return (α)
            end
        end
        else begin /* P is a min node */
            if V_MM[P_i] < β then begin
                β ← V_MM[P_i]
            end
            if α >= β then begin
                return (β)
            end
        end
    end

    /* Back up the evaluated values */
    if P is a max node then begin
        V_MM[P] ← max   V_MM[P_i]
                 1≤i≤n
    end
    else begin /* P is a min node */
        V_MM[P] ← min   V_MM[P_i]
                 1≤i≤n
    end
    return (V_MM[P])
```

Fig. 4. Pseudocode for the α-β-pruning part of the α-β^2 algorithm.

5.1 Experiments with Random Trees

In order to investigate the performance of a search algorithm, a number of game-tree models have commonly been used [12,13]. However, for OM-like algorithms we need a model including an opponent model. Iida et al. have proposed a game-tree model to measure the performance of OM search and tutoring-search algorithms [7]. On the basis of this model, we build another game-tree model including the opponent model to estimate the performance of (D,d)-OM search. As a measure of performance, we use the H value of an algorithm like we did for OM search. With this game-tree model and the H values, the performance of (D,d)-OM search is studied.

Game-Tree Model The game-tree model we use for this experiment is a uniform tree. A random score is assigned for each node in the game tree and the scores at leaf nodes are computed as the sum of numbers on the path from the root to the leaf node. This incremental model was also proposed by Newborn [14] and goes back to a scheme proposed by Knuth and Moore [11]. The max player's score for a leaf position at depth D (say P^D) is calculated as follows:

$$EV_{max}(P^D) = \sum_{k=0}^{D} r(P^k); \qquad (4)$$

the min player's score for a leaf position at depth $(d+1)$ (say P^{d+1}) is calculated as follows:

$$EV_{min}(P^{d+1}) = \sum_{k=0}^{d+1} r(P^k), \qquad (5)$$

where $-R \leq r(\cdot) \leq R$, and $r(\cdot)$ has a uniform random distribution and R is an adjustable parameter. The resulting random numbers at leaf nodes have a normal distribution. Note that the min player uses the same random score $r(\cdot)$ as the max player. It is implied that $EV_{max} = EV_{min}$ when $D = d+1$. In this case, (D,d)-OM search is identical to the minimax strategy according to Remark 2.

This game-tree model comes close to approximating the parent/child behaviour in real game trees and reflects a game tree including models for both players, in which different opponent models are simulated by various search depths d. For this game-tree model, we recognize that the strength of the min player is equal to that of the max player when $d = D-1$ and that the min player has less information from the search tree about a given position when $d < D-1$. Note that we only investigate positions for which $d \leq D-1$, since otherwise (D,d)-OM search is unreliable and should not be used.

H Value In order to estimate the performance of (D,d)-OM search we define the so-called H value (*Heuristic performance value*) for the root R by

$$H(R) = \frac{V(R, OM(D,d)) - V_{min}(R,D)}{V_{max}(R,D) - V_{min}(R,D)} \times 100 \qquad (6)$$

Here, $V(R, OM(D, d))$ represents the value at R by (D, d)-OM search. $V_{min}(R, D)$ is given by

$$V_{min}(P, D) = \min_i EV_{max}(P_i), P_i \in \text{all the leaf nodes at depth} D \qquad (7)$$

$V_{max}(P, D)$ is similarly given by

$$V_{max}(P, D) = \max_i EV_{max}(P_i), P_i \in \text{all the leaf nodes at depth} D \qquad (8)$$

The procedure indicated by (7) obtains the minimum value of the root R by looking ahead D plies and the strategy indicated by (8) analogously the maximum value. $H(R)$ then represents the normalized performance of (D, d)-OM search and can be thought of as a characteristic of the strategy. Although the value of this performance measure remains to be proven, we feel that the scaling applied by using the minimum and maximum values of the leaves sets the resulting performance in appropriate perspective.

Preliminary Results on the Performance of (D, d)-OM Search To get insight in the performance of (D, d)-OM search, several preliminary experiments were performed using the game-tree model proposed above.

In a first experiment, we observed the performance of (D, d)-OM search for various values of d. In this experiment, D is fixed at 6 and 7, and d ranges from 0 to $D - 1$. A comparison of $(6, d)$-OM search and minimax search is presented in Figure 5, while $(7, d)$-OM search and minimax search are compared in Figure 6, all with a fixed branching factor of 5. All curves shown in Figures 5 and 6 are averaged results over 100 experiments.

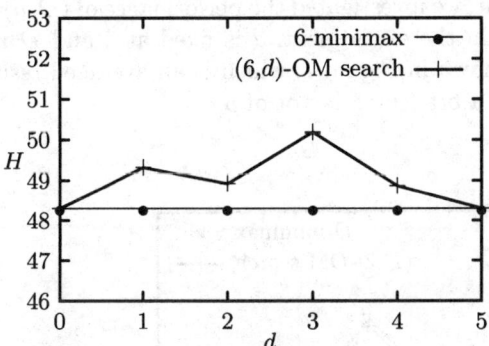

Fig. 5. $(6, d)$-OM search and minimax compared.

Figures 5 and 6 show that

- the results are in accordance with Theorem 1 and Remark 2. In particular,
  - $d = 0$ means that the opponent does not perform any search at all. The max player therefore has to rely on minimax.

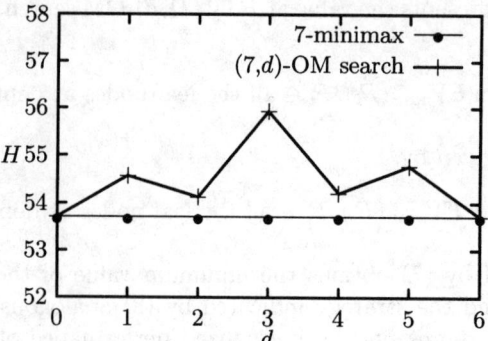

Fig. 6. $(7,d)$-OM search and minimax compared.

- when $d = 5$ in Figure 5 and $d = 6$ in Figure 6, i.e., $d = D - 1$, the min player looks ahead to the same depth in the search tree as the max player. In this case, the max player actually performs pure OM search. Since $EV_{max}(P) = EV_{min}(P)$ in our experiments, the conditions laid down in Remark 2 are fulfilled, and (D,d)-OM search is identical to minimax.

- the fluctuation in H values of (D,d)-OM search for depths d from 1 to $D-1$ hardly seems dependent on the value of d. This is explained by the fact that the ratio of mistakes of OM search does not depend on the depth of search, but only on the branching factor [6]. The results may suggest that the fluctuation in H values of (D,d)-OM search has a maximum at $d = \lfloor D/2 \rfloor$.

In a second experiment, we investigated the performance of (D,d)-OM search for various values of D. In the experiment, d is fixed at 2 and D ranges from 3 to 7. The results are shown in Figure 7, which is an averaged result over 100 experiments, again using a branching factor of 5.

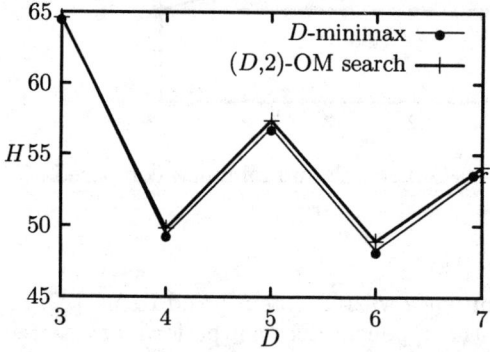

Fig. 7. $(D,2)$-OM search and minimax compared.

Figure 7 tells us that the H value of (D, d)-OM search is greater than that of D-minimax. Of course, the gain of (D, d)-OM search over D-minimax is very small, since d is fixed at 2, which means that OM search is only performed in the upper 2 plies, whereas in the remainder of the search tree minimax is performed. In addition, (D, d)-OM search and D-minimax show the same fluctuation in H values, a consequence of both using the same evaluation function.

5.2 Othello Experiments

In the subsection above, the advantage of (D, d)-OM search over D-minimax has been verified with random-tree-model simulations. However, simulating tree behaviour is fraught with pitfalls [15]. So, now let us turn to the study of effectiveness of the proposed speculative strategy in real game-playing. Due to the simple rules and relatively small branching factor, Othello is selected as a test bed. We assume that the rules of the game are known. In determining the final score of a game we adopt the convention that empty squares are not awarded to any side. The concept *net score* is used as the difference in number of stones of a finished game, e.g., in a game with final score 38-25 the first player has a net score of 13.

Experimental Design For easy comparison, we let program A with model $M_A = \langle D, EV, (D,d)\text{-OM} \rangle$ and program B with model $M_B = \langle D, EV, \text{MM} \rangle$ play against program C with model $M_C = \langle d, EV, \text{MM} \rangle$. The results of A against C compared to those of B against C then serve as a measure of the relative strengths of (D, d)-OM search and D-MM search. EV again denotes the evaluation function. To simplify the experiments, we do not consider the influence of the evaluation function for the moment, i.e., we use the same evaluation function for programs A, B and C.

In the experiments programs A and B search to the same depth D, whereas program C searches to depth d. The cases $D = d + 1$, $D = d + 2$ and $D = d + 3$ are investigated.

Performance Measure Two parameters $\overline{\Delta S}$ and R_w are defined to estimate the performance of (D, d)-OM search and D-MM search. $\overline{\Delta S}$ represents the average net score and R_w denotes the winning rate of the player. For a given player X, $\overline{\Delta S}(X)$ is given by

$$\overline{\Delta S}(X) = \frac{1}{2N} \sum_{j \in (B,W)} \sum_{i=1}^{N} \Delta S_i^j(X) \qquad (9)$$

In this formula, $\Delta S_i^B(X)$ denotes the net score obtained by player X when he plays with Black. Similarly, $\Delta S_i^W(X)$ is the analogous number for playing White, and $2N$ represents the total number of games, equally divided over games starting with Black and with White. Therefore, this performance measure offsets

Programs	Performance Measure	d			
		1	2	3	4
A vs. C	Scores	37.4/26.6	35.8/28.2	38.8/25.0	39.2/24.8
	$\Delta S(A)$	10.8	7.6	13.8	14.4
	$R_w(A)$	66%	65%	69.5%	73.5%
B vs. C	Scores	37.4/26.6	35.8/28.2	38.8/25.0	39.2/24.8
	$\Delta S(B)$	10.8	7.6	13.8	14.4
	$R_w(B)$	66%	65%	69.5%	73.5%

Table 2. The results of programs A and B vs. program C, for $D = d + 1$.

the influence caused by having the initiative, which in general is widely believed to be a decisive advantage in White's favor.

The winning rate of player X, $R_w(X)$ is defined as,

$$R_w(X) = \frac{n+m}{2N} \times 100\%, \qquad (10)$$

where n denotes the number of won games when X plays with White, and m is that when X plays with Black.

In our experiments, we let $N = 50$, i.e., a total of 100 games are played for each case.

Preliminary Results Table 2 shows the results for the case $D = d + 1$, where the average scores by 100 games are given in the format x/y, with x the number of stones obtained by the first player and y by the opponent.

From Table 2 we see that programs A and B obtain identical scores against program C, in accordance with Remark 2, i.e., that in the case $D = d+1$ (D,d)-OM search is identical to D-MM search. In addition, the results indicate that deepening search can confer some advantage. When $D = d + 1$, the average winning rate is approximately 68.5%.

Table 3 lists the results for the case $D = d+2$, showing that the performance of (D,d)-OM search then always is significantly better than that of D-MM search by a small margin.

We speculate that the edge of (D,d)-OM search over D-MM search will increase with a better evaluation function (the present one mainly just counting disks). This is an area for future research.

Table 4 gives the results for the case $D = d+3$. Again it is clear that (D,d)-OM search is stronger than D-MM search. However, when $d = 3$, although the winning rate of (D,d)-OM search is greater than that of D-MM search, the average net gain of (D,d)-OM search is surprisingly lower. We believe that this also is a result of the use of a simplified evaluation function. Comparing Tables 2-4 we also notice that the benefit of (D,d)-OM search over D-MM search grows with larger difference in search depth between the opponents. Obviously, OM

Programs	Performance Measure	d			
		1	2	3	4
A vs. C	Scores	39.9/24.1	41.7/22.3	41.2/22.8	40.2/23.8
	$\Delta S(A)$	15.8	19.4	18.4	16.4
	$R_w(A)$	75.5%	78.5%	79%	76.5%
B vs. C	Scores	37.8/26.2	39.7/24.3	40.8/22.9	39.9/24.1
	$\Delta S(B)$	11.4	15.4	17.9	15.8
	$R_w(B)$	68.5%	76%	78%	74.5%

Table 3. The results of programs A and B vs. program C, for $D = d + 2$.

Programs	Performance Measure	d		
		1	2	3
A vs. C	Scores	43.9/20	45.4/18.6	42.1/21.9
	$\Delta S(A)$	23.9	26.8	20.2
	$R_w(A)$	88%	88.5%	94%
B vs. C	Scores	41.8/22.1	43.7/20.3	44.4/19.5
	$\Delta S(B)$	19.7	23.4	24.9
	$R_w(B)$	85%	86.5%	90%

Table 4. The results of programs A and B vs. program C, for $D = d + 3$.

search is suited to profit as much as possible from defects in the evaluation function, which is precisely the reason why (D, d)-OM search was proposed. Moreover, although the margins are small we see from Tables 2-4 that (D, d)-OM search always is as good as (when $D = d + 1$) or better (when $D > d + 1$) than minimax. We feel that the significance of this observation also depends on the evaluation function in use. This will be subject of future research.

6 Applications of (D, d)-OM Search

Since (D, d)-OM search stems from grandmasters' experience, it is implied that the player using this strategy has a higher strength. Even then, a grandmaster employs only in some special cases (D, d)-OM search to get some advantage. These include the case that the opponent is really weak, and the case that the grandmaster reaches some weak position. Regarding the former, (D, d)-OM search can help the player win in fewer moves or with more gains. With respect to the latter, the grandmaster has to wait for mistakes by his opponent, in which case (D, d)-OM search can help him to change a situation.

6.1 The Requirements for Applying (D, d)-OM Search

So far, we assumed that the max player's static evaluation function EV_{max} is possibly different from the min player's one EV_{min}. However, it is very difficult to have reliable knowledge of the opponent's evaluation function to perform (D, d)-OM search. On the other hand, knowledge of the opponent's search depth (especially when the opponent is a machine) may be more reliable. We therefore restrict ourselves in this section to potential applications of (D, d)-OM search for the case $EV_{max} = EV_{min}$.

Under this assumption the requirements for applying the proposed (D, d)-OM search can be given by the following Lemma.

Lemma 1.
Let δ be the search depth difference between the max player and the min player in game-playing, i.e., $\delta = D - d$. If $\delta \geq 2$, then (D, d)-OM search can be applied.

This means that the condition $\delta \geq 2$ gives the minimum depth difference at which it is beneficial to use (D, d)-OM search over minimax in order to anticipate on the opponent's errors resulting from its limited search depth.

The detailed proof for the above lemma can be found in [5]. Furthermore, we can estimate in how many ways (D, d)-OM search can be applied. Each way of applying (D, d)-OM search is completely defined by the players' search depths D and d, where, for definiteness, $D \geq d + 2$ (from Lemma 1 and Definition 2). By simple discrete summation, we find for the number of ways, considering that the min player may, from instance to instance, choose any model with depth at most equal to d and since the max player may respond by choosing his D to match, that

$$N(D, d) = \sum_{i=1}^{d}(D - i - 1) = D \times d - \frac{1}{2}d(d+3),$$

where $N(D, d)$ denotes the number of ways of applying (D, d)-OM search.

6.2 Possible Applications

Since (D, d)-OM search is a speculative strategy, the reliability depending on the correctness of the model of the opponent, it may seem unlikely that such a strategy will be of much practical use in game-playing. However, there are several situations where such a strategy can be of significant support.

One such possible application is in building a tutoring strategy for game-playing [7]. In this case, compared with the pupil, the tutor can be considered as a grandmaster. It is essential, if tutoring is to be successful, that the tutor has a clear representation of his pupil. This statement is paramount in ranging tutoring strategies into the wider context of methods possessing a clear picture of their opponents. Tutoring strategies therefore are necessarily a special case of models possessing an opponent model. The balance in tutoring strategies is delicate: on the one hand it is essential that the tutor has a good model of his opponent. Yet it is also required that the give-away move be not so obvious as to be noticeable by the person being tutored. Thereby, with the help of (D,d)-OM search, the game is manipulated in the direction of an interesting position from which the novice may find a good or excellent move "by accident"; the novice's interest in the game may increase, stimulating his progress on the way towards becoming a strong player.

Another place of possible application is to devise a cooperative strategy for multi-agent games, such as soccer [10], 4-player variants of chess [16] and so on. In such games, (D,d)-OM search can be used by the stronger player to construct a cooperative strategy with his partner(s). Here, compared to the weaker partner(s), the stronger one is a grandmaster, who can apply (D,d)-OM search in order to model his partner(s) play [9]. One large advantage of such cooperative strategies is that it is much easier to obtain a reliable partner model than an opponent model.

7 Conclusions and Limitations

In this paper, a speculative strategy for game-playing, (D,d)-OM search, is proposed using a model of the opponent, in which difference in search depths is explicitly taken into account. The algorithm and characteristics of this search strategy are introduced. A more efficient variation, named $\alpha - \beta^2$. Experimental results with random-tree simulations and using Othello confirm its effectiveness.

Although the opponent model used by (D,d)-OM search is more flexible than that by pure OM search, it is difficult to have a reliable estimate of the search depth and evaluation function of the opponent. Mostly, the max player will only have a tentative model of his opponent, and as a consequence this will lead to a risk if the model is not in accordance with the real opponent's thinking process. Whereas preliminary experiments indicated that the applicability of OM search is greater for weaker opponents [8], more work will be needed to investigate whether this holds also for (D,d)-OM search.

Another point for future research is the recursive application of (D,d)-OM search, analogous to Carmel and Markovitch' [3] M^* algorithm. Suppose we use (4,1)-OM search. In the present implementation the algorithm uses 2-MM search to determine the Max player's values at depth 2. A better exploitation of the opponent's weakness would be to use (2,1)-OM search then. The computational costs for this extension should carefully be weighed against the benefits.

Acknowledgement

This work was supported in part by the Japanese Ministry of Education Grant-in-Aid for Scientific Research on Priority Area 732. We are grateful to the anonymous referees whose comments have resulted in numerous improvements to this paper.

References

1. G.M. Adelson-Velskiy, V.L. Arlazarov and M.V. Donskoy. Some Methods of Controlling the Tree Search in Chess Programs. *Artificial Intelligence*, 6(4):361–371, 1975.
2. M. Buro. ProbCut: An Effective Selective Extension of the Alpha-Beta Algorithm. *ICCA Journal*, 18(2):71–76, 1995.
3. D. Carmel and S. Markovitch. Pruning Algorithms for Multi-Model Adversary Search. *Artificial Intelligence*, 99(2):325–355, 1998.
4. H. Iida, J.W.H.M. Uiterwijk, and H.J. van den Herik. Opponent-Model Search. *Technical Reports in Computer Science*, CS 93-03. Department of Computer Science, Universiteit Maastricht, Maastricht, The Netherlands, 1993.
5. H. Iida, J.W.H.M. Uiterwijk, H.J. van den Herik, and I.S. Herschberg. Potential Applications of Opponent-Model Search. Part 1: The Domain of Applicablity. *ICCA Journal*, 16(4):201–208, 1993.
6. H. Iida. *Heuristic Theories on Game-Tree Search*. Ph.D. thesis, Tokyo University of Agriculture and Technology, Tokyo, Japan, 1994.
7. H. Iida, K. Handa, and J.W.H.M. Uiterwijk. Tutoring Strategies in Game-Tree Search. *ICCA Journal*, 18(4):191–204, 1995.
8. H. Iida, I. Kotani, J.W.H.M. Uiterwijk, and H.J. van den Herik. Gains and Risks of OM Search. *Advances in Computer Chess 8* (eds. H.J. van den Herik and J.W.H.M. Uiterwijk), pages 153–165. Universiteit Maastricht, Maastricht, The Netherlands, 1997.
9. H. Iida, J.W.H.M. Uiterwijk, and H.J. van den Herik. Cooperative Strategies for Pair Playing. *IJCAI-97 workshop proceedings: Using Games as an Experimental testbed for AI Research* (ed. H. Iida), pages 85–90. Nagoya, Japan, 1997.
10. H. Kitano, M. Asada, Y. Kuniyoahi, I. Noda, and E. Osawa. Robocup: The Robot World Cup Initiative. *Proceedings of the IJCAI-95 Workshop on Entertainment and AI/Life* (eds. H. Kitano, J. Bates and B. Hayes-Roth), pages 19–24. IJCAI, Montreal, Québec, 1995.
11. D.E. Knuth and R.W. Moore. An Analysis of Alpha-Beta Pruning. *Artificial Intelligence*, 6:293–326, 1975.
12. T.A. Marsland. Relative Efficiency of Alpha-Beta Implementations. *Proceedings of the 8th International Joint Conference on Artificial Intelligence (IJCAI-83)*, pages 763–766, 1983.
13. A. Muszycka and R. Shinghal. An Empirical Comparison of Pruning Strategies in Game Trees. *IEEE Transactions*, SMC-15(3):389–399, 1985.
14. M.M. Newborn. The Efficiency of the Alpha-Beta Search on Trees with Branch-dependent Terminal Node Scores. *Artificial Intelligence*, 8:137–153, 1977.
15. A. Plaat, J. Schaeffer, W. Pijls, and A. de Bruin. Best-First Fixed-Depth Minimax Algorithms. *Artificial Intelligence*, 87(1–2):255-293, 1996.
16. D.B. Pritchard. *The Encyclopedia of Chess Variants*. Games & Puzzles Publications, Godalming, Surrey, UK, 1994.

An Adversarial Planning Approach to Go

Steven Willmott[1], Julian Richardson[2], Alan Bundy[2], and John Levine[3]

[1] Laboratoire d'Intelligence Artificielle,
École Polytechnique Fédérale de Lausanne, Lausanne
willmott@lia.di.epfl.ch

[2] School of Artificial Intelligence, Division of Informatics,
University of Edinburgh
julianr@dai.ed.ac.uk

[3] Artificial Intelligence Applications Institute, Division of Informatics,
University of Edinburgh
johnl@aiai.ed.ac.uk

Abstract. Approaches to computer game playing based on (typically $\alpha - \beta$) search of the tree of possible move sequences combined with an evaluation function have been successful for many games, notably Chess. For games with large search spaces and complex positions, such as Go, these approaches are less successful and we are led to seek alternative approaches.

One such alternative is to model the goals of the players, and their strategies for achieving these goals. This approach means searching the space of possible goal expansions, typically much smaller than the space of move sequences.

In this paper we describe how adversarial hierarchical task network planning can provide a framework for goal-directed game playing, and its application to the game of Go.

1 Introduction

Most approaches to computer game playing are based on game tree search and position evaluation functions (data driven approaches). Data driven approaches are appropriate for games with small branching factors, and for which it is possible to accurately assign values to positions which indicate who is winning. While this approach has been very successful for many games including computer Chess, it has been less successful when applied to games with high branching factors and complex positions, such as Go or for games with a high degree of uncertainty such as Bridge.

An alternative to the data driven approach is goal driven search in which a single agent tries to satisfy its goals in the game. Goal driven search has been extensively explored in the AI literature, in particular as Hierarchical Task Network (HTN) planning [23, 8]. When multiple agents need to be modeled and can compete against one another this becomes *adversarial planning*. This paper presents an adversarial planning architecture for performing goal driven reasoning in games and describes its application to the game of Go.

1.1 Goals and Data

Within a specific game, move or action choices often depend upon the state of the game, the phase of the game (e.g. opening, endgame etc.), the future actions of the opponent, the ability of a player to follow up an action appropriately and many other diverse factors. It is these interacting influences on the choice and effect of moves which make games so fascinating for human players and so challenging for machines.

In computer game playing there are two main approaches to making move choices:

- **Data Driven:** At each step, rules, patterns or heuristics are applied to the game state to suggest useful moves. The resulting set of plausible actions is then evaluated using search in the tree of moves. Each move is played out in a world model followed by the possible responses of the opponent. The search continues until the leaves of the tree are reached.[1] These leaf nodes are then evaluated and used to select one of the original plausible actions as the one which leads to the most desirable (by some measure) set of leaf states.

- **Goal Driven:** During play, a goal driven system keeps a number of abstract goals in an agenda. The goals in the agenda represent the things the system would like to achieve in the short, medium and long term. To choose a move, goals are expanded into plans (which are conjunctions of goals at lower levels of abstraction) and eventually into concrete moves. Repeated decompositions form a plan for achieving the goal.

In a data driven search tree, each node represents a possible game position and has one branch for every legal move in that position. By contrast, each node in a goal driven search tree represents a *plan* for achieving the top level goal with some parts still sketchy (abstract goals) and others fixed (concrete actions), and there is one branch at the node for every way the system suggests to further refine the plan.

Which approach (goal driven or data driven) is most advantageous is heavily dependent upon the domain, in particular on the size of the data driven and goal driven search trees. In Bridge, for example, the locations of the cards are not in general known during play, which leads to a large space of possible card plays and therefore a prohibitively large data driven search tree. Frank ([10]) shows that a goal driven approach can very successfully play bridge; a relatively small number of operators is sufficient to describe all the relevant plays.

1.2 The Search Space in the Game of Go

The game of Go is considered by many to be the next great challenge for computational game playing systems. It presents new, significant and different challenges to Chess which has been long been considered the "task par excellence"

---

[1] Which nodes are the "leaves" can be variously defined by a depth cut off point, quiescence, or further domain dependent heuristics.

for AI (Berliner [2]). Go's search space is both wider and deeper than that of Chess; the size of Go's search space is estimated to be 10^{170} states (cf. Chess $\approx 10^{50}$), games last approximately 300 moves (cf. Chess ≈ 80) and the branching factor at each turn is approximately 235 states (cf. Chess ≈ 35). It is often hard to evaluate the relative strength of Go positions during play. We therefore expect that the brute-force game tree search which has been so effective for Chess will have much greater difficulty with Go.

Due to space limitations, we do not include the rules of Go in this document. A good introduction to the game can be found in, for example, [3].

1.3 Approaches to Computer Go

Although Go has received far less attention than Chess in terms of research, there have been many varied approaches to computer Go. A good summary can be found in [4]. The programs which can play the whole game most successfully, such as GO4++ and MANY FACES OF GO [9], are primarily data driven but also employ other techniques. GOGOL [6] is able to learn patterns for play and various non-symbolic techniques have been used to learn/evolve Go playing controllers (see [26] for more references). Important techniques have also been developed for playing part of the game, in particular focusing on "Life and Death" [27], [28], and using combinatorial game theory to play the endgame [13].

Despite the success of current approaches there is recognition that there is still substantial room for improvement. The advantages mentioned in §6 and earlier work [20, 12] along with the success of this approach in other domains (notably bridge [22]) suggest that a goal driven approach may be useful. It also has much psychological validity, since protocol analysis indicates that Go players consider few candidate moves, and concentrate on their own and on their opponents' purposes [19]. Finally, even in data driven approaches to computer Go, it is still necessary to consider high-level goals, for example in order to decide whether or not a satisfactory result has been achieved in life and death problems (e.g. some strings may be allowed to die but others must live) [28].

1.4 Applications of Adversarial Planning

Early attempts to use goal driven reasoning in adversarial domains include [14] and [5]. The work by Pitrat ([14]) was extended by Wilkins in [24] to produce the successful PARADISE system for Chess. More recent work includes [1] (battlefield management) and [30] (command and control). The most recent work is by Smith et. al. on bridge (described in [22] and [21]) which presents a goal driven system for bridge declarer play (TIGNUM2) good enough to beat the current top commercial computer player.

There have also been several attempts to apply this kind of technique to Go. Early Go Planners due to Sander and Davies [20] and Lehner [12] suffered from the fact that they only work in very open positions and provide high level vague plans. These early systems have difficulties in complex tactical situations, to a large extent due to the difficulty of modeling the interactions between high level goals which must be done for effective HTN planning. One of the crucial differences in the framework presented here is the use of linearisations to produce a total order planner and the use of a world model to track these interactions.

2 An Adversarial Planning Architecture

This section describes an Adversarial Planning Architecture which models goal driven reasoning for adversarial domains.[2] The goal driven approach and use of abstract plans is motivated by work on Hierarchical Task Network (HTN) planning. HTN systems were first used in NOAH [17] and INTERPLAN [23] and have since been extensively studied in the AI planning field. Erol et. al. ([8]) give a complete definition for an HTN scheme and presents UCMP which is a provably sound and complete HTN planner and provides a good template for this type of system.

2.1 Principles of HTN Planning

HTN planning is based on three types of object: *Goals*, *Operators* and *Plan Schemas*. Operators are actions which can be performed in the world (such as flicking a switch, taking a step). Goals are more abstract and express aims in the world such as "Go to the Moon", "Become Prime Minister". Schemas (also called Task Networks), specify the subgoals which must be achieved in order to satisfy the goal. For example, the following schema expresses the fact that G can be achieved by satisfying the conjunction of subgoals G_1, G_2 and G_3:

$$G : - > G_1 + G_2 + G_3$$

The G_i should be at a lower level of abstraction than G, and can generally be satisfied in any order. Operators are at the lowest level of abstraction.

Given these three types, HTN planning starts by taking a starting world state and a set of goals which form the initial abstract plan, which is then refined step by step by expanding the goals within it. Goals are expanded by selecting a schema with the chosen goal as the antecedent (the G) and replacing the instance of G in the current plan by the subgoals (the G_i) listed in the consequent of the schema. As the planning process continues, interactions, incompatibilities and conflicts may arise between combinations of goals. These "interactions" in the plan must be resolved, which can result in backtracking and (in partial order planners) ordering constraints between goals.

The process is complete when all goals have been expanded into sets of operators and all arising interactions have been resolved. The sequence of operators thereby generated should, upon execution in the initial world state, lead to the achievement of the initial goals in the world.

The extension of this idea into adversarial domains is non-trivial since plans are no longer sequences of actions but trees of contingencies which take into

---

[2] More details can be found in [25].

account the actions of opponents. The interactions in the plan are considerably more complex and serious since the goals of opponents in the world are often conflicting and since the agents are non cooperative. HTN planning for adversarial domains is computationally considerably more complex than HTN planning in standard domains.

2.2 Planning Framework

The adversarial planner presented here models two agents (named Alpha and Beta) which represent two players (*adversaries*) in a game (the framework can be generalised to more than two players). Each agent keeps an open agenda of goals which represents its current plan of action. To solve a problem in a domain, each agent is given an abstract goal (or set of goals) to achieve. The agents then attempt to find sequences of moves which satisfy their goals. Since the goals of the agents are usually contradictory and the agents must take turns in performing actions, their interaction in trying to satisfy their goals can be used to find a plan for the situation.[3]

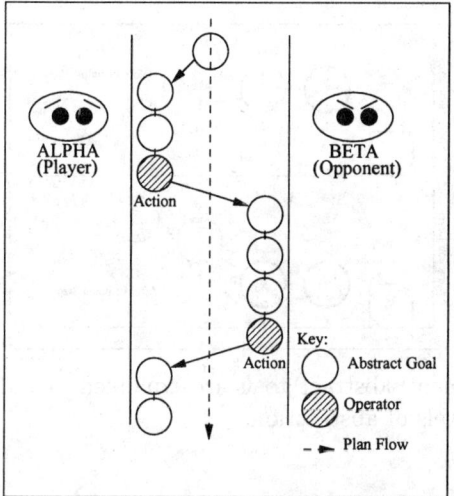

Fig. 1. Planning steps alternating between two agents.

The system allows the two agents to take control of the reasoning apparatus in turns. Once an agent has control it expands some of its abstract goals until it is able to decide upon a concrete action. The chosen action is then performed in a world model[4] before control is passed to the other agent. Figure 1 shows

---

[3] See below for how this helps choose moves.
[4] A world model is not a standard feature of HTN planners — see §2.3 and [25] for more explanation of its use.

the flow of control during the reasoning process. An agent may need to expand several abstract goals before being able to decide upon an action in the world. During this "active" period it uses its own agenda of goals and has control of the shared reasoning apparatus. Once an action is chosen, control passes to the other agent. Agent Alpha models the player who is *next to move in the game* and agent Beta the opponent, thus the planner is trying to plan for Alpha's move (Alpha takes control first).

At any one time an agent has a plan which consists of actions already taken (square boxes in figure 2) and goals at various levels of abstraction (circles in figure 2). The actions (squares) are represented in the world model, the abstract goals (circles) are held in the agenda.

A planning step involves selecting an abstract goal (such as X in figure 2) and expanding it. To do this a plan schema is selected for X which expresses how X could be achieved using a conjunction of subgoals at a lower level of abstraction. In figure 2, X is replaced in the plan by the two subgoals X1 and X2. Once expansion has reached the lowest level of abstract goals these lowest level goals need to be shown to be already true or replaced by actions which make them true.

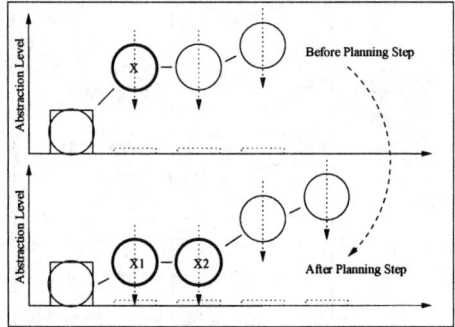

Fig. 2. Plan refinement: abstract goals are expanded to and replaced by sets of subgoals at lower levels of abstraction.

Once one of the agents (Alpha say) has achieved all of its goals (been able to perform actions in the world model which make them true) it knows that it must have satisfied its top level goals (since all its subgoals are strictly descended from these). The other agent is made aware of this and, since in general a good outcome for one agent is a bad outcome for the other, both agents are allowed to force backtracking. Agents are allowed to backtrack to any of their previous goal or expansion choices but only to their *own* decisions. Neither agent may force the other to change plans directly.

The backtracking activity explores the various interacting plans Alpha and Beta have for the situation and creates a plan tree as shown on the left of figure 3. Each choice made by an agent creates a new branch. Underlying the

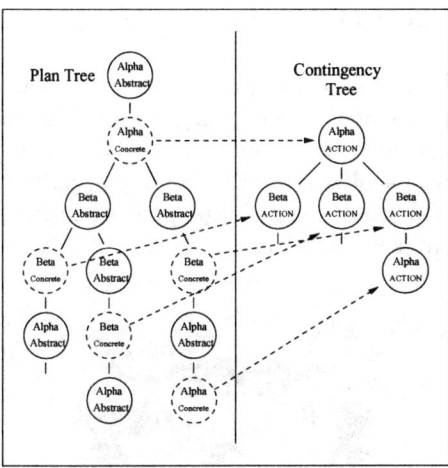

Fig. 3. The plan tree on the left is reduced to the contingency tree on the right by dropping the abstract reasoning nodes.

plan tree is the *contingency tree* which is found by removing all the abstract goal decomposition steps in the plan tree to leave only the operators/actions (shown on the right in figure 3). The final contingency tree acts as a form of proof that the first move is a good step towards achieving Alpha's goals. Hence it supports the choice of the first move in the tree.[5] In general, the final contingency tree contains only a small subset of the move tree which would be generated by considering all the available options at each turn as in a data driven approach. (See §6 below.)

2.3 Discussion

Moves in the contingency tree are *directly descended* from the goals of the two agents, and the tree structure naturally reflects the interactions between the two adversaries. Taking any branch, the moves chosen near the leaf (at the end of a move sequence) are still directly related to the same plan and aim that suggested those moves near the root (the first moves in the tree).

A key difference from standard HTN planning is that goals are expanded in time order using a linearisation. Such a linearisation turns the planner from a *partial order* into a *total order* planner and allows the use of a world model. Figure 2 illustrates how the least abstract goals/actions appear first in the sequence. As soon as goals reach the lowest level of abstraction (become operators) they can be added to the world model. This is a significant difference from standard HTN planners, its importance is discussed further in [25] and [26].

---

[5] The tree itself can also be used to respond to any of the opponents moves which are represented in it, but re-planning may be required if other moves are made. The question of how long (for how many moves) such a plan remains valid is not addressed here.

3 An Adversarial Go Planner

The planning architecture was instantiated as a Go reasoning system called GOBI both to test the architecture in a challenging domain and to investigate the usefulness of the goal driven approach for Go. GOBI consists of a set of knowledge modules which plug into the planning architecture. The knowledge modules provide the Go domain knowledge, plan schemas and goal types which the reasoner can use to solve problems.

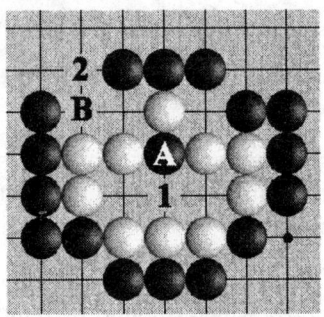

Fig. 4. Example 1: Black to play and kill the white group.

3.1 An Example Go Plan

Figure 4 shows a problem from Volume I of "Graded Go Problems for Beginners" (Yoshinori [29]). The aim is for black to move first and kill the white group of stones. The task is specified to GOBI as two abstract goals: the goal *kill-group* for agent Alpha (playing black) and the goal *save-group* for agent Beta (playing white).[6] Agent Alpha takes control, first decomposing the *kill-group* goal using one of the available plan schemas. An abstract plan for killing this group might be a conjunction of the following subgoals:[7]

- *surround-group* — stop the group from running and connecting.
- *squeeze-space* — reduce the space the group has to live.
- *prevent-eye-formation* — block any attempt by the group to make eyes.[8]

One of these subgoals is then expanded further to the next level and so on until at the lowest level in the hierarchy a move such as *play at B* is chosen to satisfy a simple goal such as *prevent-escape-at-2* (figure 4).

---

[6] Note: the goals need not be directly opposing.
[7] This abstract plan is quite intuitive. It is not obvious how a data driven system would represent the equivalent of such a plan.
[8] An eye in Go is an enclosed space where the opponent may not play - a group with two eyes is unconditionally alive.

Alpha plays this move onto the board in the world model which gives the new world state for Beta to work with. Alpha still has a set of goals at various levels of abstraction remaining in its agenda. These remaining goals represent the plan on how to follow the first move, i.e which other subgoals/actions need to be achieved to make the plan complete. To validate that this first move is good (in this case playing at B would not be), Alpha must eventually show that all these goals can be achieved no matter what Beta does. These remaining goals are kept by Alpha until after Beta's turn.

Beta now begins by expanding its single goal *save-group* in the context of the new board position (after Alpha playing at B in Figure 4). A possible plan for this goal is:

— *make-eye-space*.
— *make-eyes* (try to form two eyes).

After Beta's move (whatever it is) is played into the world model, control is returned to Alpha which then tries to satisfy the rest of its goals. The interleaving of goal expansions by the two agents continues until one is able to satisfy all of its aims (and thus implicitly its main aim). The opposing agent is informed of this and it then backtracks to explore any alternative options it has which might produce a better outcome for itself. In this way a contingency tree is generated which either proves or refutes the validity of the first move (Alpha's).

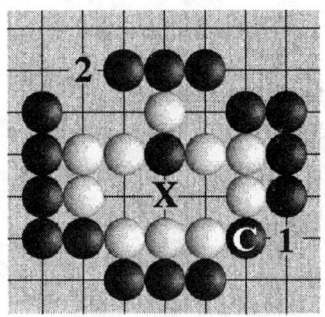

Fig. 5. GOBI plays at X and this kills the group.

For this example (figure 4) GOBI returns the move at X in figure 5, which kills the group. Among the defences tried by Beta are trying to run out at 2 and counter-attacking by playing at 1 (this puts the single black stone C under threat). Since all the moves tried by both agents must be *part of plan of action*, the number of possible moves searched is very small compared to the number of available moves in even this small problem.

3.2 Representing Go Knowledge

The planning architecture and Go knowledge modules which make up GOBI are all written in Common Lisp. Around 1400 lines of code make up the plan knowledge (i.e. schemas and goal types) GOBI has. Writing a full-board Go-playing program is a significant exercise in knowledge engineering, so to enable us to add enough depth in knowledge to do useful testing in a short time, GOBI's knowledge is focused on the area of killing and saving groups.[9] The knowledge base is made up of 45 goals at five different levels of abstraction. The average number of possible plans per goal is approximately two (thus the knowledge has a relatively low branching factor). The two highest level goals available in GOBI are *kill-group* and *save-group*, which emphasises the focus on life and death problems.

The following example goal taken from GOBI's knowledge base illustrates the structure of the knowledge the program holds (the *save-group* goal was also mentioned in the previous example). Note that the plan knowledge is not complete (for example making eyes is not the only way of following up a counter attack); more work is needed to extend the knowledge base.

```
GOAL: save-group,
LEVEL = 5,

  Plan 1 - Find Eyes:
  *make-eye-space,
  *make-eyes.

  Plan 2 - Escape Group:
  *running-sequence,
  *secure-escape.

  Plan 3 - Counter Attack String:
  *locate-vulnerable-string,
  *kill-string,
  *make-eyes.
```

In turn the *make-eye-space* goal from `Plan1` has two alternative plans:

```
GOAL: make-eye-space,
LEVEL = 4,

  Plan 1 - Ambitious Extend:
  *large-extending move,
  *consolidate-space.

  Plan 2 - Creep extension:
  *creeping-extending-move,    //A single
  *consolidate-space.          //step extension.
```

The structure of the knowledge shown here is intuitive for Go and is very different from the kind of heuristic information added into data driven approaches. The knowledge represented can be seen as an AND-OR tree with the AND component

---

[9] Note however that this does not mean GOBI is limited to enclosed problems (see §4.2).

represented in the conjunction within the plan schema and the OR component in the choice between alternative plans. Goals within plans are not evaluated at every step, for simple goals truth is established using a simple test or if an appropriate move can be made. For higher level goals truth is inferred from the fact that the goals descended from them were achieved.

3.3 The Planning Process

Currently GOBI has no goals and plans which persist *between* game moves and top level goals are given to the planner for each position to solve. In order to extend GOBI to whole-board play, one of the most important tasks is to define goals and abstract plans above the current highest level. The planner may keep these goals open (and still unexpanded) from one turn to another and indeed for long periods of the game. Earlier work [20, 12] has already shown the usefulness of planning at the higher strategic level, GOBI aims to be useful and applicable at all levels.

As described in §2 during the reasoning process a planning step involves two steps: 1) Choosing a current goal to replace in the plan and 2) Choosing an appropriate plan to replace it with. In GOBI the choice of which goal to expand next is unordered, however after choosing a goal from the agenda GOBI will try to work on it down through several levels of expansion. Repeatedly choosing one of it's descendents for expansion, this leads to some parts of the plan being expanded quickly (so they can be tracked in the world model) while others remain abstract for longer. Once a goal has been chosen, one of its associated set of plans must be selected for use. The plan schemas in GOBI are tried in a fixed order designed to try the most promising goal decompositions first. Plan preconditions can also be used to screen out plans not thought to be suitable for the current situation. If one plan leads to a dead-end, backtracking allows GOBI to try the others available for the situation.

The plans stored in the Go modules are not pre-programmed solutions and are expanded *in the context of the current game state*. The development of a plan is influenced by schema preconditions and by the choices in the world for making the lower level goals true. The failure of goals early in the plan quickly forces choice of alternative subplans for making these goals true.

The expansion of goals into plans eventually leads to goals at the lowest level of abstraction. These need to be checked for satisfaction and used to choose moves. Some example lowest level goals are: *fill-a-liberty*, *play-a-hane-move* (near *here*), *play-a-connecting-move* (between *string1* and *string2*), *play-a-placement-move*, *play-a-blocking-move* (near *here*), *play-an-extending-move* (near *here*).

The cost of move choice can be divided into three components:

1. Abstract Planning steps: these are very cheap since they consist purely of two cycles of matching items in a list (choose a goal, choose one of its plans). They can be more expensive if complex pre-conditions are used to select between plans.

2. Checking the satisfaction of low level goals: Since the goals are very focused (are these two strings connected? could black run out here?) this is also not too expensive. Checking is made easier by the fact that it can be supported by various representations of the game state — strings, groups, influence etc.

3. Using low level goals to generate moves: This is the most expensive part of the process although again the goals by this stage are very focused and limited to a small area (the *here* above). In GOBI selection is done using simple rules which define the type of move which could satisfy a goal.[10] An example set of rules is that a connecting move must:
   (a) be a liberty of (next to) *string1* and
   (b) be a liberty of (next to) *string2*.

Although the workload is divided between these three parts it is clear that this move choice method is in general more expensive for choosing a single more to try in the move tree than most data driven methods (see §6.2 for some estimates). The potential gain of using this method is in reducing the number of candidate moves in the move tree which needs to be tried. There are two important tradeoffs in the planning process:

- If goal selection and expansion mechanisms get more complex (i.e. through the extensive use of complex preconditions) their time cost will increase, however it should also mean a better use of the knowledge available.

- The simpler and more specific the lowest level goals are, the easier it is to establish their truth and choose actions with them but the more planning is needed to decide upon them.

3.4 Mixing Goal and Data Driven Approaches

Although the work here focuses on goal driven approaches to Go, it is clear that human players mix both types of reasoning. Patterns are thought to play a large part in human Go play. Thus, there is a need to consider how the two approaches can be mixed successfully.

One of the advantages of using a world model in the planner is that the changing situation of the state during planning is reflected in the world model. The changing world state may highlight interesting opportunities (or obvious problems) which arise as a result of the plan actions but were not expected effects. The architecture described above was extended to include plan *critics* which have access to the world model and watch for important situations. The critics are able to insert goals into the agendas of the agents for the planner to consider *alongside* the current track of reasoning. Thus the planner is made aware of opportunities, for example, during planning and can react to them.

---

[10] Note this could just as well have been done with local pattern matching. The planning framework poses no restriction on how this relationship between abstract goals and concrete actions is established (in general this is domain dependent).

Such opportunistic reasoning seems a plausible mechanism in terms of human play. Plans suggest moves which might lead to unexpected opportunities which in turn might lead to new plans etc.

Two critics were added to GOBI which detect groups under immediate threat of capture and insert (optional) group-saving/group-attacking goals into the agents' agendas.

4 Testing and Evaluating GOBI

Yoshinori's four-volume series [29] provides an excellent source of problems for testing Go programs. Since setting up each test problem was time-consuming, we chose (test set I) a representative sample (85 problems, approximately one third) of the problems from volume I of [29]. Problems were chosen for the test set essentially randomly, with a bias towards harder problems and excluding the simple capturing problems at the beginning of the book (which were considered too easy). A second set of tests using problems from volume II [29] and [7] was also conducted. All the tests were limited to 30 seconds of runtime on a SUN UltraSparc.

4.1 Test Results

The system successfully solved 74%[11] of the examples in test set I, which is a significant achievement given its limited knowledge and the complexity of some of the problems. GOBI has also solved considerably harder problems from volume II [29] and [7]. These tests were not comprehensive however, and used a range of hand picked examples, so the most useful indicator of performance remains the performance on problems from volume I [29].

In 98% of the correct solutions, GOBI considered most of the significant defences or responses of the opponent in its plan. This statistic is encouraging since not only were the answers correct — so was the reasoning behind them. Most of the failures were due to incomplete plan knowledge, several problems relied on groups connecting and escaping to live for example which is something GOBI currently has no plans for (GOBI has no plans for cutting groups). Another weak area was in problems which required a lot of forcing moves (e.g. ladders). GOBI has no special way of handling these and so is not able to take advantage of the problem simplification they provide (they are planned for in the normal way).

Curiously strengthening GOBI's defensive knowledge led to an improvement in attacking plans and vice versa, reflecting the fact that the better opponent model is more likely to find refutations for poor attacking plans.

4.2 Discussion

Further useful tests would involve comparing solutions with those of current programs such as GOTOOLS and MANY FACES OF GO, however we currently

---

[11] 67% when critics were disabled.

have access to neither of these. A second difficulty is that in fact comparisons would need to be made as to the number and relative cost of steps taken (search or reasoning) since our current system is a prototype and the other two systems are commercial with several years of development.

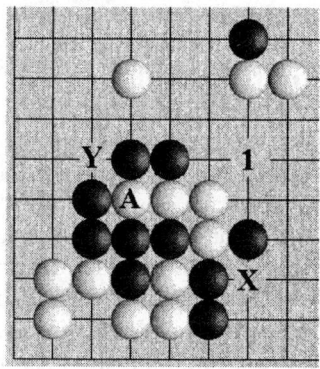

Fig. 6. Example 3: GOBI solves this more open position, killing the white string marked A by playing a net move at 1. Adding white stones at X or Y for example (or various other places) would cause GOBI to realise that the string can no longer be killed directly.

Even though most of GOBI's knowledge is about killing and saving groups it is more general than specialist life-and-death programs, knowledge for other parts of the game can be freely added and GOBI is also able to handle more open positions (which for example GOTOOLS has trouble with [28] such as the one shown in figure 6). In fact one would expect GOBI to have more of an advantage in open (though still tactical) positions where the number of possible move sequences increases rapidly. The plan knowledge in the system would then focus on only the relevant parts of the move tree.

5 Relationship to Other HTN Systems and Adversarial Planners in Go

Since what was probably the first use of the notion of goal directed search by Reitman and Wilcox [15] in 1979 there have been several attempts at applying planning approaches to Go.

Work by Ricaud [16] has similarities to that presented here although it focuses more on the use of abstraction. The GOBELIN system uses an abstract representation of the Go game state to produce plans which are then validated in the ground representation. GOBI is more flexible than this in that it shifts between several levels of abstraction (five with current knowledge) dynamically during planning rather than having a generation/validation cycle. GOBELIN probably also suffers in a similar way to [20] and [12] in that it has difficulty in representing complex interactions in the abstract representation. This difficulty may explain why GOBELIN makes the strongest plans during the opening where play is very open and there are few complex tactical situations. In his PhD thesis Hu [11] again focuses mainly on the strategic level of Go play, addressing issues of combining individual strategic and tactical goals into goals with multiple purposes.

The use of a world model to track the interactions between goals and making linearisation commitments during planning seems essential to any progress in domains with high levels of interactions between player goals. The utility of a world model has also been noted in other domains — notably Bridge [22, 21] — and makes GOBI much more applicable to tight tactical situations which can arise in Go (which was the major stumbling block in previous work [20, 12]). GOBI also uses an intuitive planning structure with the planning engine clearly separated from the knowledge.

The architecture probably owes most to the Wilkins' PARADISE Chess system [24] in the way that it uses plans to guide move choice. The use of the world model allows a similar access to the traversal of the search tree used in PARADISE.

6 Advantages of the Goal Driven Approach

The goal driven approach which is presented here has some clear advantages for Go and other similar games. Together with some of the previous work on the usefulness of planning at the strategic level of Go, GOBI shows that this approach can be used for reasoning at all levels, moving transparently between them and providing a unifying framework for Go play.

6.1 Representation and Communication of Domain Knowledge

Go knowledge in books and folklore is often expressed in a form appropriate for encoding as decompositions of abstract goals into other goals at different levels of abstraction. As reported in [18], there is a rich vocabulary which Go players can use to express their reasons for making a move. There are many popular proverbs, which convey strategies at various levels of abstraction, for example "death lies in the hane" is more tactical, whereas "don't push along the fifth line" is quite strategic. It may be easier to add this kind of knowledge to a goal driven system than to a data driven system which requires the delicate balancing of heuristics.

By following the trace of the goal decompositions one can see *why* the Go player is trying to do something — its aims and plans. Access to aims and plans is not only helpful for adding knowledge and debugging, but could be useful in the context of a Go tutoring aid. Some of the current commercial Go systems (MANY FACES OF GO for example) have teaching mechanisms but it is not clear whether these are based directly on the reasoning process of the computer player.

6.2 Search Properties

Using a goal driven approach leads to a very different search from a data driven approach, some key advantages are:

- There is no longer a need for global evaluation functions: this reduces to checking if goals can be satisfied. Evaluation functions may still be used in a limited way to carry out judgements which cannot be made on a move-by-move level, for example using an influence function to judge whether or not a group can successfully run, or using a fast $\alpha-\beta$ search to determine the life or death of enclosed groups. When an evaluation function is employed in determining whether or not a goal has been satisfied, the goal can be used to provide a focus for the evaluation function to a limited area of the board. The use of goals to focus evaluation makes sense when thinking about how humans play go — often making instinctive judgements (evaluations) of a local situation but rarely considering (evaluating) the whole board.

- Quiescence is defined automatically, thus avoiding problems such as the horizon effect in search. Since each agent has an agenda of goals, a situation can be defined as "open" (unsettled) until all goals are achieved.

- What seem like heuristically bad moves (e.g. sacrifices) are not discriminated against because the only consideration is their value to the plan.

Whilst these are important advantages a further strong point often associated with goal driven approaches is their resilience in domains with large search spaces and branching factors. Move choices are driven from above and not by the options available at each choice point. The plan knowledge therefore focuses which of the choices are directly *relevant to the current plan*. The problems GOBI was tested on gave a reduction in the number of move tree nodes visited of 10 - 100 times compared to standard $\alpha-\beta$ search. The current top Go programs are able to use a great deal of pattern and heuristic knowledge to improve on $\alpha-\beta$, however these gains are still significant.

The number of nodes visited in the move tree does not tell the whole story since GOBI requires more reasoning to choose which nodes to visit. This reasoning consists of the three components identified in §3.3 (expansion of a goal to a further abstract plan, checking if a low level goal is satisfied, choosing a move given a low level goal). For the examples in test set I (see §4) there were on average 4 planning steps per move chosen (including the steps which choose moves). Initial analysis suggests that given GOBI's current knowledge, each move tried in the game tree involves an average overhead of 2.5μ in reasoning, where μ is the average cost of modifying the internal data structures to make a move (including checking for Ko's, removing any stones killed, checking legality etc). Thus the cost of choosing a move in GOBI is about 3.5 times as much as in $\alpha-\beta$ (since $\alpha-\beta$'s main cost is adding stones to the board - i.e. μ). The computational advantages of GOBI over $\alpha-\beta$ are still clear-cut since the number of move choices (tried in the move tree) is significantly less than for $\alpha-\beta$. Note also that any system which begins to use heuristics for move choice also begins to increase the cost of each individual move choice. A more direct comparison with the systems able to apply large amounts of heuristic information to improve upon α–β would be useful but we do not have access to the relevant systems. The figures above at least illustrate how the plan knowledge in GOBI is able to cut down the search space by reasoning from above.

7 Disadvantages of Goal Driven Search

Obviously the goal driven approach is not always the best choice and has its own difficulties:

- The goal driven approach requires significant effort to encode strategies as goal decompositions. By contrast, in the data driven approach, good play can be achieved using even relatively simple evaluation functions as long as the game tree can be searched deeply enough.

- There are some types of knowledge which are hard to express in a goal/plan oriented framework, such as knowledge which is not reliant on understanding the motivation behind a move (patterns for example). It seems clear that good Go play requires both pattern (data driven) and abstract plan (goal driven) knowledge which is what leads us to try and integrate the two approaches (see §3.4).

- For games with low branching factors, shallow search trees or where near exhaustive search is possible data driven approaches have a strong advantage. It is only when searching most of the move tree is infeasible and large amounts of knowledge is needed to prune the tree that goal driven approaches can gain the upper hand.[12]

As with all knowledge based approaches, the search space is determined by the knowledge in the system, so large amounts of knowledge can dramatically impact performance. The key to making knowledge based approaches work is good organisation of that knowledge. Having many overlapping plans for example would cause serious performance problems, since each would be evaluated in turn.[13] In general there is little problem in adding "breadth" of knowledge (e.g. plans for different parts of the game) since knowledge is often obviously not applicable or leads to failure quickly. Adding more detailed plans ("depth"

---

[12] This point was illustrated by PARADISE [24] in the early eighties which despite being able to solve Chess problems requiring search up to 20 ply deep (far beyond other Chess programs of the time), still saw its knowledge based approach outstripped by the ever increasing efficiency of fast search techniques.

[13] Some moves may be chosen twice for different reasons (in different plans) which is perfectly acceptable, however if many plans overlap long sequences of move choices could be the same.

in knowledge) needs to be done carefully however since this can lead to overlaps and redundant search. The levels of abstraction used in GOBI (and in many AI planners) are key in avoiding this since knowledge can be appropriately structured and similar plans can be grouped together and expressed in a compact form.

The main problem with GOBI itself is still lack of knowledge: much more is needed before GOBI could be used as a tool to play the complete game of Go. Unfortunately adding knowledge takes a long time since it must be hand coded, most of the top programs have had knowledge added over periods of years. In this respect this approach (and many current approaches) are at a disadvantage to learning or non-symbolic approaches which can use automatic training to improve.

8 Conclusions

In this paper we have presented an adversarial planning architecture capable of reasoning about games, and an application of this architecture to Go. The planning architecture and Go reasoner reported here represent an advance on previous work for goal driven planning in Go. The system

- has a clear separation of domain knowledge from the abstract planning architecture and a clear model of the opponent in the game,

- can reason at multiple levels of abstraction simultaneously,

- can address complex tactical situations as well as high level strategic problems,

- can provide support for the integration of data driven and goal driven approaches.

We presented the advantages that a goal driven approach could have for Go and discussed the importance of mixing data and goal driven aspects when working in complex domains such as Go. GOBI as a prototype is certainly no match for current systems which play the full game of Go (and indeed cannot yet play the full game), however it does represent a step towards understanding how goal driven approaches could be applied to Go, even at a tactical level. GOBI needs more knowledge adding to be applicable in other areas of the game. A further important step is to incorporate higher level, persistent goals which allow GOBI to play the whole game of Go. We also aim to further develop structures for mixing data and goal driven approaches.

Go has several strong domain features which make goal driven approaches applicable (very large search spaces, clear layers of abstraction in domain descriptions — stones, strings, groups etc, and a wealth of knowledge similar in structure to abstract plans) GOBI represents a further step towards turning this theoretical possibility into a reality. The work described in this paper again shows that Go is an excellent test bed for AI research. There has been very little work on adversarial planning in recent years — the challenge of Go really motivated this work.

9 Acknowledgements

The first author is very grateful for financial support from EPSRC under grant number 96417312. The second and third authors are supported by EPSRC grant GR/L/11724. The fourth author is partly funded by DARPA/AFRL contract F30602-95-1-0022. Special thanks also go to Martin Müller for suggestions regarding test data and to Neil Firth for numerous instructive games of Go. We are grateful to the anonymous referees whose comments have resulted in numerous improvements to this paper.

References

[1] C. Applegate, C. Elsaesser, and D. Sanborn. An Architecture for Adversarial Planning. *IEEE Transactions on Systems, Man and Cybernetics*, 20(1):186–294, 1990.
[2] H. J. Berliner. A chronology of computer chess and its literarture. *Artificial Intelligence*, 10:201–214, 1978.
[3] R. Bozulich. *Second Book of Go*. The Ishi Press Inc, 1987.
[4] J. Burmeister and J. Wiles. An Introduction to the Computer Go field and Associated Internet Resources. Technical report, The University of Queensland, January 1997. Available online at: http://www/psy.uq.edu.au/~jay/go/go_page.html.
[5] J. G. Carbonell. Counterplanning: A Strategy Based Model of Adversarial Planning in Real World Situations. *Artificial Intelligence*, 16(1):295–329, 1981.
[6] T. Cazenave. *Système d'Apprentisage par Auto-Observation. Application au jeu de Go*. PhD thesis, L'Université Paris 6, 1996.
[7] J. Davies. *Life and Death*. The Ishi Press Inc, 1978.
[8] K. Erol, D. Nau, and J. Hendler. UMCP: A Sound and Complete Planning Procedure for Hierarchical Task-Network Planning. *AIPS-94*, June 1994.
[9] D. Fotland. Knowledge Representation in The Many Faces of Go. Technical report, American Go Association, 1993. Available online at: ftp://bsdserver.ucsf.edu/Go/comp/mfg.Z .
[10] I. Frank. *Search and Planning under Incomplete Information : a study using Bridge card play*. PhD thesis, Department of Artificial Intelligence - University of Edinburgh., 1996.
[11] S. Hu. *Multipurpose Adversary Planning in The Game of Go*. PhD thesis, George Mason University, 1995.
[12] P. Lehner. Strategic Planning in Go. In M. A. Bramer, editor, *Computer Game Playing: Theory and Practice*, pages 167–176. Ellis Horwood, 1983.
[13] M. Müller. *Computer Go as a Sum of Local Games: An application of Combinatorial Game Theory*. PhD thesis, Swiss Federal Institute of Technology, Zurich, 1995.
[14] J. Pitrat. A Chess Program which uses Plans. *Artificial Intelligence*, 8(1):275–321, 1977.

[15] W. Reitman and B. Wilcox. The Structure and Performance of the Intrim 2.0 Program. In *International Joint Conference on Artificial Intelligence*, pages 711–719, 1979.
[16] P. Ricaud. A Model of Strategy for the Game of Go Using Abstraction Mechanisms. *International Joint Conference on Artificial Intelligence*, pages 678–683, 1997.
[17] E. D. Sacerdoti. *A Structure for Plans and Behaviour*. Elsevier, 1977.
[18] Yasuki Saito and Atsushi Yoshikawa. Do Go players think in words? — interim report of the analysis of Go player's protocols. In Hitoshi Matsubara, editor, *Proceedings of the Second Game Programming Workshop*. Computer Shogi Association, 1995.
[19] Yasuki Saito and Atsushi Yoshikawa. An analysis of strong Go-players' protocols. In Hitoshi Matsubara, editor, *Proceedings of the Third Game Programming Workshop*. Computer Shogi Association, 1996.
[20] P. T. Sander and D. J. M. Davies. A strategic approach to the game of Go. In M. A. Bramer, editor, *Computer Game Playing: Theory and Practice*, pages 152–166. Ellis Horwood, 1983.
[21] S. J. J. Smith, D. S. Nau, and T. A. Throop. A Planning Approach to Declarer Play in Contract Bridge. *Computational Intelligence*, 12(1), 1996.
[22] S. J. J. Smith, D. S. Nau, and T. A. Throop. Total-Order Multi-Agent Task-Network Planning for Contract Bridge. *Proceedings AAAI-96*, pages 108–113, 1996.
[23] A. Tate. Generating Project Networks. In *Proceedings: 5th International Joint Conference on Artificial Intelligence*. 1977.
[24] D. E. Wilkins. Using Patterns and Plans in Chess. *Artificial Intelligence*, 14(1):165–203, 1980.
[25] S. Willmott. Adversarial Planning and the Game of Go. Master's thesis, Department of Artificial Intelligence, University of Edinburgh, September 1997.
[26] S. Willmott, J. Richardson, A. Bundy, and J. Levine. An Adversarial Planning Approach to Go (long version). Technical report, Department of Artificial Intelligence, University of Edinburgh, 1998. Forthcoming, Serial number TBA.
[27] T. Wolf. The program GoTools and its computer-generated tsume go database. In *Proceedings of the First Game Programming Workshop*, pages 84–96. Computer Shogi Association, 1994.
[28] T. Wolf. About problems in generalizing a Tsumego program to open positions. In Hitoshi Matsubara, editor, *Proceedings of the Third Game Programming Workshop*. Computer Shogi Association, 1996.
[29] K. Yoshinori. *Graded Go Problems for beginners (Volumes I - IV)*. The Ishi Press Inc, 1985.
[30] P. R. Young and P. Lehner. Applications of a Theory of Automated Adversarial Planning to Command and Control. *IEEE Transactions on Systems, Man and Cybernetics*, 16(6):186–294, 1990.

First Results from Using Temporal Difference Learning in Shogi

Donald F. Beal[1] and Martin C. Smith[2]

[1,2] Department of Computer Science, Queen Mary and Westfield College,
University of London, Mile End Road, London E1 4NS, England.
[1] Don.Beal@dcs.qmw.ac.uk
[2] martins@dcs.qmw.ac.uk

Abstract. This paper describes first results from the application of Temporal Difference learning [1] to shogi. We report on experiments to determine whether sensible values for shogi pieces can be obtained in the same manner as for western chess pieces [2]. The learning is obtained entirely from randomised self-play, without access to any form of expert knowledge. The piece values are used in a simple search program that chooses shogi moves from a shallow lookahead, using pieces values to evaluate the leaves, with a random tie-break at the top level. Temporal difference learning is used to adjust the piece values over the course of a series of games. The method is successful in learning values that perform well in matches against hand-crafted values.
Keywords: Learning, Shogi, Temporal Difference, Minimax, Search, Game-playing.

1 Introduction

This paper describes results from the application of Temporal Difference learning [1] to learning the relative values of shogi pieces. The learning is obtained entirely from randomised self-play, without access to any form of expert knowledge. The piece values are used to evaluate the leaves of a minimax search tree, and temporal difference learning is used to adjust these values over the course of a series of games.

A combination of machine-learning methods, including TD learning, have earlier been used to learn chess piece values for use by a program performing a 1-ply search only [3], and also for coarse-grained piece values [4]. TD learning has also been used successfully [5] by Baxter, Tridgell and Weaver to improve the weights of a complex chess evaluation function consisting of positional terms as well as piece values, when playing against knowledgeable opponents. However, they found it necessary to provide piece weights as initial knowledge to obtain good performance.

The focus of this work is on learning from self-play alone, with no knowledge input, as this is of greater potential value for problems where existing expertise is not available, or where the computer program may be able to go beyond the level of existing knowledge.

Previous experiments in learning from self-play with no initial knowledge [2] have found the Temporal Difference (TD) learning method alone to be highly successful in learning chess piece values that performed well in a computer program. Shogi is significantly different from chess, as detailed in section 2 below, and there are no standardised piece values to compare against. Although human shogi players avoid ascribing fixed values to shogi pieces, they do generally agree that rooks and bishops are the most powerful pieces, and that pawns have least value.

Our investigation was designed to discover whether the same TD technique would perform as satisfactorily in shogi as it had in chess, and whether it would yield sensible values for shogi pieces.

We performed experiments to learn suitable values for thirteen adjustable weights (seven main and six promoted shogi piece types). The experiments were conducted with a variety of learning and search parameters, and random seeds, and the consistency of results was surveyed. To demonstrate that the values learnt are reasonable, we played matches between four versions of the program, one using piece values learnt during the experiments, two using values used obtained from other shogi programs and a fourth using values estimated by a human programmer.

2 Shogi

Shogi is the Japanese name for the Japanese version of chess. It belongs to the same family of games as western chess and Chinese-chess (Xiangqi). Throughout this paper we refer to western chess as just *chess*, and Japanese chess as *shogi*. An introduction to the rules and some basic strategies of shogi is given by Fairbairn [6] and Leggett [7]. Matsubara, Iida & Grimbergen [8] suggest shogi is an appropriate target for current game-playing research, and discuss the similarities and differences between shogi and chess. In shogi, captured pieces are not eliminated from the game, but kept *in hand* by the capturing player, and may later be returned (*dropped*) on almost any vacant square. This greatly increases the branching factor of the game tree, and makes the game less amenable to full-width searching techniques. An additional feature of shogi is that all pieces apart from the king and gold are eligible for promotion once they reach the *promotion zone* (the last three ranks of the board). Pawns, Lances, Knights and Silvers may all promote to Golds upon entering the promotion zone, whereas Rooks and Bishops promote to more powerful pieces.

Choosing suitable values for shogi pieces is a problem for game programmers, as shogi experts prefer not to allocate values to the pieces. Sensible values for chess pieces are fairly widely known, but there is no generally-agreed standardized set of values for shogi pieces. Hence shogi programmers have more need for machine learning to generate material values for use in evaluation functions.

3 Temporal Difference Learning

Temporal difference (TD) learning methods apply to multi-step prediction problems. Each prediction is a single number, derived from a formula using adjustable weights, for which the derivatives with respect to changes in weights are computable. Each pair of temporally successive predictions gives rise to a recommendation for weight changes. Sutton [1] shows that TD methods make more efficient use of their experience than conventional prediction-learning methods, converging faster and producing more accurate predictions.

Weight adjustments are made according to:

$$\Delta w_t = \alpha (P_{t+1} - P_t) \sum_{k=1}^{t} \lambda^{t-k} \nabla_w P_k \tag{1}$$

where P is a series of temporally successive predictions, w the set of adjustable weights, α is a parameter controlling the learning rate, and $\nabla_w P_k$ is the vector of partial derivatives of P_t with respect of w. The recency parameter, λ, allows for an exponential weighting with recency of predictions occurring k steps in the past. TD(λ) learning enabled Tesauro's backgammon program to reach master level [9], [10].

The process may be applied to any initial set of weights. Learning performance depends on λ and α, which have to be chosen appropriately for the domain. In principle, TD(λ) weight adjustments may be made after each move, or at any arbitrary interval, but for game-playing tasks, the end of every game is a convenient point to actually alter the evaluation weights.

3.1 Obtaining Prediction Probabilities from Evaluation Scores

The evaluation score from the position at the end of the principal variation (the *principal position*) is "backed up" to the root of the search, and is regarded as a prediction of the final outcome of the game, to be compared with future values by the temporal difference method. The TD method requires a probability of winning, rather than a material score, so the values returned by the search are converted by use of a standard sigmoid squashing function. Thus the prediction of probability P, of winning from a given position is determined by:

$$P = \frac{1}{1 + e^{-v}} \tag{2}$$

where v = the 'evaluation value' of the position.

This sigmoid function has the advantage that it has a simple derivative:

$$\frac{dP}{dv} = P(1-P) \tag{3}$$

The derivative appears in the classical supervised-learning procedure on which TD(λ) is based – the adjustments are proportional to the derivative so that weights which have little effect on the prediction are adjusted less than weights to which the prediction is more sensitive.

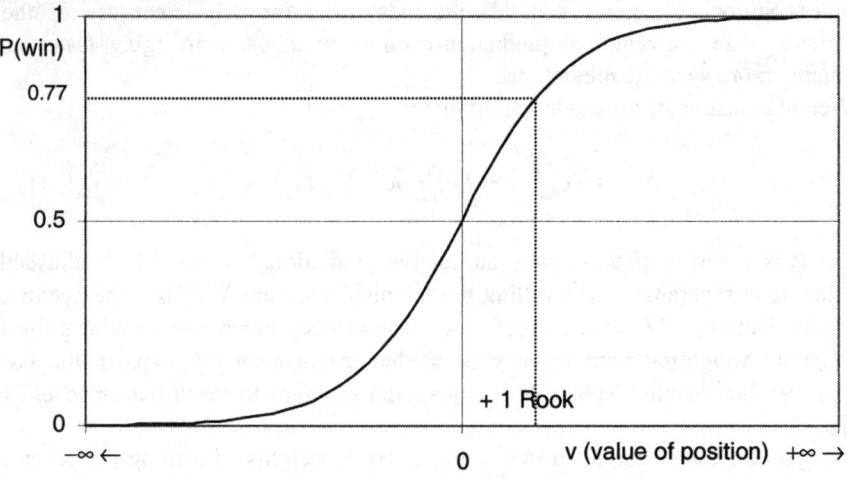

Fig. 1. Conversion of position value into prediction probability

The figure shows the conversion of position value into prediction probability. The example score of +1 Rook (using the learnt value of Rook = 1.21 from Run3) is converted into a probability of winning of 0.77.

Note that in Shogi, unlike chess, the value of a piece is not the same as the change in the material balance when a piece is captured. For example, when capturing an opponent's promoted rook the change in material balance needs to take into account both the loss of the promoted rook to the opponent, and also the gaining of a rook in hand for the capturing side. The +1 rook score shown in figure 1 represents the value of a rook, and does not represent the effect of a rook capture, which would change the material balance by twice that value.

4 The Shogi-Playing Search Engine

The experiments used a search engine derived from a conventional chess program, with an iteratively-deepened search, alpha-beta pruning and a captures and promotions only quiescence search at the horizon [11]. To prevent undue search effort being expended in the quiescence search, it was limited to six plies. Null-move pruning [12],[13] was used to reduce the size of the search tree, and the search was made more efficient by the use of a transposition table. The evaluation function applied at the leaves of the quiescence search consisted of the material score only, with the move choice at the root being made randomly from the materially-equal moves.

The thirteen piece values being learnt were used by the evaluation function. The *material balance* for a position was calculated as the sum of all the values of the side to move's pieces (including pieces in hand), minus the sum of all the values of the opponent's pieces.

5 The Experiments

We performed many learning runs to explore the behaviour of the TD method using a variety of learning rates and values for λ. All games were played using the shogi engine described in section 4, with a three-ply-deep main search. To prevent the same games from being repeated, the move lists were randomised. A random choice is thus made from all tactically equal moves, and it has the added benefit of ensuring a wide range of different types of position are encountered.

During each game a record is kept of the value returned by the search after each move, and the corresponding principal position. These values are converted into prediction probabilities by the squashing function, and then equation (1) is used to determine adjustments to the weights at the end of each game.

In the experiments reported here we used a value for λ of 0.95, and a variable value of α that decreased during each learning run, from 0.05 to 0.002. At the start of each run we initialised all weights to 1, so that no game-specific knowledge was being provided via the initial weights.

The experiments learn values for the pieces entirely from randomised self-play. This method has the advantage that it requires no play against well informed opponents, nor are games played by experts supplied. The piece weights are learnt "from scratch", and do not need to initialised to sensible values. The only shogi-specific knowledge provided is the rules of the game. Whilst each learning run consists of several thousand games, this represents a relatively short amount of machine time, and the entire run can be completed without any external interaction.

6 Results

We present results from five separate learning runs of 6000 games each. The learning runs were identical except that a different random number seed was used in each one, ensuring that completely different games were played in each. We shall refer to these learning runs as *Run1* through *Run5*.

6.1 Weight Traces

Figure 2 shows the weight traces for un-promoted pieces for a typical learning run (Run3) of 6000 games. A decaying learning rate was used for the first half of the run, decreasing from 0.05 to 0.002. Once the learning rate reached 0.002, it remained constant for the remainder of the run. Very similar results were achieved using a fixed learning rate of 0.002, but the runs required more games to achieve stable values.

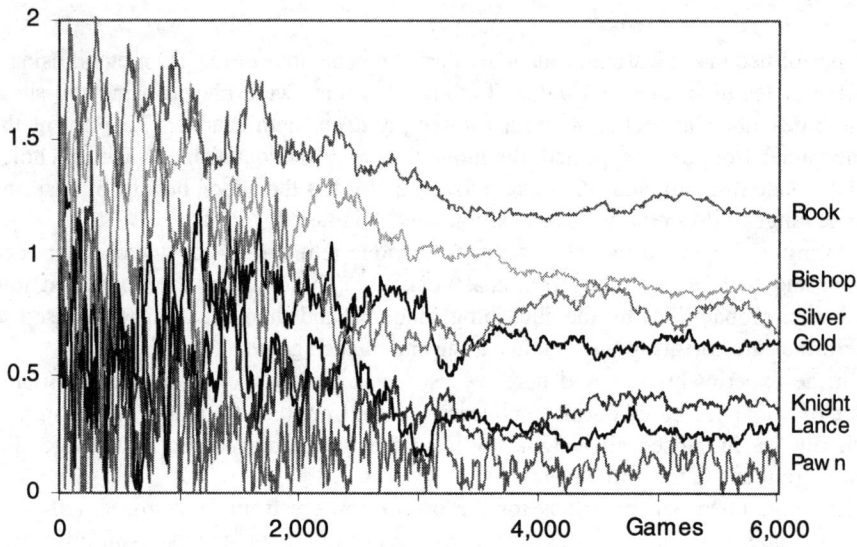

Fig. 2. Typical weight traces (main pieces)

From figure 2 we can see that the relative ordering of the main pieces has been decided after about 4000 games, and that pieces remain in that relative order for the remainder of the run. During the last 2000 games there is still considerable drift in the values. Some random drift is to be expected as a result of the random component included in the move choice. We averaged the values over the last 2000 games in order to obtain values for testing against other weight sets.

Figure 3 shows the weight traces for promoted pieces from Run3. Comparing figures 2 and 3 we can see that the promoted piece traces appear more stable than the main piece traces. This is because adjustments to the promoted piece types occur less frequently during the course of a game. Indeed, some games may not contain a single instance of a given promoted piece type. There is no trace for Gold, because they do not promote.

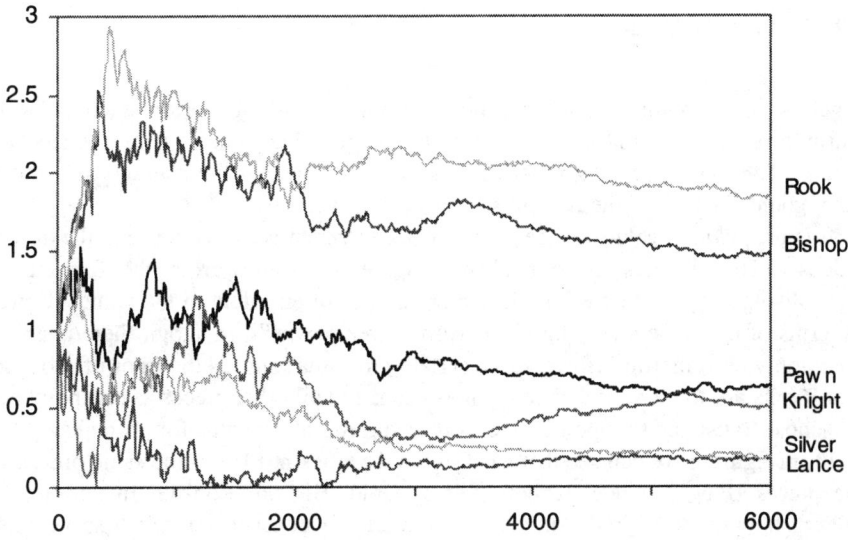

Fig. 3. Typical weight traces (promoted pieces)

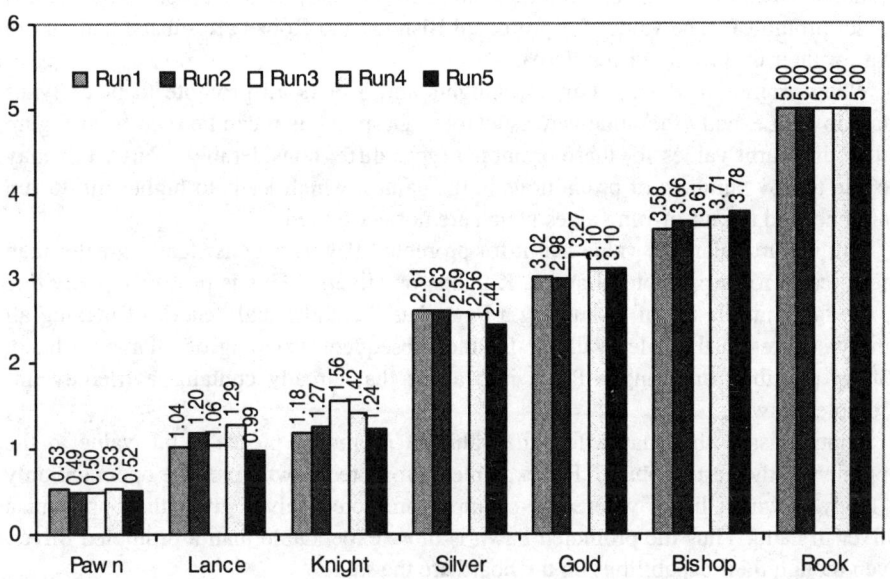

Fig. 4. Normalised learnt values for 5 runs (main pieces)

6.2 Main Piece Values

Figure 4 shows relative values for the seven main piece types, from each of the five learning runs. To avoid fluctuations in the weights due to noise from the stochastic nature of the game-playing process, these values represent the average over the last 2000 games in each of the five runs.

It is the *relative* values of the pieces that governs move selection, not the *absolute* values. This enables us to readily compare the values from the five runs by normalising them so that Rook=5. (In chess, one often refers to the values of pieces in terms of pawns, e.g., "A knight is worth three pawns". In shogi, there is no such commonly used metric. However, in certain rare situations [3] the rules of shogi state that Rooks are to be scored as five points each, and all other pieces as one point each. We chose to use the five-point rook score as our reference value for normalising.)

From figure 4 we can see that each of the five runs has learnt the same ordering of the pieces (Pawn, Lance, Knight, Silver, Gold, Bishop, Rook). In addition, the relative magnitude of the learnt values is remarkably consistent across the five runs.

6.3 Promoted Piece Values

Figure 5 shows the normalised relative values for the six promoted piece types (Golds do not promote). The values for promoted Bishops and Rooks are substantially more that for their un-promoted counterparts.

When promoted, Pawns, Lances, Knights and Silvers all promote to piece types that move in exactly the same way as a Gold. Despite this it can be seen from Figure 5 that the learnt values for these promoted types differ considerably. Partly this may be due to low numbers of promotions in the games, which leads to higher run-to-run variance, and to end-of-run values which are not yet settled.

But, in particular, the value learnt for promoted Pawns is consistently greater than those learnt for a promoted Lances, Knights, or Silvers. This is probably partly due to the fact that the act of promoting a Pawn has the additional benefit of making all empty squares in that file available for the subsequent dropping of a Pawn in hand. (Shogi prohibits dropping a Pawn into a file that already contains a friendly un-promoted Pawn.)

Another issue that may affect the value of promoted pieces is the value to the opponent if they are captured. For example a promoted Pawn gives the opponent only a lowly Pawn in hand, whereas a captured promoted Silver gives the opponent a Silver in hand. Thus the promoted Pawn is more expendable than a promoted Silver, even though their capabilities on the board are the same.

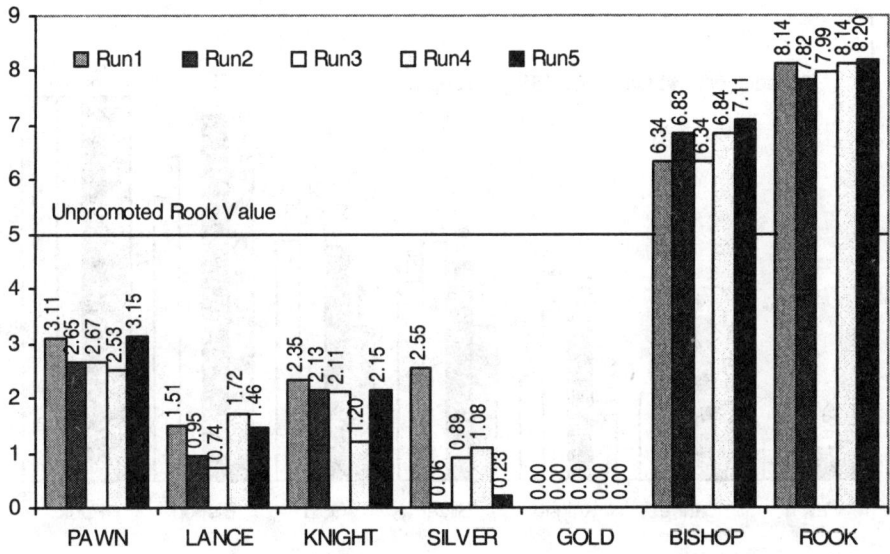

Fig. 5. Normalised learnt values for 5 runs (promoted pieces)

6.4 Matches to Test the Learnt Values

To test the effectiveness of the learnt values in our domain, a number of matches were played between identical search engines using various different piece values. The search engines used were the same as used in the learning experiments, but the piece weights were fixed to a given set of piece values, and not adjusted during the match.

Each match consisted of 2000 games, alternating Black and White (Sente and Gote). Games that ended in mate were scored as 1 point for the winning side. Games that were unfinished after 600 ply (300 moves each) were scored as ½ a point for both sides.

We ran two set of matches. The first was effectively a mini-tournament to compare the average values from all five learning runs with values obtained from other sources. The second set compared each of the five sets of learnt values with the best of the values from other sources.

The *Beginner* piece values were decided by a shogi beginner (but experienced game programmer), guided by the advice given by Leggett [7].

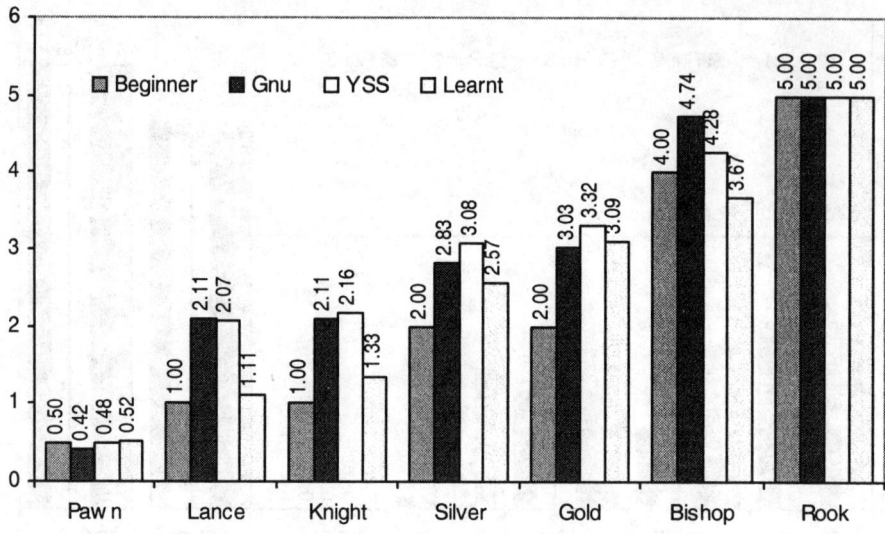

Fig. 6. Value sets tested in match play (main pieces)

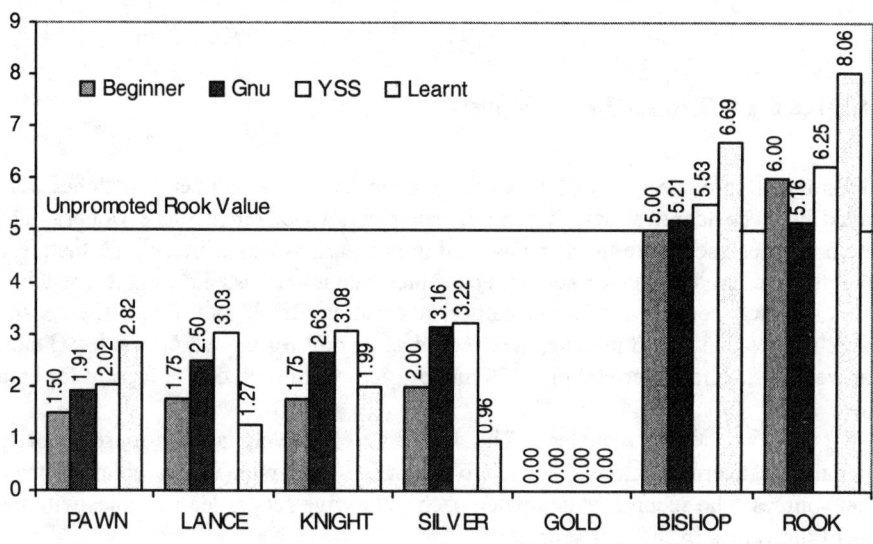

Fig. 7. Values sets tested in match play (promoted pieces)

The *Gnu-derived* piece values were derived from those used by the widely available program *Gnu Shogi* [14]. This program uses four different sets of piece values, depending on the stage of the game, as determined by various heuristics. The *Gnu-derived* values are the average of these four sets.

The *YSS* piece values are those published on the WWW by Hiroshi Yamashita [15], author of *YSS 7.0*, winner of the 7th World Computer Shogi Championship in 1997.

The evaluation functions of both Gnu Shogi and YSS also contain more sophisticated positional terms, e.g. king safety. In both programs, piece values are fundamental and typically the largest component of the overall evaluation score for a position. (Positional factors may also reward material possession indirectly. We ignored this secondary effect for these matches.)

The *Learnt* piece values are the average of the values presented in figures 4 and 5.

Figures 6 and 7 show the pieces values used in the matches, again normalised to Rook=5.

Table 1. Match results

Side 1		Side 2	Games	Win	Loss	Draw
Learnt	vs.	Beginner	2,000	1,206	718	76
Learnt	vs.	Gnu	2,000	1,170	766	64
Learnt	vs.	YSS	2,000	1,071	871	58
YSS	vs.	Beginner	2,000	1,113	835	52
YSS	vs.	Gnu	2,000	1,146	784	70
Gnu	vs.	Beginner	2,000	1,018	911	71

Table 2. Mini-tournament cross-table

	Learnt	YSS	Gnu	Beginner
Learnt	x	55%	60%	62%
YSS	45%	x	59%	57%
Gnu	40%	41%	x	53%
Beginner	38%	43%	47%	x

Table 1 gives the details of the matches played in the mini-tournament, and table 2 shows the cross-table of results. The *Learnt* values performed better than any of the other value sets under our test conditions, scoring 55%, 60% and 62% against the *YSS, Gnu-derived*, and *Beginner* value sets respectively.

The Learnt values were the average of the five learning runs. To verify that each of the individual learning runs learnt reasonable weights, each was pitted in a match against the YSS values, which performed the best of the three non-learnt sets. Table 3 shows the results from these matches, and shows that each of the five learning runs produced values that beat the YSS values under out test conditions.

Table 3. Individual learning run match results against YSS values.

	Games	Win	Loss	Draw	Percent
Run1	2,000	1,062	871	67	55%
Run2	2,000	1,044	888	68	54%
Run3	2,000	1,009	915	76	52%
Run4	2,000	1,070	852	78	55%
Run5	2,000	1,004	925	71	52%

7 Conclusions

The shogi piece values were learnt from self-play without any domain-specific knowledge being supplied. Although shogi experts are traditionally reluctant to assign values to the pieces, we believe that our learnt values would be recognised by human experts as reasonable for use in a shogi program.

It should be noted that these experiments have learnt material values within a material-only evaluation function. We would expect the material values learnt to be somewhat different if the evaluation function included positional scoring terms. Also, our results are obtained using a specific set of search parameters (depth, selectivity, quiescence details, etc). These may influence the optimum values, although we would expect changes to search parameters to have less of an effect on learnt values than additional evaluation terms. The method could be applied to any other set of search parameters, and other search engines. It is also applicable to learning an appropriate weight for positional evaluation terms, and we expect it to be useful in learning weights for more sophisticated evaluation functions in both chess and shogi.

8 References

1. Sutton, R. S.: Learning to Predict by the Methods of Temporal Differences. Machine Learning 3 (1988) 9-44
2. Beal, D.F. and Smith, M.C.: Learning Piece Values Using Temporal Differences International Computer Chess Association Journal, Vol. 20, No. 3 (1997) 147-151
3. Levinson, R. and Snyder, R.: Adaptive Pattern Oriented Chess. Proceedings of AAAI-91, Morgan-Kaufman (1991) 601-605
4. Christensen, J. and Korf, R.: A Unified Theory of Heuristic Evaluation Functions and its Application to Learning.. AAAI-86, Morgan-Kaufman (1986) 148-152
5. Baxter, J., Tridgell, A. and Weaver, L.: KnightCap: A chess program that learns by combining TD(lambda) with game-tree search. In: Machine Learning, Proceedings of the Fifteenth International Conference (ICML '98), Madison (1998) 28-36
6. Fairbairn, J.: Shogi for Beginners. Ishi Press International (1989)
7. Leggett, T.: Shogi: Japan's Game of Strategy. Charles E. Tuttle Company [Reprinted in 1993, first published in 1966]

8. Matsubara, H., Iida, H. and Grimbergen, R.: Natural Developments in Game Research: From Chess to Shogi to Go International Computer Chess Association Journal, Vol. 19, No. 2 (1996) 103-112
9. Tesauro, G.: Practical Issues in Temporal Difference Learning. Machine Learning 8 (1988) 9-44
10. Tesauro, G.: TD-Gammon, a Self-Teaching Backgammon Program, achieves Master Level Play. Neural Computation, Vol. 6, No. 2 (1994) 215-219
11. Marsland, T.A.: Computer Chess and Search. In: Shapiro, S. (ed.) Encyclopaedia of Artificial Intelligence. 2nd edn. J. Wiley & Sons (1992)
12. Beal, D.F.: Experiments with the Null Move. In: Beal, D.F. (ed.) Advances in Computer Chess 5. Elsevier Science Publishers (1989) 65-79
13. Donninger, C.: Null Move and Deep Search: Selective Search Heuristics for Obtuse Chess Programs. International Computer Chess Association Journal, Vol. 16, No. 3 (1993) 137-143
14. Mutz, M.: Gnu Shogi v1.2p03. Available from many sources, including ftp://ftp.uni-passau.de/pub/local/shogi (1994)
15. Yamashita, H.: YSS: About the Data Structures and the Algorithm. Published on the WWW at http://plaza15.mbn.or.jp/~yss (1997)

From Simple Features to Sophisticated Evaluation Functions

Michael Buro

NEC Research Institute
4 Independence Way
Princeton NJ 08540, USA

Abstract. This paper discusses a practical framework for the semi–automatic construction of evaluation-functions for games. Based on a structured evaluation function representation, a procedure for exploring the feature space is presented that is able to discover new features in a computationally feasible way. Besides the theoretical aspects, related practical issues such as the generation of training positions, feature selection, and weight fitting in large linear systems are discussed. Finally, we present experimental results for Othello, which demonstrate the potential of the described approach.

Keywords: automatic feature construction, GLEM, Othello

1 Introduction

Many AI systems use evaluation functions for guiding search tasks. In the context of strategy games they usually map game positions into the real numbers for estimating the winning chance for the player to move. Decades of research has shown how hard a problem evaluation function construction is, even when focusing on particular games. In order to simplify the construction task, the notion of *evaluation features* was introduced. The underlying assumption is that there exist reasonable approximations of the perfect evaluation function in the form of combinations of a few distinct numerical properties of the position — called features. Given this, evaluation functions can be constructed in two phases by 1) selecting features and 2) combining them.

Selecting features is one of the most important and difficult sub–tasks in the construction of a game playing program. It requires both domain specific knowledge and programming skills because of the well known tradeoff between speed and knowledge in game–tree search. A couple of years ago, the authors of the best game playing programs still picked not only features but also their weights in course of a tedious optimization process. This is somewhat surprising, since already in [7] proposed ways for automatically tuning weights. While selecting features is difficult for a machine, fitting even a large number of weights given a set of training positions is not. Research focused on the latter topic produced TD–Gammon, a world–class backgammon–program [8, 9], and contributed to Deep Blue's victory over Kasparov in 1997 [4].

In this article we go a step further towards the ultimate goal of automatic evaluation function construction. First, a generalized linear evaluation model is presented. It restricts evaluation features to boolean combinations of given atomic functions. The model parameters can be tailored such that an automatic feature space exploration becomes feasible. The following sections cover all aspects of evaluation function construction — from generating training positions over feature selection to weight estimation — with respect to the new model and emphasis on efficient implementation. Finally, we show how the presented techniques can be applied to the game of Othello and discuss the new approach with regard to related work.

2 Evaluation Model

We first give a definition of the evaluation model we are proposing and discuss its properties. In what follows, $\mathcal{P}$ denotes the set of all legal game positions[1], and $\mathbb{R}$ the set of real numbers. Let A be a finite set of integer valued — so called *atomic* — features and $R_A := \{ (f(\cdot) = k) \mid f \in A,\ k \text{ is an integer}\}$ the set of relations over A that compare feature values with integer constants. *Configurations* are conjunctions of relations in R_A. For a position $p \in \mathcal{P}$ and a configuration $c = r_1 \wedge \ldots \wedge r_l$ we define

$$\text{val}(c(p)) := \begin{cases} 1, & \text{if } r_1(p) \wedge \ldots \wedge r_l(p) = \text{true} \\ 0, & \text{otherwise} \end{cases}.$$

A configuration c is called *active* in a position p, iff $c(p) = \text{true}$.

With this notation we can now define the *Generalized Linear Evaluation Model* — GLEM($\mathcal{P}, A, g$) for short. Evaluation functions in this model have the following form:

$$e(p) = g\Big(\sum_{i=1}^{n} w_i \cdot \text{val}(c_i(p))\Big), \qquad (1)$$

where $c_1, \ldots, c_n$ are configurations over R_A, $w_1, \ldots, w_n \in \mathbb{R}$ are weights, and $g : \mathbb{R} \to \mathbb{R}$ is an increasing and differentiable link function.

The weights are subject to the usual least–squares optimization. That is, given a set of configurations $c_1, \ldots, c_n$, a link function g, and a sequence of scored training positions $((p_i, r_i) \mid i = 1 \ldots N)$, the weights are chosen such that the total squared error

$$E(w) := \sum_{i=1}^{N} (r_i - e_w(p_i))^2.$$

is minimized. This model has several desirable properties:

---

[1] W.l.o.g. it is assumed that game positions in $\mathcal{P}$ are normalized in such a way that a fixed player is to move.

- Atomic features are the building blocks of more sophisticated ones. This, in principle, allows the automated discovery of new important features by systematic combination.
- If necessary, complex features can be added to A. Thus, "atomic" is not necessarily a synonym for "simple".
- When evaluating a position, features are combined linearly. This keeps the time overhead low. Actually, not even a multiplication with the weight is necessary since val($c_i(p)$) is either 0 or 1.
- Non–linear effects can be approximated by using configurations that consist of several relations.
- In order to deal with saturation an increasing non–linear link function, such as $g(x) = 1/(1 + exp(-x))$, can be used without increasing the run time during minimax search. There is no need to compute g, because $g(x_1) > g(x_2) \iff x_1 > x_2$.
- The simple linear core of the evaluation function allows an efficient approximation of optimal weights, even for large systems. In the application reported later, more than a million weights were fitted to a training set consisting of eleven million scored positions in a reasonable period of time.

At this point GLEM should be moved into the right perspective: in the stated form it is neither a new revolutionary evaluation approach, nor does it ease the task of automatic evaluation function exploration. This is because the model is built upon well known linear evaluation functions and does not impose a severe restriction on the structure of functions it includes. E.g., for any atomic feature set A, which is capable of distinguishing any two different positions (including game history if the game result depends on it) via conjunctions over R_A, GLEM covers *all* evaluation functions over $\mathcal{P}$. A trivial example for such a complete atomic feature set for board games without position repetition is

$$A = \left\{ f_s \mid \begin{array}{l} f_s(p) = \text{contents of square } s \text{ in position } p, \\ s \text{ is a square} \end{array} \right\},$$

where the contents of a square is considered to be an integer value.

However, GLEM allows one to define a hierarchy of submodels in a natural way, which reflects different levels of computational complexity and the expressive power of the covered evaluation functions. By restricting the size of A, the number of configurations, or their structure, an automated search for new features becomes feasible. In the application discussed later, evaluation functions based on GLEM outperformed the best known functions so far. In this respect, GLEM breaks new ground.

Good evaluation functions accurately estimate the winning chances in positions visited during game–tree search and are optimized for speed. Therefore, the following topics have to be borne in mind when using scored training positions for tuning configuration weights:

- The training positions have to be representative of the positions that will be evaluated later in actual game–tree search.

- Training positions must be scored accurately.
- The selected configurations and their combination must have the expressive power to explain the data reasonably well while avoiding over-fitting. Given the flat evaluation function representation in GLEM, meeting this condition may require a large number of configurations. Their automatic construction is therefore of great interest.
- Evaluation speed is important.
- While computing weights is an off-line process, its memory and time consumption should still be subject to optimization. The reason is that in the feature selection phase usually many evaluation function versions have to be compared. Moreover, without optimization the current solver might not be able to handle the number of features one would like to use.

In the following sections these topics are discussed in detail in the context of GLEM.

3 Training Positions

A theory of how to generate good training sets in the context of evaluation function tuning has not been developed yet. In this section practical ideas are discussed which may become the seed for further investigations.

Training positions can be generated and scored in several ways. If the considered game has a long tradition and is quite popular, many games may be available in electronic format. The simplest scoring procedure assigns the final game result (depending on the side to move) to all positions occurring in a game. Obviously, this ad hoc procedure has limitations, since it does not ensure accurate scoring. Selecting games between good players alleviates this problem. But this approach leads to high-quality games, in which hardly any catastrophe takes place, such as losing material in chess or a corner in Othello without compensation. The reason for this is obvious: good players know the important evaluation features and keep them mostly balanced in their games. What we (and machines) can learn from such games are the finer points of play, which make the difference between good and the best players. However, an evaluation function must also be aware of the most important features. Thus, our training set should also contain games in which at one point a player makes a serious mistake that is rigorously exploited by the opponent. In summary, a reasonable strategy for generating training positions from a game database is to select games played by at least one good player and to score game positions according to the final game result. This procedure is efficient and its output can serve as the basis for tuning the first evaluation function version.

Besides the still present potential mis-scoring problem, the question arises, whether the so generated training set is representative to positions encountered in game-tree search. This question is of importance, since the weight fit for a linear evaluation function is influenced by the correlation among features in the training set. The answer obviously depends on the type of game-tree search we are conducting: in a highly selective search evaluated positions are in the vicinity of principal variations, whereas in brute–force searches many ridiculous positions are evaluated, which one would never encounter in actual games. It seems natural to let the search algorithm generate the training positions by itself. For instance, starting searches with positions from played games, a random subset of evaluated positions can be saved in a file and serve as the training set after scoring. In this way, the generated positions are surely a representative sample of the positions encountered in game–tree searches. It remains to assign accurate scores to the positions. This task can be accomplished again by game–tree searches, which normally return more reliable results than the evaluation function itself. In particular, in many games endgame positions can be evaluated perfectly — or at least more accurately than middle–game or opening positions — in a reasonable amount of time. In this case, a game–stage dependent evaluation function can be improved iteratively by first tuning the endgame weights. Thereafter, training positions from the previous game stage are evaluated by a game–tree search, which utilizes the just tuned evaluation function, and so on. The next step would be to generate even positions and those with a narrow advantage for one side. Similar to considering games between good players mentioned above, these positions are useful for tuning weights of minor features or revealing possible tradeoffs between major features (e.g. material vs. king safety in chess or corner possession vs. mobility in Othello).

If training positions are selected randomly during minimax–based searches, one soon discovers that the winning chance in such positions is biased towards the player to move. This phenomenon is easy to explain, given the fact that in typical positions the majority of searched moves lose. Its undesirable effect on fitted weights is an artificial bonus for the player to move. This, in turn, leads to unstable evaluations, which compromise comparing evaluations backed–up from depths of odd difference during selective search. Because the proposed generation procedure labels positions with search results, a simple cure for this problem is to add the principal variation successor positions to the training set after labelling them with the negated search result.

4 Selecting Configurations

GLEM proposes a new perspective on how to look at evaluation features. In the classical approach a couple of complex features are combined linearly. Weights were mostly hand–tuned. Later, the study of neural networks opened up a practical way of combining features non–linearly. Application of the well known gradient descent procedure (in this context called "back–propagation") makes it possible to automatically tune a large number of network parameters. A prominent and very successful example is Tesauro's backgammon network which, in its strongest version, makes use of hand–crafted features in addition to a raw board representation. GLEM uses a different approach. Instead of modelling non–linear effects by applying parameterized analytical functions to features, GLEM handles non–linearities *directly* by assigning values to boolean feature combinations, called configurations. In this way, distinct cases can be handled naturally, without the detour over non-linear analytical functions. The design of neural networks corresponds to configuration selection in GLEM, which is the topic of this section. After stating basic requirements for the atomic features, we will present an algorithm for generating configurations by analyzing training positions, and discuss several optimizations.

4.1 Atomic Features

Atomic features are the building blocks for configurations. As the scope of automatic configuration selection is limited by its time and space complexity, choosing the right abstraction level for atomic features is crucial. In Othello, configurations based upon the raw board representation are sufficient for building good evaluation functions — as we shall see later. The reason is that many relevant features in this game can be expressed by local board configurations of small cardinality. Other games may require a greater abstraction level. For instance, the relation "piece A attacks piece B" in chess has a long description length when using raw board representation languages. Since many important features, such as forks and pins, are based on those attack features, they certainly should be included in the atomic feature set. In general, candidates for atomic features are common parts of relevant features, that — combined in novel ways — may lead to new important features. Obviously, this selection task is beyond current program abilities.

Not all atomic features have to be useful for building other features. Limitations of the configuration generator may suggest the inclusion of complex features that can not be expressed or well approximated by restricted combinations of other members of the atomic feature set.

Moreover, GLEM generalizes the classical use of features — $w \cdot f(p)$ — because $(w \cdot k)$ in

$$w \cdot f(p) = \sum_k (w \cdot k) \cdot \text{val}(f(p) = k).$$

specializes the weight of $\text{val}(f(p) = k)$. This generalization is only meaningful if f has a small range. In case one likes to incorporate a feature f having a large range, GLEM can be easily extended by allowing summation terms of the form $w \cdot f(p)$.

4.2 Generating Configurations

In a balanced evaluation function design the number of features can be increased up to a point where either 1) adding additional knowledge is compensated for by a decreased evaluation speed or 2) over-fitting becomes a problem. Since configurations can be computed quickly, once the atomic features have been evaluated, GLEM encourages to use many configurations rather than a few complex features. Our chief concern is therefore over-fitting.

We will first present an algorithm for generating a configuration set that does not suffer from over-fitting. Thereafter, we will discuss how to deal with a possibly unacceptably long run time for the configuration generator, for weight fitting, or for the configuration value look-up during game-tree search.

Configurations have to cover positions that occur in game-tree search while avoiding over-fitting when optimizing weights. Both requirements can be met by using a large set of training positions — generated as described in the previous section — and selecting configurations that match a sufficiently large number of these positions. Fig. 1 shows a straight forward algorithm for this task. Given a set of atomic features A, training positions E, and a minimal match count n, it computes all *valid* configurations over A that occur in at least n positions in E. Beginning with all valid configurations of length one, the algorithm iteratively builds larger configurations by specializing previously generated configurations, until the matches count drops below n. The algorithm certainly halts, since the set of valid configurations is finite. Its correctness can be shown by induction using the fact, that for $k > 1$, valid configurations of length k have valid sub-configurations of length $k - 1$.

The run time of the algorithm is $O(|C| \cdot |R_A|^2 \cdot |E|)$, where C is the computed set of valid configurations and E the set of training examples. The most time-consuming part is computing the match counts in the inner loop. Since in the beginning the number of checked configurations grows exponentially in

Function GenConf

Input: atomic feature set A, training position set E, minimal match count n
Output: configurations over A that are active in at least n positions of E

$R := \{\{f(\cdot) = k\} \mid f \in A,\ k \in \text{range}(f),\ \#\text{match}(\{f(\cdot) = k\}, E) \geq n\}$
$C := R$; collects all valid configurations
$N := R$; set of configurations created in previous iteration

```
     while N ≠ ∅ do
        M := ∅                        ; set of valid configurations in current iteration
(*)     foreach c ∈ N, d ∈ R do
           e := c ∪ {d}               ; specialize configuration c
           if #match(e, E) ≥ n then
              M := M ∪ {e}            ; append if valid
           endif
        endfor
        N := M                        ; next configurations to specialize
        C := C ∪ N                    ; add valid configurations
     endwhile
     return C
```

Fig. 1. Pseudo code for generating the set of configurations that occur in at least n training positions. The function iteratively specializes configurations, which are implemented as sets of relations, until the number of matching positions ($\#\text{match}(e, E)$) drops below n.

each iteration, it is crucial to optimize the match computations, especially if the number of positions is large. The following optimizations speed up a naive implementation considerably:

- Due to the commutativity of $\wedge$, valid configurations of length k may have several valid subconfigurations of length $k-1$. This observation suggests that we should check whether a given specialization has been tested before in the current iteration, in order to avoid repeated match computations. An even better solution is to generate specializations in an ordered fashion by defining a total order over R and replacing line (*) by

$$\textbf{foreach } c \in N,\ d \in R \text{ with } d > \max_{d' \in c} d' \textbf{ do}$$

It is not hard to show that after applying this time–saving modification the algorithm still generates all valid configurations.
- A naive algorithm for deciding #match$(e, E) \geq n$ evaluates the relations in e for every member of E. The computation time of this algorithm can be reduced by preprocessing and parallelizing computations. The idea is to compute, for each $r \in R$, a sequence of bits $(b_i)_{i=1}^{\#E}$ defined by $b_i := \text{val}(r(p_i))$, where $p_i \in E$ is the i-th training position. After this preprocessing step, the actual features and positions are no longer needed. The match count computation reduces to and–combining the bit sequences of the involved relations and counting set bits in the result sequence. Modern CPUs allow

Function MatchHeuristic

Input: configuration e, chunk size s, random partition $E_1, ..., E_m$ of position set E as described in the text, confidence level $t > 0$
Output: true, if #match$(e, E) \geq n$ is likely; false, otherwise

$q := n/\#E$; match count fraction aimed for
$d := 0$; number of elements checked
$u := 0$; current match count
for $i := 1$ **to** $m - 1$ **do**
$\quad u := u + \#\text{match}(e, E_i)$; update counts
$\quad d := d + s$
$\quad$ **if** $u \geq dq + t\sqrt{dq(1-q)}$ **then**
$\quad\quad$ **return true** ; #match$(e, E) \geq n$ is likely
$\quad$ **endif**
$\quad$ **if** $u < dq - t\sqrt{dq(1-q)}$ **then**
$\quad\quad$ **return false** ; #match$(e, E) < n$ is likely
$\quad$ **endif**
endfor
return $u + \#\text{match}(e, E_m) \geq n$

Fig. 2. A fast procedure for testing the hypothesis #match$(e, E) \geq n$

a very efficient implementation of the and–part by handling 32 or even 64 bits in parallel. Iterating $x := x \wedge (x-1)$, which clears the rightmost one in the binary representation of x, allows us to count set bits quickly. In this application table–based techniques for counting bits are inferior because the number of set bits is decreasing rapidly due to specialization.
– Replacing the condition $\#\mathrm{match}(e, E) \geq n$ by a sequential statistical test procedure speeds up the computation further. This optimization can be motivated by an intuitive example: if among the first 100 randomly selected bits of 1000 there is only a single one, it is very unlikely that the total number of ones exceeds 500. More formally, we propose the following heuristic function, which quickly checks whether $\#\mathrm{match}(e, E) \geq n$ holds with a prescribed likelihood. In a preprocessing step, E is randomly partitioned into chunks $E_1, ..., E_m$ of size s (E_m might have less elements). For a given configuration e, the function then iteratively computes the match counts for increasing subsets beginning with E_1. If the match count fraction at one point significantly differs from the one we aim for, the function returns the likely truth value of $\#\mathrm{match}(e, E) \geq n$ early. The pseudo code implementation shown in Fig. 2 makes use of the fact that the expected number of ones in a sequence of d randomly generated bits is dq, if $\mathrm{Prob}\{1\} = q$, while its standard deviation is $\sqrt{dq(1-q)}$. The behaviour of this function is controlled by confidence level t. For large values of t, hardly any break condition will be met — the function will be slow, and almost always return the correct result. If t is small, the function is quick, but it also returns unreliable results. Experiments can tell how to choose t depending on the speed/reliability one likes to achieve.

4.3 Finding Active Configurations

During weight fitting and position evaluation the set of active configurations has to be computed quickly for a large number of positions. For this purpose, we represent the set of all configurations over R_A by a DAG G. Nodes in G correspond to configurations, and arcs mark direct specializations. A detailed example is shown in Fig. 3a. The just described selection algorithm computes all configurations that occur at least n–times in a set of training positions. This set of valid configurations induces a sub–DAG G' of G. Given a position, all active configurations can be found by a depth–first search in G' starting at its root. During search, all visited configurations are marked and their active status is determined. The search stops in nodes that have been visited before or have been found inactive. This algorithm quickly finds all active configurations. However, the only relevant active configurations for evaluation purposes are those without active specializations, because generalizations are redundant. It is easy to extend the described algorithm accordingly by restricting its output to leaves of the active configuration sub–DAG. Fig. 4 illustrates the entire procedure.

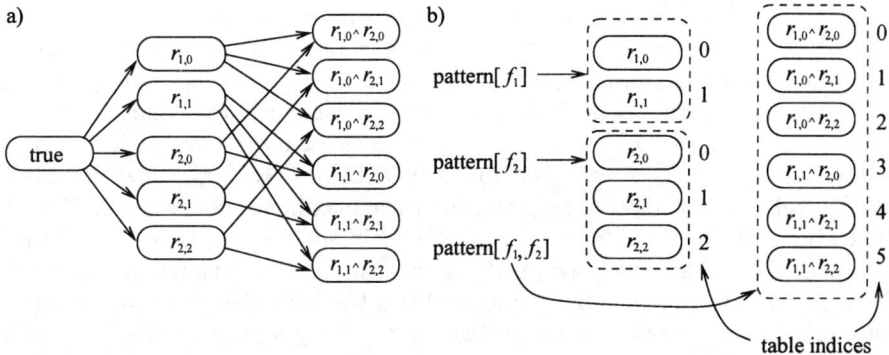

Fig. 3. a) Configuration DAG for two features f_1, f_2 with range$(f_1) = \{0,1\}$ and range$(f_2) = \{0,1,2\}$. $r_{i,k}$ denotes the relation $f_i(\cdot) = k$. b) Configurations belonging to patterns over f_1 and f_2.

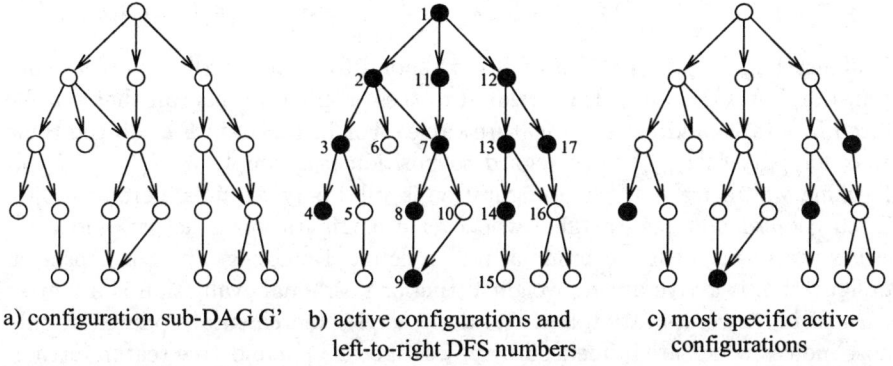

a) configuration sub-DAG G'   b) active configurations and   c) most specific active
                                 left-to-right DFS numbers   configurations

Fig. 4. Finding the most specific active configurations by depth–first search in the configuration DAG

4.4 Reducing Complexity: Patterns

So far, our focus has been on efficient ways for generating configurations and computing active configurations. Despite the optimization efforts, GenConf may still not be able to generate all valid configurations due to time or space limitations. Furthermore, a large number of generated configurations might prevent an efficient position evaluation, because too many configurations are active, or the configuration data needs too much memory.

One solution to these problems is to increase the minimal match count n, until the number of generated configurations is manageable. This approach, however, narrows the evaluation function's view by focusing it on the most common phenomena. A compromise is to generate all valid configurations choosing n high enough to avoid over–fitting, and to reduce their number afterwards by looking at their statistical significance with regard to winning chance prediction.[2] Another option for reducing the number of configurations is to limit their size or to choose subsets of the atomic feature set as the base for generating configurations.

Finally, considering sets of mutual exclusive configurations helps to reduce the number of active configurations in order to speed up the evaluation considerably. Let G be the complete configuration DAG for $\{f_1, ..., f_m\} \subset A$ (Fig. 3a), and let $r_{\min}$ and $r_{\max}$ denote the minimum/maximum range cardinality of the features. Then the number of nodes in G is bounded by $(1 + r_{\min})^m$ and $(1 + r_{\max})^m$, and for any position the number of active configurations is 2^m.[3] Thus, in case of complete configuration DAGs the DFS algorithm presented in the last subsection seems to waste time by searching a large number of nodes before it eventually returns the single active configuration we are interested in. This observation motivates looking for a more efficient data structure. For $\{f_1, ..., f_m\} \subset A$ we collect all possible most specific configurations in a set called pattern$[f_1, ..., f_m]$, i.e.

$$\text{pattern}[f_1, ..., f_m] := \{r_{1,l_1} \wedge ... \wedge r_{m,l_m} \mid r_{i,l_i} = (f_i(\cdot) = l_i),\ l_i \in \text{range}(f_i)\}$$

Configurations in pattern$[f_1, ..., f_m]$ correspond to leaves of the complete configuration DAG (Fig. 3b). Data related to these configurations can therefore be stored in a table addressed by feature values. For instance, in Fig. 3b the table index for pattern$[f_1, f_2]$ with regard to position p is simply $3 \cdot f_1(p) + f_2(p)$. Checking whether a pattern configuration is valid only requires incrementing a match counter stored in a table whenever a configuration is active, and comparing the result with the minimal match count. Detecting whether a pattern configuration is active during weight fitting or positional evaluation is a matter of a fast index computation and one table access. Incremental updates of only those indices which are influenced by moves speeds up game–tree search further. In summary, the flat table is the data structure of choice for storing information regarding small and medium sized complete configuration sets. The fast access encourages to restrict configuration sets to patterns.

Large patterns require a more memory efficient representation. In order to avoid over–fitting, we are still only interested in configurations that match several training positions. Consequently, large patterns are sparse. Fig. 5 outlines a very fast and — to our knowledge — novel technique for accessing sparse data which trades memory for speed. It is based on representing valid configurations as index tuples (i_1, i_2). For a given position and pattern, i_1 and i_2 are computed by splitting the pattern's feature set into two parts and performing the index calculations described above separately for each subset. Both indices are then used for accessing a hash–table, in which data regarding configuration (i_1, i_2)

---

[2] The general problem of deciding the relevance of variables in a multivariate regression model in advance is hard. Nevertheless, simple statistics like the feature's correlation with the training position scores can serve as a reasonable first approximation.

[3] These numbers can be derived by adding lower/upper bounds for the number of nodes/active configurations for each depth and applying the identity $\sum_{i=0}^{m} \binom{m}{i} x^i = (1 + x)^m$.

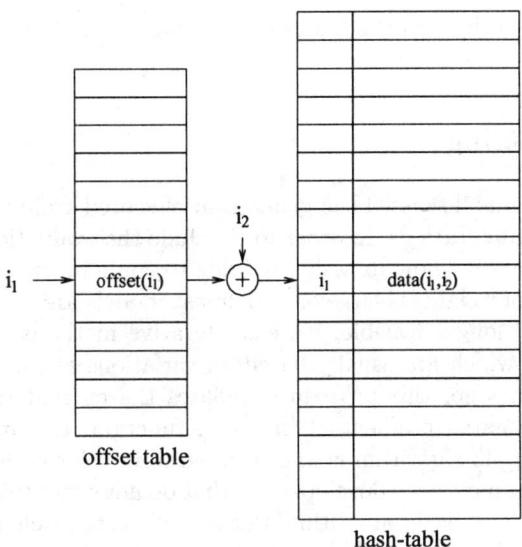

Fig. 5. Fast sparse data access. Data regarding a configuration represented by two indices i_1 and i_2 can be accessed quickly in two steps.

is stored. First, an offset is looked–up in a table using index i_1. Then, this offset, incremented by i_2, is used to access the hash–table. For the algorithm to be correct, 1) unique hash–table entries have to be assigned to valid index tuples, and 2) invalid index tuples must be detected. The first condition can be met by choosing suitable offsets and a sufficiently large hash–table. In practice, the following greedy algorithm for constructing collision–free hash–tables has produced reasonable results: beginning with the most frequent i_1–values, offsets are assigned to them in first–fit manner. That is, whenever a collision occurs when attempting to occupy the hash–table entry offset$(i_1) + i_2$, all i_1 entries claimed so far are erased and offset(i_1) is incremented before restarting. The hash–table size must be greater than the sum of the maximal offset and maximal *possible* value of i_2, in order to avoid accesses beyond table end. A simple way for meeting condition 2) is to add the lock i_1 to hash entries for all valid tuples (i_1, i_2) and to reject tuples (i_1, i_2), for which the lock stored in the accessed hash entry does not match i_1. Locks of unused hash entries must be initialized with a value different from any *possible* i_1 (e.g. -1). Finally, offsets for all i_1, which are not the first component of any valid index tuple, can be safely set to 0, since all locks in the hash–table are different from those i_1 values.

Patterns may outperform configuration sets constructed by GenConf due to a much faster generation and evaluation of configurations. However, patterns suffer from their limited scope because patterns may miss essential generalizations. This observation suggests building a hierarchy of patterns in order to quickly cover both general and specific position aspects. Since this approach also increases the evaluation time, experiments have to tell, which is the better strategy for a given application.

5 Weight Fitting

The previous sections discussed the generation of scored training positions and the selection of configurations. In order to conclude the evaluation function construction, we must show how to assign weights to configurations.

If the number of weights is large or non–linear models are used, direct weight computation is no longer feasible. Instead, iterative methods have to be used for weight fitting, which are usually based on variations of the gradient decent procedure. In each step, this procedure updates the current weight vector in direction of the negated gradient of the error function. If features are highly correlated, this simple algorithm is known to converge slowly. Faster conjugate gradient algorithms have been developed [6], that do not suffer from this problem. However, because the basic algorithm works sufficiently well in practice and is easier to implement, its application will be discussed in more detail in the remainder of this section.

5.1 Basic Considerations

In games, the purpose of evaluation functions is to estimate the winning chance for the player to move. This goal can be accomplished literally by constructing functions that map positions into $[0,1]$. Alternatively, the game may provide a numerical scoring of terminal positions reflecting the win size. In this case, a reasonable evaluation objective is to estimate the final game score. In either case, experiments should be conducted to find a suitable link function g. The most commonly used candidates are the identity function and sigmoid functions of the form $g(x) = 2C/(1+exp(-x)) - C$. For instance, for modeling the winning chance an S–shaped link function $g : \mathbb{R} \to [0,1]$ can be used in order to deal with saturation. In this regard, $g(x) = 1/(1 + exp(-x))$ is of special interest, because the weight fitting process benefits from a quickly computable derivative of g, which in this case is $g(x)(1 - g(x))$. A straight forward scoring scheme for terminal positions in this model assigns 0.9 to won positions, 0.5 to draws, and 0.1 to lost positions for the player to move. It is important to realize that an optimal weight vector may not exist if the extreme values 1.0 and 0.0 are chosen.

Given a sequence of scored training positions $((p_i, r_i))_{i=1}^{N}$ the objective is to find a weight vector $\boldsymbol{w}_0$ which minimizes the error function

$$E(\boldsymbol{w}) = \frac{1}{N} \sum_{k=1}^{N} \Delta_k(\boldsymbol{w})^2,$$

where

$$\Delta_k(\boldsymbol{w}) := r_k - g\Big(\sum_{i=1}^{n} w_i h_{i,k}\Big) \text{ and } h_{i,k} := \text{val}(c_i(p_k)).$$

Starting with an initial guess $w^{(0)}$, in each step the basic gradient descent procedure updates the weight vector according to

$$\boldsymbol{\delta}^{(t)} = -\alpha \cdot (\mathbf{grad}_{\boldsymbol{w}} E)(\boldsymbol{w}^{(t)})\ ^4$$

$$\boldsymbol{w}^{(t+1)} = \boldsymbol{w}^{(t)} + \boldsymbol{\delta}^{(t)}.$$

$\alpha > 0$ is the step size and $\mathbf{grad}_{\boldsymbol{w}} E$ is the vector consisting of E's partial derivatives $\frac{\partial E}{\partial w_i}$. This update scheme changes the weights in direction of the error function's steepest descent and is widely used for training artificial neural networks.

In this application, the partial derivatives have a simple form due to GLEM's flat evaluation structure:

$$\frac{\partial E}{\partial w_i}(\boldsymbol{w}) = -\frac{2}{N} \sum_{k=1}^{N} g'\left(\sum_{i=1}^{n} w_i h_{i,k}\right) \Delta_k(\boldsymbol{w}) h_{i,k}. \qquad (2)$$

If g is the identity function, this expression reduces to

$$\frac{\partial E}{\partial w_i}(\boldsymbol{w}) = -\frac{2}{N} \sum_{k=1}^{N} \Delta_k(\boldsymbol{w}) h_{i,k}.$$

Thus, steepest descent updates for all weights can be computed efficiently in a single pass through the training data. It is worth noting, that the computation of (2) can be arranged in such a way that its run time depends on the number of $h_{i,k}$ different from 0, rather than on N. Especially when using patterns, the savings thus achieved are significant.

Since the configuration match count may vary by large factors, the described update step changes weights at very different speeds. This is undesirable, because at one point the iteration process has to be stopped, and by then, weights of rare but important configurations might not have reached a proper level yet. A simple way to deal with this problem is to normalize the updates by dividing the sum by the number of $h_{i,k} \neq 0$ instead of N.

5.2 Position Type Dependent Weights

The evaluation of configurations may depend on the game stage or, more generally, on the particular type of the position. For instance, centralizing the king in chess openings is considered suicide, whereas his activation is crucial in many endgames. It may therefore be worthwhile to partition the training set according to position type, and to select configurations and fit weights separately for each set. In order to avoid big evaluation jumps when crossing type boundaries, which can cause undesired artifacts in game–tree search, it is helpful to define fine grained position types and to smooth evaluations across adjacent types. Fitting

---

[4] adding $\beta \cdot \boldsymbol{\delta}^{(t-1)}$ — known as "momentum" — can improve the convergence in case of correlated features.

weights for many position types, however, requires a large number of training positions, provided the minimal match count is maintained in order to eliminate over–fitting. Globally lowering the match count is therefore not an option. Instead, a more local view can help to reduce the number of needed positions. One suggestion when fitting weights for a particular position type, is to consider the training positions from adjacent types as well. This method increases the number of positions for any single position type and weights are smoothed automatically. The second option is to fit position type dependent weights in a more flexible manner. For this purpose, valid configurations are generated by considering *all* training positions. The weight fitting process then decides, how to compute the configuration weights separately for each type of position. For any type, for which the particular configuration match count is sufficiently high (say ≥ 20), it is safe to fit the according weight as described in the previous subsection. If the count is small (say ≤ 4), over–fitting is likely and the configuration should be treated as if there is no information available, i.e. the weight is set to 0. Cases in between can be handled by merging adjacent position types, until the total match number allows a robust weight fit. Here, the alternatives are to have only a single weight for all involved types or, if there are enough positions available, to fit a parameterized weight model. An example for such a model is $w(k) = a \cdot k + b$, $k_0 \leq k \leq k_1$, which states a linear relationship between the weight and the position type k — coded as an integer — in $[k_0, ..., k_1]$. Of course, this kind of model is only meaningful for position types that can be totally ordered, such as opening, middle–game, and endgame. Incorporating the update of parameters a and b in the gradient descent procedure is not hard.

This technique allows a flexible and robust fitting of position type dependent weights. After generating training positions and selecting configurations, this concludes the evaluation function construction.

6 Application: Othello

The presented general framework for the construction of evaluation functions has been inspired by the work on our Othello program LOGISTELLO. Besides the progress in selective search and automated opening book construction, the application of the techniques discussed has contributed to the considerable playing strength of this program. LOGISTELLO is able to beat the best human Othello players handily, even when running only on ordinary hardware [2]. The details of LOGISTELLO's evaluation function already have been discussed in [1]. We will therefore only give a short overview and concentrate on its recent improvement, which is based on the sparse pattern approach presented above.

Othello is a popular Japanese board game, played by two players on an 8x8–board using 64 two–colored discs. Moves consist of placing one disc on an empty square and turning all bracketed opponent's discs over. Fig. 5 shows an example. The game ends when neither player has a legal move, in which case the player with the most discs on the board has won.

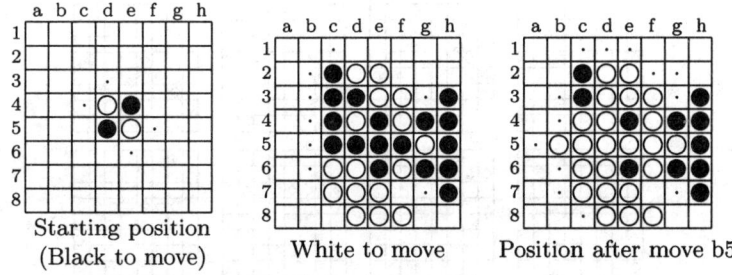

Fig. 6. Example positions. Legal moves are marked with a dot.

The most important concepts in Othello are disc stability, mobility, and parity. In particular:

- Stable discs can not be flipped by the opponent. Therefore, they directly contribute to the final score. The most prominent stable discs are occupied corners, which can be used as anchors for creating more stable discs.
- Having fewer move options than the opponent is dangerous, because it increases the chance of losing a corner in the near future.
- Making the last move in an Othello game is advantageous, since it increases one's own disc count while decreasing the number of opponent's discs. Parity generalizes this observation by considering last move opportunities for every empty board region.

In [1] it has been shown, that all of these features can be quickly approximated by pattern configurations built upon a raw board representation. The chosen patterns are shown in Fig. 7. Horizontal, vertical, and diagonal lines of length ≥ 4 are included for covering mobility. The remaining patterns deal with the important corner regions and edges. The evaluation function distinguishes 13 game stages, depending on the number of discs on the board. Applying the techniques described in the previous sections, about eleven million scored training positions were generated to fit approximately 1.5 million weights. This figure takes weight sharing among symmetrical configurations into account. Starting with $w^{(0)} = \mathbf{0}$, the weight fitting procedure took a Pentium II/333 CPU about 30 hours to reach an acceptable accuracy level after 250 iterations. Equipped with an evaluation function very similar to that we have just described, LOGISTELLO beat the human Othello World–champion 6–0 in August 1997 [2]. After four years of successful tournament play, LOGISTELLO ended its career in October 1997 with a straight 22–win victory in its last computer Othello tournament.

Recently, the incorporation of larger patterns has improved the evaluation performance. In the current implementation, configuration weights are represented as 16 bit integers. Storing weights for 10–square patterns in 13 flat tables thus requires $3^{10} \cdot 2 \cdot 13 \approx 1.5$ million bytes. Using the same approach for storing weights for much larger patterns is therefore out of the question. The first experiments with several sparse data access schemes based on binary search were

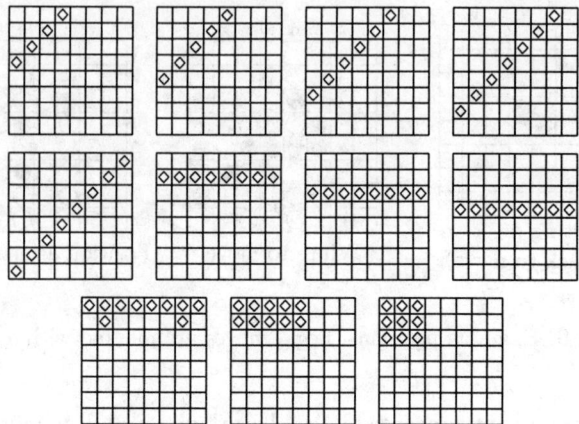

Fig. 7. LOGISTELLO's previous pattern set. Patterns that can be obtained by rotating and mirroring the board have been omitted. Each diamond represents an atomic feature f with range $\{0, 1, 2\}$. $f(p)$ is defined by the particular square contents (e.g. white disc $\mapsto 0$, empty $\mapsto 1$, black disc $\mapsto 2$).

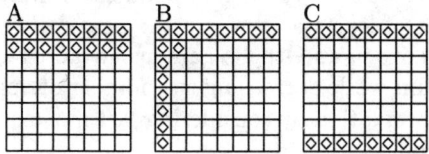

Fig. 8. Large patterns tested. For each of these patterns the simplified pattern version of GenConf generated about 88,000 valid configurations (#E ≈ 11 million, $n = 75$). All configuration sets fit in hash–tables with about 310 thousand entries.

disappointing. Increasing the program's knowledge by adding the patterns shown in Fig. 8 could not compensate for a slowdown of about 45%. Only after utilizing the fast hash–table access scheme and adding just one of the three features, the program achieved its best performance so far. Table 1 summarizes the results of all tournaments that have been played to evaluate each version. All games were played by brute–force versions of LOGISTELLO running on Pentium II/333 PCs. On this hardware LOGISTELLO achieves a middle–game speed of approximately 270K nodes/sec when the patterns shown in Fig. 7 are used. This speed enables the program to look 12–14 ply ahead in the opening and middle part of ten minutes games.

The patterns presented in Fig. 8 were chosen based on both game and evaluation speed considerations. Human players frequently make use of their abilities to evaluate large disc formations which are not covered by the basic patterns. Of special interest are edge interactions and 2 × 8–corner configurations. On the other hand, it is preferable to add patterns for which the index computation can make use of already determined indices. The chosen 16–square patterns meet this

Table 1. Tournament results. LOGISTELLO using the basic patterns played 434–game tournaments against several versions that — in addition — employed the large patterns shown in Fig. 8. The results indicate that speed matters. The strongest versions are those that only use either pattern A or B. They beat the previous version significantly, although they are 11% slower. When playing at equal strength the best version only needs to search about 2/3 of the nodes — as the results of the time–handicap tournaments indicate.

opponent	time/game (minutes)	#nodes (fraction)	opponent results wins	draws	losses	winning percentage
A	10-10	0.89	213	58	163	55.8
B	10-10	0.89	211	60	163	55.5
AB	10-10	0.83	203	60	171	53.7
ABC	10-10	0.8	211	49	174	54.3
A	6-10	0.51	172	59	203	46.4
A	7-10	0.62	183	55	196	48.5
A	8-10	0.71	195	63	176	52.2

preference. Nevertheless, the results show, that the combined knowledge coded in the new patterns does not compensate for the speed drop. This finding indicates that a significant improvement of a sequential program may not be possible by adding further patterns based on the raw board representation. However, a more effective atomic features might exist which in combination outperform the current evaluation function.

7 Summary and Discussion

In this paper a practical framework for the semi–automatic construction of evaluation functions has been presented. Based on a generalized linear evaluation model — called GLEM — efficient procedures have been developed for generating training positions, exploring the feature space, and fitting feature weights. Rather than combining a few features by using complicated non–linear functions, we propose to construct evaluation functions by combining many — possibly more than hundred thousand — features, which are boolean combinations of atomic relations. This approach allows us to model non–linear effects directly, without the detour over analytic functions, and opens up practical ways for generating features automatically. GLEM allows the program author to concentrate on the part of evaluation function construction, where humans excel: the discovery of fundamental positional features by *reasoning* about the game. GLEM simplifies this task because the exact feature formulation is no longer needed. The system is able to approximate complex features by combining atomic fragments. In this way, it is now possible for the programmer to speculate about feature building blocks and to leave the creation of actually used features as well as assigning weights to them to the system. One example for this strategy has been presented in this paper: the observation that configurations can approximate important Othello concepts combined with the "mechanical" analysis of millions of training positions has produced an expert program capable of beating any human player. An interesting fact is that the game knowledge encoded by the set of over a million configuration weights goes far beyond the features we intended the system to approximate in the first place [1]. This result encourages the application of GLEM to other games or even to search or decision problems in other domains. Attractive candidates are chess and Go since both games are very popular and well analyzed. And yet, for chess, hardware roughly equivalent to four thousand ordinary PCs is currently needed[5] to compete with the human World–champion. For Go the status is even worse because brute–force search is not feasible due to the large branching factor. Since a good evaluation function is not known either amateurs are still able to beat the best Go programs. It is our opinion that the key to better chess and Go programs lies in improved evaluation functions. A starting point is the analysis of known features with regard to their approximation by weighted configurations as proposed by GLEM.

The automatic construction of features has been studied by several authors. Utgoff [10] proposes a general evaluation function learner, called ELF, which combines the processes of constructing boolean feature combinations and weight fitting. This approach has been shown to be effective in small artificial problems, but could not convince in its application to checkers. The main problem of ELF is its low speed. Taking into account the large number of features needed for an adequate evaluation in complex domains, and the resulting considerable effort for optimizing weights, it seems hopeless to combine feature construction and weight fitting. Other approaches for constructing features or adapting the combination function while fitting weights (e.g. MORPH [5], meiosis networks [3], node splitting [11]), face similar complexity problems. Our solution is to separate these tasks in order to speed–up the process and to give many opportunities for optimization.

References

[1] M. Buro. Experiments with Multi–Probcut and a new high–quality evaluation function for Othello. *Workshop on Game-Tree Search, NEC Research Institute*, 1997.

[2] M. Buro. The Othello match of the year: Takeshi Murakami vs. Logistello. *ICCA Journal*, 20(3):189–193, 1997.

[3] S.J. Hanson. Meiosis networks. *Advances in Neural Information Processing Systems*, pages 553–541, 1990.

[4] F. Hsu, S. Anantharaman, M.S. Campbell, and A. Nowatzyk. Deep Thought. In T.A. Marsland and J. Schaeffer, editors, *Computer, Chess, and Cognition*, pages 55–78. Springer Verlag, 1990.

---

[5] Deep Blue searched around 200 million nodes per second in the 1997 match with Kasparov. Assuming a speed–up of four gained by using special purpose evaluation hardware and a speed of 200K nodes/sec of a state–of–the–art PC chess program leads to the given speed factor estimate.

[5] R.A. Levinson and R. Snyder. Adaptive pattern–oriented chess. In L. Birnbaum and G. Collins, editors, *Proceedings of the 8th International Workshop on Machine Learning*, pages 85–89, 1991.

[6] W.H. Press, S.A. Teukolsky, W.T. Vetterling, and B.P. Flannery. *Numerical Recipes, 2nd edition*. Cambridge University Press, 1992.

[7] A.L. Samuel. Some studies in machine learning using the game of checkers. *IBM Journal of Research and Development*, 3(3):211–229, 1959.

[8] G. Tesauro. TD–Gammon, a self–teaching backgammon program, reaches master–level play. *Neural Computation*, 6(2):215–219, 1994.

[9] G. Tesauro. Temporal difference learning and TD–Gammon. *Communications of the ACM*, 38(3):58–68, 1995.

[10] P.E. Utgoff. Constructive function approximation. Technical Report 97–4, Univ. of Mass., 1997.

[11] M. Wynne-Jones. Node splitting: A constructive algorithm for feed–forward neural networks. *Neural Computing and Applications*, 1(1):17–22, 1993.

A Two-Step Model of Pattern Acquisition: Application to Tsume-Go

Takuya Kojima and Atsushi Yoshikawa

NTT Basic Research Labs.
3-1 Morinosato Wakamiya, Atsugi,
Kanagawa 243-0198, JAPAN
{kojima, yosikawa}@rudolph.brl.ntt.co.jp

Abstract. It has been said to be very useful for Go playing systems to have knowledge. We focus on pattern level knowledge and propose a new model of pattern acquisition based on our cognitive experiments. The model consists of two steps: pattern acquisition step, using only positive examples, and pattern refinement step, using both positive and negative examples. The latter step acquires precise conditions to apply and/or the way of conflict resolution. This model has advantages in computational time and precise control for conflict resolution. One algorithm is given for each step, and each algorithm can change independently, it is possible to compare algorithms with this model. Three algorithms are introduced for the first step and two for the second step. Patterns acquired by this model are applied to Tsume-Go problems (life and death problems) and the performance between six conditions are compared. In the best condition, the percentage of correct answers is about 31%. This result equals the achievement of one dan human players. It is also shown that the patterns enhance search techniques when the search space is very large.
Keywords: knowledge acquisition, evolutionary learning, pattern acquisition, Go, Tsume-Go

1 Introduction

1.1 Purpose

Studies on games have mainly focused on search techniques, and these techniques support systems playing most games — such as chess, checkers, and Othello — to perform as well as human experts. In these games, full-width search to some depth is very common. Full-width search is, however, very difficult to apply in games such as Go and Shogi (Japanese chess), whose branching factors are about 250 and 80, respectively, as compared with the 35 of chess. Therefore, selection of moves is indispensable. One promising method for selecting moves is to use knowledge.

We have classified Go knowledge into two major levels based on cognitive studies: pattern level knowledge and language level knowledge [11]. Pattern level knowledge includes patterns and sequences of moves. A pattern is a rule whose condition part is a partial board configuration, and whose action part is a move.

A sequence of moves is a rule suggesting several moves. It is said that they are very useful for human players but are missing in current computer systems [10] [1]. Language level knowledge is a verbal rule consisting of Go terms used by human players.

In this study, we focus on pattern level knowledge, especially patterns. Although patterns have often been used in Go playing systems [2], most of them have been entered by programmers and it is very hard to input enough patterns to replicate the "skill" of human players. Instead, some studies have examined automatic pattern acquisition [13,1,8]. However, when the patterns are used in the actual games, too many patterns are matched, and mechanisms to resolve the conflicts among patterns are needed. These mechanisms are computationally expensive and hard to implement. Therefore, we propose, in this paper, a new model of acquiring patterns that also learns the conflict resolution mechanisms. The reason why too many patterns are matched in a situation is because the patterns contain only a partial board configuration as the matching condition. We assume that patterns with additional information – such as detailed descriptions about when the patterns are applied or the priority of patterns – can substitute for the conflict resolution mechanisms. This process can be considered as a refinement of the rules. Thus we propose a model of pattern acquisition where patterns are acquired and the acquired patterns are refined, as explained in Subsection 1.2.

The effectiveness of these refined patterns is investigated by solving Tsume-Go problems (life and death problems of Go), because performance is easy to measure and compare. This paper compares the performance of our model to that of human players.

Most systems solve Tsume-Go by searching almost all possible moves without using knowledge to narrow the search space. GoTools [14], which is one of the best Tsume-Go solvers, uses patterns only at the end node of the search tree. We narrow the search space with patterns by finding the first move. In general, you can narrow more if you narrow earlier, so narrowing the first move narrows the most. We use patterns to select candidates of the first move of sequences of Tsume-Go answer moves. A related work is that of Sasaki [12], wherein Tsume-Go problems are solved using Neural Networks without search.

1.2 A 2-Step Model of Pattern Acquisition

Our cognitive studies [17] have shown that *kyu* level (beginner level) amateur players have patterns whose condition is only a part of board configuration, and that *dan* level (expert level) amateur players have patterns with precise conditions when the rules that can be described by Go terms can be applied.

---

[1] Sequences of moves in Go are important and helpful for humans and computers, as well as in chess. One reason is that they are helpful for "answer moves", your response to an opponent's move. Another reason is that in Go there are many sequences which cannot be stopped by halves. For example, a sequence that yields good results if completed may lead to bad results if stopped by halves.

We propose the following assumption: when players are weak, they acquire patterns by reading books or watching good players' moves. However when they actually apply the acquired rules, they sometimes fail. This is because the conditions of the acquired patterns are inappropriate and they sometimes apply a rule in the wrong situation. Another reason is that they do not know the priority of rules and they apply less important rules. As a consequence of trial and error, they gradually learn more precise rule conditions or priority.

Therefore, we propose the following two-step model to acquire proper patterns. The first step is to mainly acquire patterns. This is the pattern acquisition step. In the second step, some additional information on the patterns, such as the conditions under which the rules can be applied or priority to resolve conflicts among rules, are added in order to refine the rules. This can be considered as the pattern refinement step.

This is a concept-level description of the model. When the model is implemented, we should create an algorithm for each step. In this paper, the first algorithm acquires patterns whose conditions are only partial board configurations. It only requires positive examples. The second algorithm, however, requires positive and negative examples, because they are learned as a consequence of trial and error. These algorithms can change independently.

1.3 Procedures of This Paper

This model needs two algorithms: one is for the first step and the other is for the second step. We introduce three algorithms for the first step in Section 2, and two algorithms for the second step in Section 4. We carry out six experiments (2×3) to compare these algorithms by solving Tsume-Go problems in Section 6. In Section 5, human performance against the Tsume-Go problems is shown in a comparison to that of our system.

2 Pattern Acquisition Algorithm

This section introduces three algorithms for the first step of pattern acquisition. The first algorithm acquires flexible patterns whose shape and size are widely varied, and is called "Flexible Algorithm" in this paper. Other two algorithms acquire patterns of fixed size and shape. One of them, called "Fixed Algorithm", always acquires patterns in much the same size and shape. This type of patterns are often used in the previous studies [13]. The other, called "Semi-Fixed Algorithm", acquires almost fixed patterns but the size and shape of acquired rules are slightly varied between patterns. These three algorithms are explained in this section.

2.1 Flexible Algorithm

The first algorithm acquires flexible rules. This algorithm was initially proposed to acquire flexible patterns from game records [8]. Although Tsume-Go problems and their answers can be considered game records of a few moves (usually one or three moves), the small number of Tsume-Go problems causes overfitting [2]. Thus a new mechanism is implemented to avoid overfitting. This subsection briefly explains the Flexible Algorithm. For details, see [7,9].

Overview of the Algorithm The algorithm's aim is to acquire useful rules, herein *individuals*, which match the given training data, herein *food*. Each rule takes the form of a production rule, consisting of an IF-part (consisting of conditions) and a THEN-part (consisting of an action), and has an *activation value*. There are no rules in the initial state.

Rules matching a training datum get food, and their activation values increase. If there are no rules matching a given datum, a new rule matching the datum with only one condition is created. The number of rules with only one condition thus increases at an early stage.

After being fed the rules with activation value over a threshold are *split* into two rules: the original one and a more complex one. More complex rules are thus created by splitting, and the number of rules increases.

Every rule consumes food at each step, and decreases one activation value. Rules whose activation value is 0 *die*. Thus, rules that are too complex and get food too rarely, die. As a result, all the rules are expected to get food at almost the same frequency. The procedure of this algorithm is shown below.

Algorithm 1.1 Pattern Acquisition Algorithm
```
1    step ← 1
     while step ≤ the number of iterations
2        choose a random game record from the game database
3        move ← 1
         while move ≤ the number of moves in the game
4            a training datum ← move-th move
5            if no rule matches the datum
                 then create a new rule
                 else feed matched rules
             for all rules
6                if activation of a fed rule > threshold
                     then split the rule
7                activation of a rule ← activation of a rule − 1
8                if activation of a rule = 0
                     then the rule dies
             end for
9            move ← move + 1
10       step ← step +1
     end while
end while
```

---

[2] When the volume of training data is small, less general rules that suit only the training data tend to be acquired. This is usually called *overfitting* in machine learning.

Details of the Algorithm

Rules Each rule takes the form of a production rule consisting of an IF-part and a THEN-part.

A rule is described in relative terms so that it can be executed by either player. Eight equal board configurations (4 rotations (90 degrees) times 2 reflections) are converted to one of the configurations. Rules are described as follows:

IF exist($[x_1, y_1], obj_1$) ∧ ...∧ exist($[x_n, y_n], obj_n$) **THEN** play($[0,0]$),

where $[x_i, y_i]$ represents relative coordinates from the action place [3], whose coordinate is always $[0,0]$, and obj_i is one of the four objects shown below:

1. SAME: the same color stone as the action stone [4]
2. DIFF: a stone with different color from the action stone
3. EDGE: an edge of the board [5]
4. $-n$: a previous move [6] ($n \in N, n > 0$)

The following rule is an example: "**IF** exist($[-1,-1]$, -1) ∧ exist($[0,-1]$, SAME) ∧ exist($[-2,-1]$, DIFF) ∧ exist($[0,-5]$, EDGE) **THEN** play($[0,0]$)". This rule is shown in Figure 1. The active player is always shown as Black in this paper. The action stone and the move are presented as the stone with the largest number. This stone is "2" in this example. This rule means that **if** the previous move (shown as "1") is played to point [-1,-1] (relative to the action place) and a stone of the *same* color as the action stone (Black) exists at point [0,-1] and a stone of a *different* color from the action stone (White) exists at point [-2,-1] and an *edge* exists at point [0,-5], **then** put a stone on [0,0].

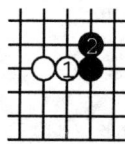

Fig. 1. An example of rules: **IF** exist($[-1, -1], -1$)∧ exist($[0,-1]$,SAME) ∧ exist($[-2,-1]$,DIFF) ∧ exist($[0,-5]$,EDGE) **THEN** play($[0,0]$)

---

[3] The place where a stone is placed according to the rule.
[4] The stone to be placed according to the rule.
[5] An edge is located just outside the board, not inside the board.
[6] "-3" represents the move played three moves before. This is used to acquire sequences of moves.

Feeding Rules Among matched rules, those that do not have rules more specific than themselves are *fed*. An activation value equal to Parameter $FOOD$ is equally shared among the fed rules. The following is an example.

Suppose that the following five rules matched a given training datum.

1. **IF** C_1 **THEN** A_1
2. **IF** $C_1 \wedge C_2$ **THEN** A_1
3. **IF** $C_2 \wedge C_3$ **THEN** A_1
4. **IF** $C_4 \wedge C_5$ **THEN** A_1
5. **IF** $C_2 \wedge C_3 \wedge C_4$ **THEN** A_1

Rules 1 and 3 are not fed because they have more specific rules than themselves (Rules 2 and 5, respectively). The others (Rules 2, 4, and 5) share food, and each gets one third of the food.

As a consequence, the more general the rules are, the more frequently they are matched: however, because of this feeding method, the probability that they are fed falls although they are matched. As a result, too general rules are not acquired because they often match with food, but rarely get food as long as more specific rules exist.

Splitting and Creating Rules A fed rule whose activation value is greater than a certain threshold value is split into two rules: one, called *parent*, is the original rule and the other, called *child*, is created by adding a new condition to the IF-part of the original rule. For example, rule, **IF** $C_1 \wedge C_2$ **THEN** A_1, is split into the original rule and a newly created rule, **IF** $C_1 \wedge C_2 \wedge C_3$ **THEN** A_1. The new condition, C_3, is chosen at random from among the objects on the current board (stones, edges of the board, and the previous moves).

A child extracts a certain amount of the activation value from its parent. As a result, after the split, the activation value of the rule changes as listed in Table 1. The total activation value, however, does not change after the split.

Table 1. Change of activation value after splitting.

	before	after
parent	P_ACT	P_ACT − INI_ACT
child		INI_ACT

If no rule is matched, a new rule is created which has only one predicate chosen randomly from among the objects on the current board. A newly created rule that is the same as a rule already in the rule set is deleted. There is thus no duplication of rules.

Avoidance of Overfitting: Memory The total number of training data of Tsume-Go is much smaller than that of real games, because the number of moves (training data) in one Tsume-Go problem is usually 1, 3 or 5 and is much smaller than that in game records (about 207 [3]); moreover, there are fewer Tsume-Go problems than game records. Since the small number of training data usually causes overfitting, a new mechanism, *memory*, is introduced to avoid this problem.

The memory avoids acquiring rules used by only a few training data. When memory is used, IDs [7] of the last M training data which a rule matches are recorded in the rule. When a rule matches a training datum whose ID is recorded, it is not fed. By this mechanism, a rule is not fed by training data by which it is fed last M times. M is usually set to 2 or 3.

2.2 Fixed Algorithm

The second algorithm acquires fixed patterns. The shape of the acquired patterns is the same, but the size can change. A pattern includes all the squares within d Manhattan distance from the center and their adjacent 8 squares. The IF-part is all the squares except the center, and the THEN-part is the center square, so when a pattern is matched, the move is to the center square. Figure 2 shows the shapes of patterns. In the figure, the numbers indicate the Manhattan distance from the center square, 'a' means their adjacent squares, and '1' means the center square. A square is occupied by one of the following five objects: SAME, DIFF, EDGE, EMPTY(the square is not occupied by any stone), and OB(the square is out of the board). Note that empty squares are explicitly described in this algorithm. In this paper, d ranges from 0 to 3. The number of conditions according to d is as listed in Table 2.

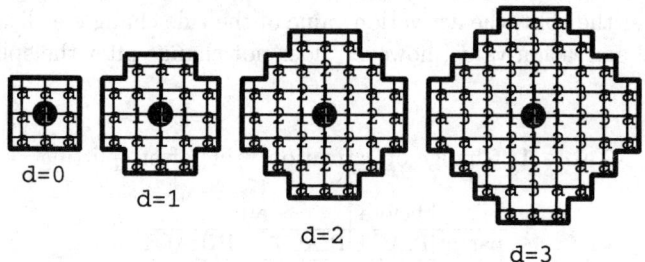

Fig. 2. Shapes of patterns acquired by Fixed Algorithm.

---

[7] ID of training datum is both game ID and move number. Training data of the same game but different moves have different IDs.

Table 2. Then number of conditions in a rule.

d	0	1	2	3
The number of conditions	8	20	36	56

Patterns are acquired by a simple mechanism. When a training datum is given, a pattern for each d, whose center is the training move, is simply stored into a database. A total four patterns ($d = 0, 1, 2, 3$) are stored. Eight rotations are considered and only one of them is stored and the others will not be stored.

When the patterns are used (for refinement explained in Section 4 or for solving Tsume-Go explained in Section 6), only the pattern with the largest d is used, which is considered as "fed" in Subsection 3. For example, if two patterns with $d = 0$ and $d = 2$ match a board configuration, only the pattern with $d = 2$ is considered as "fed".

2.3 Semi-fixed Algorithm

The third algorithm acquires semi-fixed patterns. Patterns acquired by this algorithm are almost the same as those acquired by the Fixed Algorithm. The difference is that it mainly considers empty squares. In our cognitive experiment [15] using an eye mark recorder, experts mainly solve Tsume-Go problems by seeing empty squares and stones adjacent to the empty squares. This mechanism is introduced in the Semi-Fixed Algorithm. The Manhattan distance is how many steps you can walk vertically and horizontally from the center. In this algorithm you walk only on empty squares. That is, when you count d, only empty squares are counted and squares occupied by a stone are not counted. The counted squares and squares adjacent to the counted ones are considered in a rule. As a result, the number of objects in the IF-part varies. When there is no stone within d Manhattan distance, the rule is the same as that acquired by the Fixed Algorithm. An example is given in Figure 3.

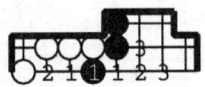

Fig. 3. An example of rules acquired by Semi-Fixed Algorithm. $d = 3$.

Mechanism of pattern acquisition is the same as that in the Fixed Algorithm.

3 Results of Pattern Acquisition

3.1 Methods

The three algorithms explained in Section 2 were used to acquire patterns from Tsume-Go problems. The training examples were 1,039 Tsume-Go problems and their answers (total 3,993 moves, the average number of training data per problem is 3.8) [8]

Flexible Algorithm Training sets were chosen as follows. One problem was randomly chosen and all the correct moves of the problem were taken as training data from the first move to the last move. After all the moves were used as training data, another problem was randomly chosen. This procedure was repeated. The time step is a single move in this simulation. The parameters are shown in Table 3.

Table 3. Parameters used in acquiring Tsume-Go knowledge.

Name of Parameters	Value
INI_ACT	100
$FOOD$	1000
$threshold$	2000
M (memory)	2 or 3
the number of iterations	1,000,000

Fixed and Semi-fixed Algorithm All training problems were used once. For each training datum, patterns with $d = 0,1,2,$ and 3 were acquired. How many time patterns were matched was recorded and patterns that were matched more than once were stored.

3.2 Results

Flexible Algorithm Figure 4 lists examples of acquired rules. The number of acquired rules is listed in Table 4. Some rules, called *reliable rules* [9], were selected for future evaluation as follows. After the acquisition process finished, all training data were given once again as training data, and we checked how many times the

---

[8] Both White and Black moves are used as training data. Tsume-Go problems are like "next move problems" in chess. In the answer moves (training data), only a small number of moves are given but you have to think much deeper to answer a correct move.

[9] Rules fed over 100 times are called reliable rules.

acquired rules were fed. Rules fed over M (memory) times were selected, because rules fed less than M times will eventually die due to the memory mechanism. 10% of the reliable rules were randomly selected to be evaluated by two human expert players [10].

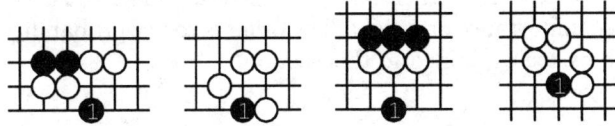

Fig. 4. Examples of acquired rules of Tsume-Go.

Table 4. The number of acquired rules from Tsume-Go.

M	No. of acquired rules	No. of reliable rules	No. of evaluated rules
2	896	649	65
3	885	569	57

The results of the experts' evaluation of the acquired rules are listed in Table 5. They show that the quality of acquired rules is very high.

Table 5. The results of expert evaluation of Tsume-Go rules.

expert	M	good	average	bad	total
A	2	21 (32%)	29 (45%)	15 (23%)	65 (100%)
	3	22 (39%)	15 (26%)	20 (35%)	57 (100%)
D	2	37 (57%)	17 (26%)	11 (17%)	65 (100%)
	3	35 (62%)	19 (33%)	3 (5%)	57 (100%)

Fixed and Semi-fixed Algorithm 1252 rules were acquired by the Fixed Algorithm and 1512 rules by the Semi-Fixed Algorithm. Since these rules are fixed and hard to evaluated by human experts, they were not evaluated.

---

[10] Both are strong amateur players, 5 dan and 6 dan.

4 Rule Refinement Algorithm

Our cognitive studies [17] show that the patterns which players stronger than 2 dan have contain conditions that use Go terms and that those which 1 dan players have contain few such terms. We assume that 1 dan players use only a simple priority of patterns. Therefore, we assign numbers to rules in two ways. One is to assign importance or weight. The other is to use probability of accuracy. These algorithms are explained in this section.

4.1 Assignment of Weights to Rules

In this algorithm, weights, indicating how important the rules are, are automatically assigned to the rules acquired in the first step.

Methods A weight is an integer. In the initial state, all rules are given the same weight, 200. When the value is larger that this, it takes more time to assign the weights. When the value is too small, most rules tend to have similar values and discrimination does not work properly. The initial value of weights, 200, was decided after some preliminary experiments.

The points of Move m are calculated by summing all the weights of rules that are fed by Move m. After all the points of possible moves are calculated, the moves are ranked in descending order of points. A move that duplicates a training datum is called a correct move.

Initially given weights are automatically adjusted in the following way. The value of one is subtracted from the weights of all the rules fed by moves ranked higher than the correct move, and the value of S/n is added to the weights of all the rules fed by the correct move, where S is the total subtracted values and n is the number of rules fed by the correct move.

The adjustment of the weights does not change the total amount of the weights of all rules, because the total subtraction of weights is S, and the total addition is $S/n \times n = S$.

We use Table 6 as an example to explain the adjustment. Each $Move_i$ is a candidate move, ranked as the i-th rank by the summation of all the current weights of rules, R_{ij}, which are fed by $Move_i$. Numbers in the table indicate the weights of each rule. In this example the correct move is in the fourth rank; thus, all the weights of the rules fed by Move 1, 2, and 3, which are all negative examples, are decreased by one and the total subtraction is 8. The number of rules fed by the correct move is two, thus $8/2 = 4$ is added to each rule. As a result, the correct move will be in the third rank. Repeating these procedures is expected to make the weights more appropriate.

This realizes automatic adjustment of the degree of change because a larger amount is given to the rules fed by the move when the rank of the correct move is low, and a smaller amount is assigned when the rank of the correct move is high.

Table 6. Adjustment of priority values.

candidate moves		Priority values before adjustment	Priority values after adjustment
$Move_1$		$R_{11} = 230, R_{12} = 210, R_{13} = 180$	$R_{11} = 229, R_{12} = 209, R_{13} = 179$
$Move_2$		$R_{21} = 200, R_{22} = 180, R_{23} = 170$	$R_{21} = 199, R_{22} = 179, R_{23} = 169$
$Move_3$		$R_{31} = 200, R_{32} = 180$	$R_{31} = 199, R_{32} = 179$
$Move_4$	Answer	$R_{41} = 195, R_{42} = 180$	$R_{41} = 199, R_{42} = 184$

4.2 Rules with the Same Weights

In order to confirm the effect of the weight assignment explained in Section 4.1, Tsume-Go problems were solved with the same weights.

Procedures for Solving Tsume-Go When the system is given a problem, it calculates points of every possible move in the manner shown below. The moves getting more points are considered more promising ones. The moves are ranked in descending order of points. The rank of the correct move is registered.

Calculation of Points P_m, the point score of Move m, is calculated as follows. Taking Move m as a training example, the number of matched rules to the example is denoted as NR_m. The number of objects appearing in the condition parts of the matched rules (an object appearing in multiple rules is counted as one) is denoted as NO_m. The point score of Move m, P_m is calculated as follows:

$$P_m = 10 \times NO_m + NR_m$$

NR_m ranges from 0 to 30 and its average is about 10. The rank of the move is thus determined mostly by NO_m and minor adjustment is done by NR_m.

Take Figure 5 as an example of calculating the point score of a move.

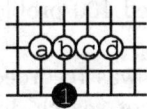

Fig. 5. Examples of calculation of points of moves.

Suppose that the following two rules are matched to Move 1.

if a ∧ b then Move 1
if b ∧ c ∧ d then Move 1

NR_1 is 2, and NO_1 is 4 (**a, b, c** and **d**), thus, P_1 is 42.

4.3 Probability of Rule Accuracy

This algorithm calculates the probability of the accuracy of each rule. When only the IF-part is matched and the THEN-part is not matched, it is called a "match", which is a negative example. When both the IF-part and the THEN-part are matched, it is called a "hit", which is a positive example. The probability of accuracy (p) for a rule, i, is calculated as follows: a_i = No. of hit / No. of match.

The probability of a move, $m, (A_m)$ is calculated as follows: $A_m = 1 - \prod_{i=1}^{n}(1-a_i)$, where n is the number of hit rules. When $n=1, A_m = a_1$. When Tsume-Go problems are solved, probability of all moves is calculated and the highest one is selected.

5 Human Performance in Tsume-Go

The performance of human players, described in this section, was determined to allow a comparison with that of the proposed system. Detailed procedures for the cognitive experiment are explained in [16,15].

5.1 Methods

Subjects were given a Tsume-Go problem and replied with the first move of the solution within three or four seconds. [11] The subjects were allowed multiple replies. There were two kinds of experiments. In one, the problems were shown on paper for three seconds and the solutions were written on paper. In the other, the problems were shown on a computer display for four seconds and replies were made by mouse actions. [12] The results of the experiments using a mouse are marked "*" below. The subjects were nine amateurs from 4 kyu to 6 dan as determined by standard tests. Three kinds of problems were used as test problems: basic problems [6], problems for 3 dan [5], and problems for 5 dan [4]. [13] Each kind of problem contained 100 problems, and a total of 300 problems were solved.

The eye path of the subjects was recorded by an eye tracker. The records indicated that the subjects did not search during the experiments. Therefore, the performances of human subjects can be compared with those of a system which does not search.

---

[11] Solving Tsume-Go normally means showing a sequence of moves and the result such as White dead, Ko, or Black alive. In this experiment only the first move of the correct sequence of moves was requested.

[12] Using a mouse needs a bit more time than writing, so one second was added. Training in mouse usage was carried out before the experiment.

[13] The names of the problems, such as "basic", "for 3 dan", do not exactly represent the difficulty of the problems, and they are simply taken from the title of the books.

5.2 Results

Table 7 lists the correct rate of each subject. The correct rate is the sum of the reciprocals of the number of replies containing a correct answer [14].

Table 7. Correct rate of Tsume-Go within a few seconds. (%)

strength	basic	for 3 dan	for 5 dan	average
6 dan(1)*	63	68	43	58.0
6 dan(2)*	53.5	59.5	38.3	50.4
4 dan	58	68	38	54.7
3 dan	27.0	30.5	21.0	26.2
1 dan(1)	40.2	36.8	20.3	32.4
1 dan(2)	30.8	40.3	22.3	31.1
2 kyu(1)*	21.5	18.3	16.0	18.6
2 kyu(2)	12.3	15.8	11.8	13.3
4 kyu	13.5	12.5	10.8	12.3

The value of the correlation coefficient between the subjects' strength [15] and the correct rate of Tsume-Go within a few seconds is 0.923, which is very high. The results show that the task of replying within a few seconds is directly related to the subjects' strength and that the task is a proper one to evaluate the strength of players. They also show that these problems are rather difficult.

6 Solving Tsume-Go

6.1 Methods

Patterns were acquired in the first step and then refined in the second step. Using the refined patterns, Tsume-Go problems were solved. We tested all six combinations of the three algorithms for the first step and two for the second step.

The same 300 problems that presented to the human subjects in Subsection 5.1 were used. These problems were not part of the training problems.

6.2 Overall Results

The correct rates of solving Tsume-Go for the six combinations are listed in Table 8. The results show that as the first step algorithm, Flexible Algorithm is

---

[14] When one of the three replies is a correct answer, 1/3rd of a point was added to the correct rate. 6 dan (A) and 4 dan subjects always made one reply. The correct rate is thus the same as the number of correct replies.

[15] 2 kyu is -1 dan, and 4 kyu is -3 dan.

much better than either Semi-Fixed Algorithm or Fixed Algorithm. Semi-Fixed Algorithm seems a bit better than Fixed Algorithm but the difference is not so clear.

As for the second step, Probability Algorithm seems a bit better than Weights Algorithm with Fixed and Semi-Fixed Algorithms, but Weights Algorithm is better for Flexible Algorithm. This is a very interesting result. It is said that combinations of First step and Second step are very important.

Table 8. Correct rates of solving Tsume-Go for six conditions.

		first step		
		Flexible	Semi-Fixed	Fixed
second	weights	31.0	13.3	11.0
step	probability	25.0	15.0	13.3

The time it takes to solve problems is very short, less than 10 seconds for solving all the 300 problems for any combination using an Ultra Sparc Station.[16]

6.3 Results of Flexible Rules with Weights

Using Flexible Algorithm and Weights Algorithm produces the best performance. In this section the result is shown in detail. Table 9 lists the detailed results of solving Tsume-Go using the acquired rules with weights. The number indicates the total percentage of correct moves for candidates. For example, solving *Basic* Tsume-Go, the correct percentage of the *1st* rank answer of the system is *36%*, and the correct percentage of the *1st and 2nd* rank answers by the system is *51%*.

Table 9. Correct percentage of solving Tsume-Go using acquired rules with weights. (%)

Rank	Basic	3 dan	5 dan	Average
1st	36 (36)	31 (31)	26 (26)	31 (31)
2nd	15 (51)	26 (57)	20 (46)	20 (51)
3rd	12 (63)	10 (67)	9 (55)	11 (62)
4th	10 (73)	7 (74)	14 (69)	10 (72)
5th	6 (79)	6 (80)	8 (77)	7 (79)

---

[16] Almost the same speed as a Pentium II 400 MHz.

By comparing the average correct rate of the first-ranked move of the system (31%) and the average correct rates of human players (Table 7), the performance of this system almost equals that of 1 dan human players.

In order to know the effect of weight assignment, Tsume-Go was solved by rules, all of which were assigned the same weight as explained in Section 4.2. Table 10 lists the results.

By comparing the average correct rate of the first-rank move of the system (19%) and the average correct rates of human players (Table 7), the performance of this system almost equals that of a 2 kyu human player.

Comparing these results with the previous experiments in Table 9 shows that the performance is much improved by assigning weights to rules in the second step.

Table 10. Results of solving Tsume-Go by Acquired Patterns. (%)

Rank	Basic	3 dan	5 dan	Average
1st	20 (20)	21 (21)	15 (15)	19 (19)
2nd	14 (34)	12 (33)	13 (28)	13 (32)
3rd	8 (42)	8 (41)	12 (40)	9 (41)
4th	10 (52)	9 (50)	9 (49)	9 (50)
5th	10 (62)	11 (61)	14 (63)	12 (62)

6.4 Performance for the Difference Search Space

In order to see whether the performance of the system differs according to the size of search space, the 300 test problems were divided into two groups according to the number of candidate moves: the problems of one group had 10 or more candidate moves (165 problems) and those of the other group had less than 10 candidate moves (135 problems). While the percentage of correct answers of the system for the former set was 29.1%, that for the latter set was 33.3%. This result shows that the system is effective in selecting the first move even if the problem has a lot of candidates. Hence this system will help search-based Tsume-Go solvers when the problems have many candidates, a situation which is very difficult for current search-based solvers.

6.5 Effects of Selecting Candidates of the First Move

In order to know how effective it is to select first move candidates in Tsume-Go, a preliminary experiment was carried out. For some problems which our system chooses the correct first move, the time of solving a problem by GoTools was compared to that of solving a position where the first move is added to the problem. The result is that the time taken by GoTools is reduced by 80%. For one problem, the reduction was 92.6% (from 25266.3 to 1870.8 seconds). This preliminary experiment implies the effectiveness of selecting first move candidates.

7 Discussion

7.1 Two-Step Learning

We identified that two steps were needed to acquire patterns, and separated the step of acquiring patterns from the step of adding additional conditions or information to the patterns. The meaning of this is as follows. In existing systems, conflict resolution mechanisms are implemented by hand, which is very hard. The proposed two-step model includes learning mechanisms. Some advantages of this two-step model are explained below.

One advantage is that more precise control is possible in this model. When conflicts are resolved without additional information to patterns (without refining patterns) – for example complexity is the only conflict resolution mechanism available and the most complex rules are always winners among matched rules – this resolution approach is applied to all rules. On the other hand, in the two-step model, the conflict resolution mechanisms can control each rule differently. The experiments in this paper showed this advantage. When Tsume-Go problems are solved by one general rule without refinement, the percentage of correct answers is about 19% (same as a 2 kyu human player). On the other hand, when weights are assigned to patterns (more precise control), the percentage of correct answers becomes 31% (same as a 1 dan human player).

Another advantage is that the steps are independent of each other and the effect of an algorithm in one step is easy to measure. This makes it easy to compare algorithms for the first step and those for the second step. This comparison was carried out in Section 6.

Two algorithms implemented in the model in this paper have the following features: the first algorithm requires only positive examples, whereas the second algorithm requires both positive and negative examples. An advantage in this implementation is a computational one. If both steps are calculated in one step, both positive examples and negative examples are needed. In Go, for a training move, the move is only a positive example, whereas all the other possible moves are negative examples, the number of which is about 200 on average. Thus, using negative examples is computationally much more expensive than using only positive examples. The first step uses only positive examples. After acquiring promising patterns in the first step, negative examples are used in the second step in this study. When the two steps are integrated in preliminary experiments, it takes too much time to acquire as many patterns as this study. This is because negative examples are used when a huge number of non-promising rules are evaluated. The experiments confirmed that two-step learning is a practical way of using negative examples.

The performance recorded in solving Tsume-Go in this paper is not so important. This model is a framework for developing and comparing algorithms for each step. It is important that this model allows algorithms to be compared. The performance of solving Tsume-Go in this paper is the starting point. We need to develop better algorithms for this model and compare the results achieved in order to acquire better patterns.

7.2 First Step

Comparison with Other Algorithms In previous studies, knowledge that involved fixed sizes and shapes was acquired. For example, one system [13] acquires knowledge as shown in Figure 6, which almost duplicates the patterns acquired by Fixed Algorithm. As shown in Table 8, Flexible Algorithm, offers much better performance than Fixed Algorithm.

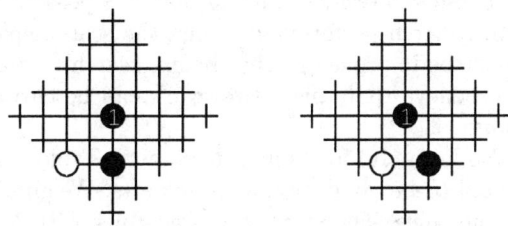

Fig. 6. Examples of knowledge acquired by Sei's system [13].

Effects of the Number of Rules The number of patterns may affect their performance. When the number of rules increases, the performance may become better or worse when the number of rules is too large [17]. The algorithm implemented as the first step in this paper can change the number of acquired rules by changing the parameter $FOOD$. Since the effect of the number of rules is easy to examine in this model, future work will be to solve Tsume-Go using a different number of rules.

In this simulation, patterns with weights are acquired, which are expected to be the same knowledge representation as that of a 1 dan human player. The performance of this simulation is as good as that of a 1 dan human player. By increasing the number of rules until performance saturates, we will determine whether using patterns with weights can exceed the performance of a 1 dan level human player.

---

[17] When the number of rules increases, the time it takes to match rules will also increase. It is, however, not so expensive for computers nowadays; it takes a Pentium II 300 MHz machine about 1 second to play a move using 15,000 rules for a 19 × 19 board game. If the performance becomes even worse after adding too many rules, you can improve the performance in many ways, such as by adding Go terms or purposes to rules.

7.3 Second Step

This paper considered only two algorithms for the second step. Many other algorithms, such as Neural Networks, should be investigated. This is a future work.

7.4 Usefulness of Patterns

Patterns have been thought useful, but how and when they are useful has not been investigated. How patterns are useful is discussed in this section.

Patterns may improve search performance when the search space is very large. The results in Subsection 6.4 show that patterns are useful in solving Tsume-Go even when the search space is large. The result in Subsection 6.5 shows that selecting first move candidates in Tsume-Go may improve the speed of the search techniques. These two results indicate that it is possible to conclude that patterns can improve search performance when the search space is very large. The correct rate must be improved further before search techniques can be put into actual use. We believe that this future work can be carried out using the proposed two-step model.

Patterns may also be useful for human beginners. Table 5 shows that many patterns are evaluated to be good by human experts. Weights of rules indicate the importance of the rules. Therefore, acquired rules with large weights may be useful for human beginners. It may be a very interesting experiment to investigate whether beginners can improve their performance by memorizing the patterns.

Toward More Human-like Knowledge This simulation used patterns; that is, only stones and the edges of the board are used to describe rules. Cognitive studies on the human knowledge of Tsume-Go [17,16] have revealed the following. A 2 kyu player has only rules of board configurations, which is the same as this system. A 3 dan player and a 6 dan player have "hybrid patterns", whose conditions are described by Go terms. In order to improve the quality of rules, acquiring rules with Go terms seems to be a hopeful way. For example, a term representing the number of liberty of stones may be a good candidate. Since the first step algorithm is flexible, once Go terms are defined, rules with the terms can be acquired. This would be a very interesting future work.

7.5 Application to Game Playing Systems

We believe that the proposed two-step model can be applied to game playing systems. Each algorithm in the first step and the second step should be reconsidered for application to the game playing systems. The first step may need major modification. In actual games, the importance of language level descriptions, such as purpose and concept, may increase. On the other hand, the second step, conflict resolution step, may need only minor modification. We assume, however, that the model itself does not need to be changed. Note that this discussion becomes possible because of this two-step model.

8 Conclusions

A two-step model of pattern acquisition was proposed in this paper based on our cognitive experiments. This model has advantages in short computational time over one-step models. Another advantage is that more precise control over conflict resolution is possible.

Tsume-Go problems were solved using patterns acquired in this model. The usefulness of the patterns acquired by this model was also investigated. First, the performance of the system is as good as 1 dan human players. Second, patterns are effective even when the search space is large, thus patterns can improve search technique performance when the search space is very large. Third, acquired patterns with large weights may be useful for human beginners.

Acknowledgement

The authors would like to thank Dr. Kazuhiro Ueda for his helpful comments and discussions. Support for this research is gratefully acknowledged from Dr. Ken-ichiro Ishii, the executive manager of the Information Science Research Laboratory in NTT Basic Research Laboratories. Members of the Mental Process Research Group gave us invaluable comments on this paper.

References

1. T. Cazenave. Automatic acquisition of tactical go rules. In *Game Programming Workshop in Japan '96*, pages 10–19, 1996.
2. D. Fotland. Knowledge representation in the Many Faces of Go. Available from ftp://igs.nuri.net/Go/computer/mfg.Z, 1993.
3. I. D. House. *Igo Data enjoy Books '60 ~ '95*. Nihon Ki-in, 1996. in Japanese.
4. Y. Ishida. *Godan Toppa-no Tsume-Go 100*. Tsuchiya Shoten, 1989.
5. Y. Ishida. *Sandan Chosen-no Tsume-Go 100*. Tsuchiya Shoten, 1989.
6. Y. Ishida. *Kihon Tsume-Go 100 Dai*. Nihon Bunge-sha, 1994.
7. T. Kojima, K. Ueda, and S. Nagano. An evolutionary algorithm extended by ecological analogy and its application to the game of Go. In *Proceedings of the Fifteenth International Joint Conference on Artificial Intelligence (IJCAI-97)*, pages 684–689. Morgan Kaufmann, 1997.
8. T. Kojima, K. Ueda, and S. Nagano. Flexible acquisition of various types of knowledge from game records: Application to the game of go. In *Proceedings of the Using Games as an Experimental Testbed for AI Research*, IJCAI-97 Workshop, pages 51–57, 1997.
9. T. Kojima, K. Ueda, and S. Nagano. Flexible acquisition of various types of knowledge from game records: Application to the game of go. In V. den Herik, editor, *Using Games as An Experimental Testbed for AI Research*. Drukkerij van Spijk B.V. Velno, in press.
10. Y. Saito. *Cognitive Scientific Study of Go*. PhD thesis, The University of Tokyo, 1996. in Japanese.
11. Y. Saito and A. Yoshikawa. An analysis of strong Go-players' protocols. In *Game Programming Workshop in Japan '96*, pages 66–75, 1996.

12. N. Sasaki and Y. Sawada. Neural networks for solving Go games. In *Proc. of The 2nd R.I.E.C. International Symposium on Design and Architecture of Information Processing Systems Based on The Brain Information Principles*, pages 241–244. 1998.
13. S. Sei and T. Kawashima. The experiment of creating move from "local pattern" knowledge in Go program. In *Game Programming Workshop in Japan '94*, pages 97–104, 1994. in Japanese.
14. T. Wolf. The program GoTools and its computer-generated tsume go database. In *Game Programming Workshop in Japan '94*, pages 84–96, 1994.
15. A. Yoshikawa and Y. Saito. Perception in tsume-go under 3 second time pressure. In *Game Programming Workshop in Japan '95*, pages 105–112, 1995. in Japanese.
16. A. Yoshikawa and Y. Saito. Perception in tsumego under 4 seconds time pressure. In *Proceedings of the 18th Annual Conference of the Cognitive Science Society*, page 868, 1996.
17. A. Yoshikawa and Y. Saito. The difference of the knowledge for solving tsume-go problem according to the skill. In *Game Programming Workshop in Japan '97*, pages 87–95, 1997.

A Neural Network Program of Tsume-Go

Nobusuke Sasaki [*,1], Yasuji Sawada[1], and Jin Yoshimura[2]

[1] Research Institute of Electrical Communication, Tohoku University
2-1-1 Katahira, Aobaku, Sendai 980-8577, Japan
{sasaki, sawada}@sawada.riec.tohoku.ac.jp
[2] Department of Systems Engineering, Shizuoka University
3-5-1 Johoku Hamamatsu 432-8561, Japan
jin@sys.eng.shizuoka.ac.jp

Abstract. Go is a difficult game to make a computer program because of the space complexity. Therefore, it is important to explore another approach that does not rely on search algorithms only. In this paper, we focus on tsume-go problems (local Go problems) that have a unique solution. A three-layer neural network program has been developed to find a solution at a given position of tsume-go problems, where the attacker is to kill the defender's territory on a 9×9 board. The network consists of 162 neurons for the input layer, 300 neurons for the middle layer, and 81 neurons for the output layer. We let the network learn the current stone patterns and, hence, process a direct answer. The network learns 2000 patterns of tsume-go by the back-propagation method. Within 500 repeats, the network learns 2000 patterns correctly. We tested the network ability: the top three selected moves contain about 60% correct answers, and the top five, about 70% for unknown problems at 500 repeats of learning. We compare the rate of correct answers by the network with that of human players who replied in a few seconds only. The ability of the network is roughly equivalent to 1-dan strength of human player.
Application of neural networks for a computer program of tsume-go (and also Go) combined with a pattern classifier might provide a prospective approach to create a strong Go-playing program.

Keywords: neural network, tsume-go, back-propagation, unique solution

1 Introduction

Computer technologies have been developed rapidly in recent years and the calculating ability of computers has surpassed in some respects that of human beings. It is widely known that the best computer chess program recently defeated the world human champion. The superb ability of conventional computers has considerably contributed to this result.

However, in some games like Go, computer programs are largely inferior to moderate human players. Currently the best Go program is still weaker than the

---

[*] At present affiliated to the Satellite Venture Business Laboratory, Shizuoka University, 3-5-1 Johoku, Hamamatsu, 432-8561 Japan.

average human player. One of the main reason for this is that the search tree of a Go program is significantly larger than that of a chess program.

In chess programs, an almost full-width search can be carried out. However, it is impossible to make a Go program to look ahead the entire tree because the size of the tree is extremely large. Therefore, it is important to explore an alternative approach that does not rely on searching only.

Tsume-go is a local problem of Go for which the attacker tries to kill a group/territory of the defender in a given setting. Among tsume-go programs, the best computer program is supposed to be Wolf's program(GoTools) [5]. GoTools uses the conventional search algorithm, and is limited on the search ability. When stones of tsume-go problems are placed in a relatively large area, GoTools tends to spend a significantly long time to solve them; in some cases, it cannot solve the problem at all because the search tree is too large. Thus, even in tsume-go, we need to explore an approach other than the conventional search algorithm. In this paper we consider the application of neural networks as an alternative approach for tsume-go and Go-playing programming.

It is known that neural networks have a high ability of pattern matching and generalization. If a neural network learns many patterns of the stones in Go, the network would automatically reply to any unknown input pattern. Therefore, the network can play an individual move by learning and generalization. In addition, the neural network can output the answers immediately for a given input pattern. The time the network spends simply depends on the size of networks, and it is minimal compare to the conventional search programs. Thus we have two advantages to use neural networks for Go and tsume-go programming: an ability to find a solution for unknown problems and a fast response.

There are some studies on the application of neural networks for Go programming. Enzenberger applied a neural network for calculation of an evaluation function[2]. In his program, the network uses the transformational features of stones like strings and groups of stones. Thus it does not evaluate the current input patterns directly. Furthermore, his program NeuroGo is much weaker than the conventional Go programs. The performance level of NeuroGo may correspond to the medium playing level 8 (out of 20) of the conventional program "The Many Faces of Go" [3] played on a 9×9 board. Richards et al. also used the evolving neural network for Go programming[4]. Their networks evolve by using how well they play the game, instead of individual moves. Thus the current use of neural networks does not focus on individual moves.

In Go and tsume-go, however, individual moves highly depend on the current stone patterns. In Go, a position that needs a unique solution appears very often in "local" games and in the endgames. In tsume-go problems, usually there is only one solution.

In this paper, we describe a neural network program for tsume-go. The network learns the current stone patterns and find good candidates. By learning the best moves in many tsume-go patterns, the network should be able to find a solution in tsume-go and the Go endgames.

2 The Application of Neural Networks

2.1 The Feature of Neural Networks

The neural network has been widely used in Artificial Intelligence research. Neurons of artificial neural network models have a nonlinear input-output characteristic like real neurons. We show the typical characteristics of a neuron in Figure 1. Hereafter, we use the term "neuron" as a neuron of neural network models.

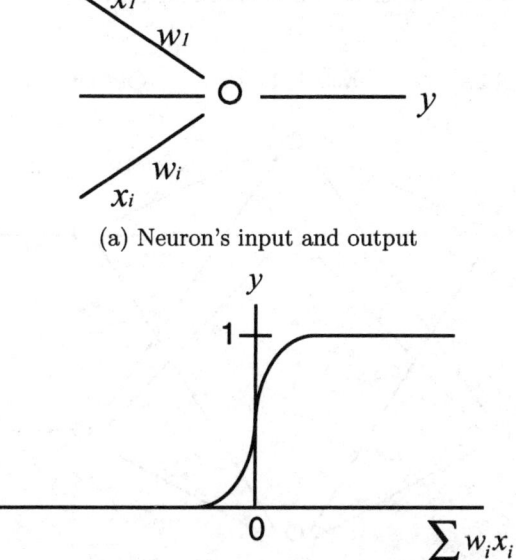

(a) Neuron's input and output

(b) Continuous type input-output characteristic of neuron(threshold = 0)

Fig. 1. The typical characteristics of a neuron.

Neurons have multiple inputs from others. Each input connection has a weight(w_i) that indicates the strength of the connection. It is allowed that w_i has a minus value. The product of input value x_i and weight w_i is the effective input of each connection.

Neurons have two states, depending on the values of effective inputs to a neuron. Usually, the input-output characteristic of neuron is expressed as either a step function or a continuous function that have sharp gain near the threshold value (Figure 1 (b) and Equation 1 [1]). We use a continuous function for the

---

[1] In general, the characteristic of biological neurons is expressed that they have two states. Though neurons of neural networks might have the middle-range value near the threshold input, we usually say that neurons have two states because of the analogy with biological neurons.

input-output characteristic of neurons in this paper. If the sum of effective input values is higher than the threshold value, the neuron is activated, and if the sum is lower, the neuron is inactivated. When the threshold value is 0, output value y is calculated by Equation 1.

$$y = \frac{1}{2}\left(\tanh\left(\sum w_i x_i\right) + 1\right) \tag{1}$$

In this paper, we use the back-propagation learning method, that is a popular supervised learning method for neural networks[1]. A multilayer feedforward neural network is used in the back-propagation learning. Let us show, in Figure 2, an example of a three-layer neural network structure.

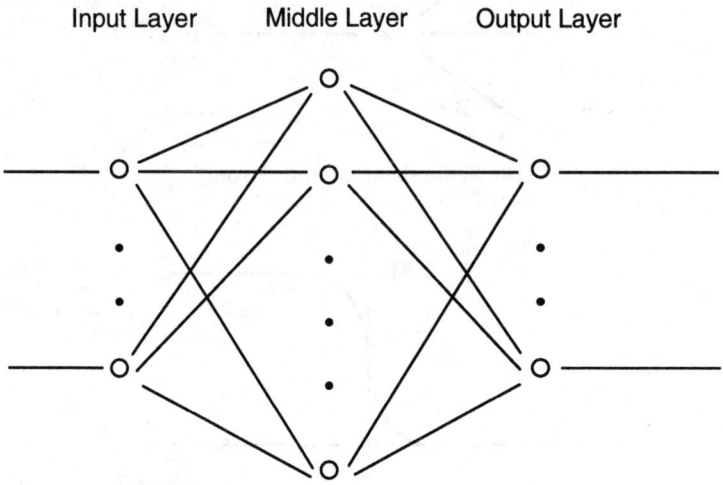

Fig. 2. An example of a three layer neural network structure.

Each connection has a weight(w_i) that indicates the strength of connection. The network can reply the specified output patterns for the specified input patterns by adjusting w_i. Adjustment of w_i means the learning of neural networks. The supervised data consists of problems for the input layer and answers for the output layer. When the problem data is input to input layer neurons, the network replies with signals from the output layer. We expect some error between the supervisor's answer and the network's answer. In the back-propagation method, w_i is updated following Equation 2, and the error values should be reduced by updating:

$$w_i(t+1) = w_i(t) - eps\frac{\partial E}{\partial w_i}x_i. \tag{2}$$

Here, E denotes the sum of squared error values between the supervisor's answer and the network's output. eps is the learning coefficient that determines the speed of learning. As a result of repeated w_i updates, the network's outputs approach to the supervisor's values.

Many patterns can be learned by a neural network. The number of patterns they can memorize depends on the structures and parameters of a network, e.g., the number of neurons and layers. Neural networks that have learned by the back-propagation method, have the abilities of classification and generalization. We use these abilities of neural networks to produce the next step in tsume-go problems.

2.2 Structure of Our Program

A three-layer feedforward neural network is used in the simulation. Each neuron in the former layer has a connection to all the neurons of the next layer. The structure of the neural network and its corresponding board position is shown in Figure 3.

Neurons of the input layer correspond to the location on the board in tsume-go problems. The board size is 9×9, about one-fourth large of the standard Go board. Therefore, all tsume-go problems used in this paper is confined in the 9×9 area (total: 81 lattice points). The number of necessary neurons in a network depends on the board size. Two neurons in the input layer are used for each lattice point on the board. One input neuron indicates whether or not to be a black stone at that position, and the other neuron indicates whether or not to be a white stone. Altogether $162 = 9 \times 9 \times 2$ neurons are present in the input layer of the current network. When a stone is in a position, the input value of a neuron that corresponds to the position and the stone color is 1. When no stone is present in the position, the input values of the two neurons are 0. In the output layer, only one neuron is allocated to each board position in the current network. We only ask the answer for the next move. Therefore, the number of neurons in the output layer is $81 = 9 \times 9$.

The number of middle layer neurons is an important factor that affects the learning ability of the network. Here, we set 300 neurons in the middle layer. We discuss the relationships between the number of middle layer neurons and the network's ability of learning in section 3.3 below.

When stone positions are input into the neurons of the input layer, the output of the network produces an answer. We interpret that the location indicated by the highest activated neuron in the output layer is the move to select as an answer.

2.3 Learning Procedures

The neural network learns tsume-go patterns by the back-propagation method. All the problems that the network learns in the current work are classified as "Kurosen-Shiroshi". Two players are distinguished as the first player (black stones) and the second player (white stones). Here, the first player is the attacker who is to kill the territory of the defender (the second player). The problem is designed that white stones should be killed in the end if the attacker selects the

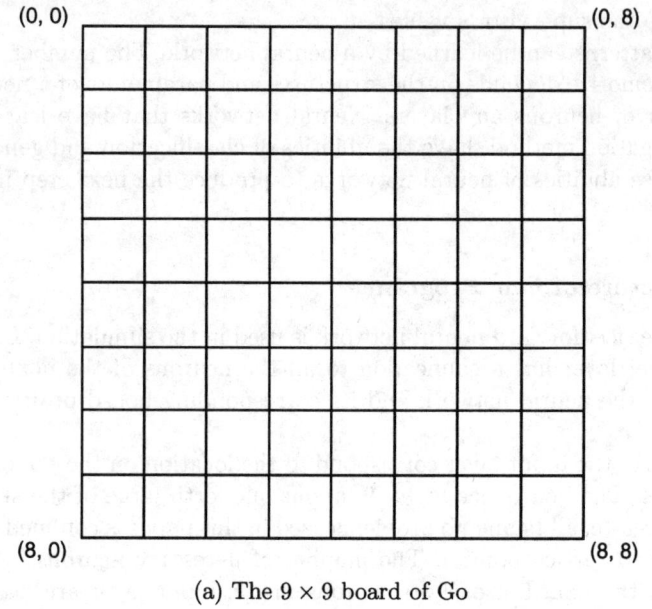

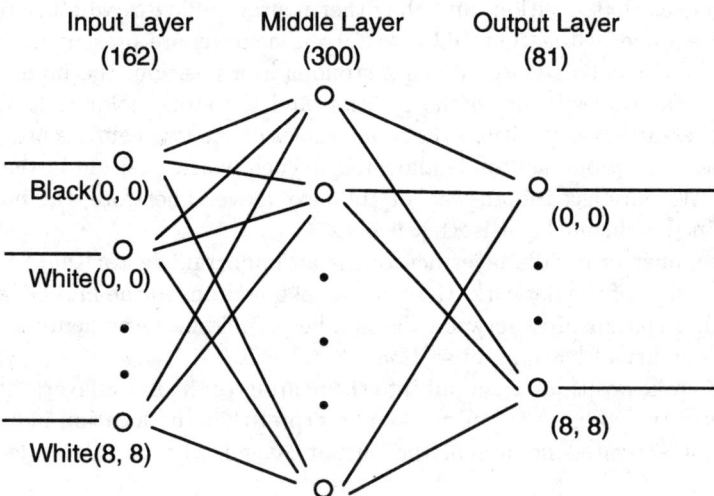

(b) Each input layer neuron indicates a position in question where the attacker is to move and a stone color, and each output layer neuron indicates a move of the attacker for the position considered.

Fig. 3. The board position and the structure of the neural network.

optimal moves. We collected problems as a test set from several tsume-go problem books ([6] and others). The answer to a tsume-go problem often consists of several sequential moves. We used every move of the attacker in a problem as a single learning datum.

It is generally known that the third move (i.e., the second move of the attacker) of tsume-go solution is easier than the first move; the fifth move is easier than the third and first moves; and so on. However, these problems contain various levels of difficulties. The first move to a given problem can be much easier than the fifth move to another problem. Therefore, we ignored these differences of moves. Thus, all attacker's moves in a problem are used as a data set. We collected the total of 3388 data of stone positions together with its solution, i.e, the attacker's move. We prepared five different sets of learning data from these data (Data set A to E). Each data set consists of 2000 moves of the attacker for learning, and 1000 moves for testing as unknown problems. We built five networks of the same structure for the five different data sets. Learning was done by the following procedure:

1. Input all the stone locations of a position in the input layer. If there is a stone, we set a neuron that indicates this location and its stone color to 1, and all other neurons are set to 0.
2. The only neuron activity level in the output layer that is recognized as a correct answer is set to 0.9, while those levels recognized as a wrong answer is set to 0.1.
3. The weight of network connections are updated by the back-propagation method (one round).
4. After the network experiences all 2000 patterns once, the network is then exposed to these patterns as a second time (2nd round). By updating (learning) many times (rounds), the neuron activities should converge to either 0.9 for a correct answer (only one neuron), or 0.1 for wrong answers (all the rest neurons).

3 Results

3.1 Learning Results

During and after learning, we examined the ability of the neural network to answer for "known" and "unknown" problems. "Known" problems are the ones used in learning (2000 patterns), and "unknown" problems are those not used in learning (1000 patterns). In the case of known problems, the rate of correct answers means the number of correct answers divided by the number of the total answers (2000). The answers by the network were evaluated in the following manner:

1. We input the stone positions in the input layer and pick up the highest activated neuron in the output layer. This position is considered the move selected.

2. When a stone is already at the selected position, we pick up the next highest activated neuron for the answer, and so on.
3. If the network's answer corresponds to a correct answer, we judge that the network replied with a correct answer.

Table 1 shows the numbers and percentages of correct answers of the neural networks. After learning a few hundred rounds, the network produces correct answers for all the problems. Figure 4 shows the rate of correct answer for one of 5 data set. Note that the horizontal axis, learning steps, is in the logarithmic scale. The rate of correct answer increases almost linearly and converges to the unity at around one hundred rounds. Thus our results demonstrate that the neural networks have an ability to learn at least a couple thousand stone movements correctly within a few hundred-round learning.

Table 1. The rate of correct answers for known problems (2000 patterns) in a three-layer networks.

Learning (Round)	correct(%) data A	B	C	D	E
0	6(0.30)	8(0.40)	8(0.40)	7(0.35)	6(0.30)
1×10^0	147(7.35)	105(5.25)	101(5.05)	87(4.35)	77(3.85)
5×10^0	847(42.35)	874(43.70)	844(42.20)	882(44.10)	916(45.80)
1×10^1	1233(61.65)	1265(63.25)	1309(65.45)	1290(64.50)	1312(65.60)
5×10^1	1978(98.90)	1970(98.50)	1985(99.25)	1974(98.70)	1966(98.30)
1×10^2	1995(99.75)	1996(99.80)	1997(99.85)	1995(99.75)	1995(99.75)
5×10^2	2000(100)	2000(100)	2000(100)	2000(100)	2000(100)
1×10^3	2000(100)	2000(100)	2000(100)	2000(100)	2000(100)

3.2 Answers to Unknown Problems

Here we examine the ability of our neural networks to answer 1000 unknown problems. We follow the same evaluation procedure as in the case of the known problems. The five highest activated neurons are used to select plausible candidates, instead of the only one highest activated neuron that is used for known problems.

The numbers and percentages of correct answers at 500 learning steps are shown in Table 2. The rates of correct answer are similar among all data sets (data A to E). The cumulative rate of correct answer improves significantly up to the third answer. But the rate of improvements becomes less significant for the forth and fifth answers. These results demonstrate that the trained networks have some ability to choose correct moves for unknown patterns.

It is also shown that the first few answers are highly likely to include the correct answers. The first and two cumulative rates of correct answer for unknown

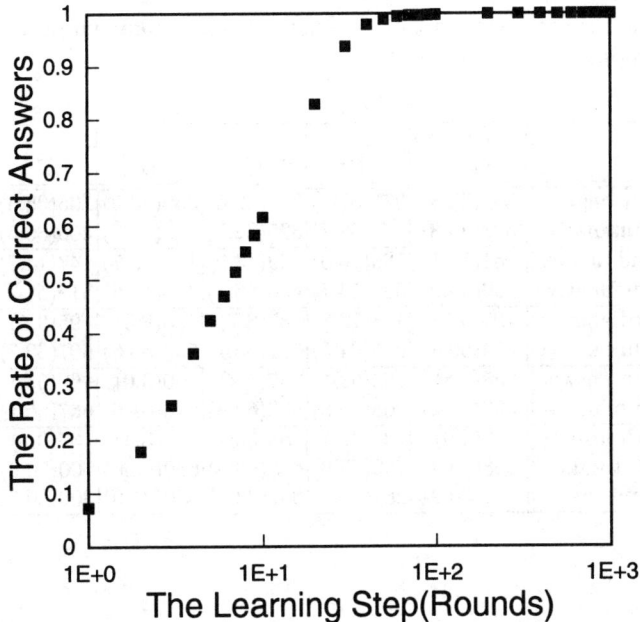

Fig. 4. The rate of correct answers of a network for known problems (data A).

problems (1000 patterns) are shown for one of 5 data set (Figure 5). The rate of correct answer increases almost linearly and at around a hundred rounds it converges to some level depending on the number of selected answers. After 500 learning steps, the cumulative rate of the top 3 answers becomes approximately 55-65%, and that of the top 5 answers, 65-75%. Thus a few hundred learning steps are enough for the network to achieve the best performance for unknown problems, as in the case of the known problems.

These results suggested that the neural network acquired the characteristic pattern knowledge of tsume-go by learning some amounts of given patterns of tsume-go problems. The networks replied correct answers to a certain degree for unknown problems by classifying such characteristic patterns of movement.

3.3 The Sizes of Network

How many neurons are necessary and/or sufficient for the middle layer of the network to produce the correct answers for tsume-go problems? We examined the relationship between the number of middle layer neurons and the learning ability of the networks for tsume-go problems. The number of neurons in the middle layer is varied from 25 to 300. Right after 1000 rounds learning, we examined the answering ability of the networks for known problems (Figure 6).

The results indicate that at least 100 or 150 neurons are necessary in the middle layer to learn 2000 patterns of tsume-go, and 150-200 neurons are suffi-

Table 2. The rate of correct answers of networks for unknown problems (at 500 learning rounds).

	correct(%) data A	B	C	D	E
1st answer	338(33.8)	377(37.7)	325(32.5)	339(33.9)	365(36.5)
Cumulative	338(33.8)	377(37.7)	325(32.5)	339(33.9)	365(36.5)
2nd answer	161(16.1)	160(16.0)	139(13.9)	137(13.7)	147(14.7)
Cumulative	499(49.9)	537(53.7)	464(46.4)	476(47.6)	512(51.2)
3rd answer	95(9.5)	108(10.8)	87(8.7)	80(8.0)	95(9.5)
Cumulative	594(59.4)	645(64.5)	551(55.1)	556(55.6)	607(60.7)
4th answer	58(5.8)	63(6.3)	57(5.7)	60(6.0)	60(6.0)
Cumulative	652(65.2)	708(70.8)	608(60.8)	616(61.6)	667(66.7)
5th answer	46(4.6)	31(3.1)	55(5.5)	34(3.4)	33(3.3)
total	698(69.8)	739(73.9)	663(66.3)	650(65.0)	700(70.0)
mistake	302(30.2)	261(26.1)	337(33.7)	350(35.0)	300(30.0)

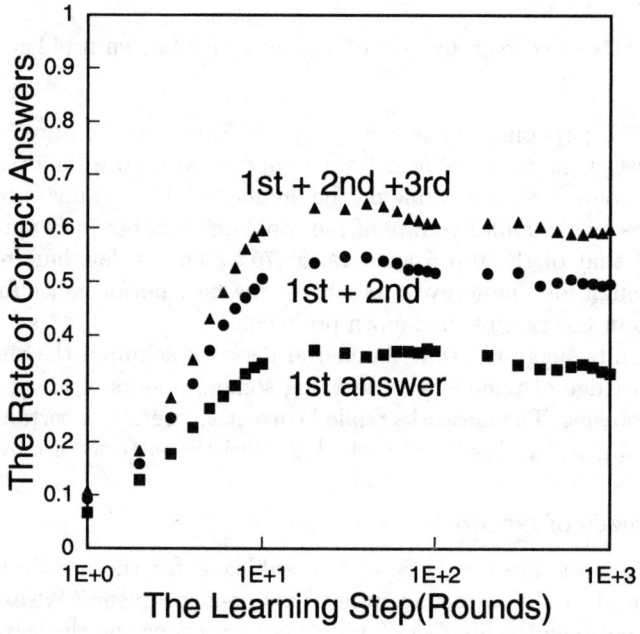

Fig. 5. The rate of correct answers of the network for unknown problems (data A).

cient. These results assures 300 neurons used in all the tests in this paper are sufficient to gain the best performance of the neural networks.

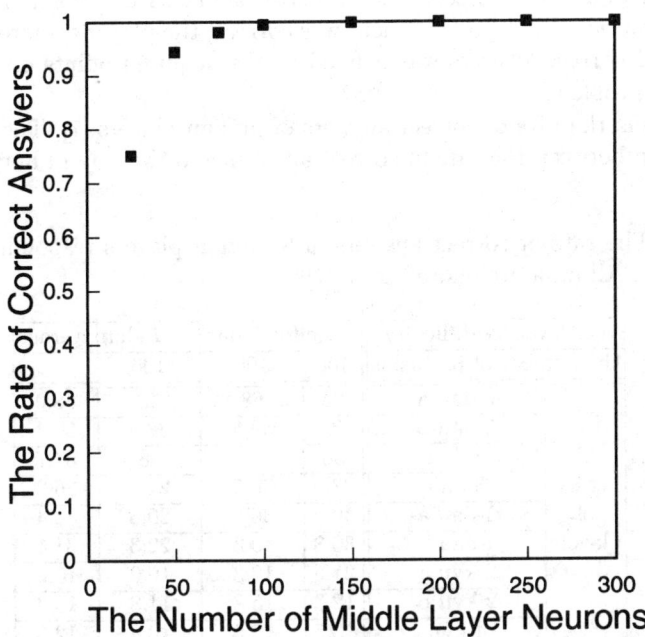

Fig. 6. The relationship between the number of middle-layer neurons and the learning ability (at 1000 learning rounds).

4 Discussion

4.1 The Comparison with Human Ability

The ability of the neural networks is compared with that of human players. When human players are forced to answer in a short time, they seek answers from the impression of the stone patterns, because they do not have time to search and evaluate sequential moves. Similarly the neural network quickly reaches to answers based solely on stone patterns. Therefore, the network performance is comparable to that of human players in such a quick response,

Yoshikawa and others studied the ability of human players by showing them tsume-go problems in a few seconds [7]. Their experiment was carried out under the conditions and methods described below.

Nine human players (subjects) saw a problem and answered a move to play in a few seconds. Problems were collected from three tsume-go books of three differing levels of difficulty (basic, 3-dan and 5-dan) [8]-[10]. Each tsume-go book contains 100 problems; the total of 300 problems are used for the experiment. The players were allowed to reply multiple answers. If a player replied the only one answer and if it was correct, the player acquired one point. If the player replied N answers and if one of them was correct, the player acquired $\frac{1}{N}$ points. The rate of correct answers was defined as the acquired points divided by the number of problems.

Table 3 is the rate of correct answers of human players [7]. There is a clear relationship between the rate of correct answers and the skill of human players.

Table 3. The rate of correct answers of 9 human players responding in a few seconds(%). All data are quoted from [7].

the levels of difficulty		basic	for 3-dan	for 5-dan	average
the number of problems		100	100	100	
skills of players	6-dan a	63	68	43	58.0
	6-dan b	53.3	59.5	38.3	50.4
	4-dan	58	68	38	54.7
	3-dan	27.0	30.5	21.0	26.2
	1-dan a	40.2	36.8	20.3	32.4
	1-dan b	30.8	40.3	22.3	31.1
	2-kyu a	21.5	18.3	16.0	18.6
	2-kyu b	12.3	15.8	11.8	13.3
	4-kyu	13.5	12.5	10.8	12.3

The ability of the neural network is compared with the skill of human players based on the rate of correct answer in the following manner. It is preferable to use the exact same problems used for human players to test the network ability. However, the current network is set to learn only "Kurosen-Shiroshi" problems on a 9×9 board. Therefore, we selected the problems to be tested from all the problems tested for human skills. The number of the unknown problems used for the network tests was 61 for the book of basic, 43 for the book of 3-dan, and 47 for the book of 5-dan.

The rate of correct answers of the network after 500 learning rounds is shown in Table 4. Here "1st" means the network's 1st answer. Similarly, 2nd, 3rd, 4th and 5th mean the network's n-th answer. "Cumulative" of n-th is the sum of 1st to n-th.

The network cannot decide how many answers to select. Therefore, we have to decide how many answers to choose. In "score 1", the first answer is only used as a sole answer for each problem. In "score 2", the first and second answers are used for each problem; the sum of the two scores are divided half. This is equivalent to the case when a human player responds with two answers for each problem. Both score 1 and score 2 are calculated by the same rule for calculating human scores in [7].

Therefore, we can roughly compare the data presented in Table 3 and the scores 1 and 2 in Table 4. Based on the individual data and averages (see Table 3 and 4), the ability of the current network is roughly equivalent to human 1-dan skill.

Table 4. The rate of correct answers of the neural network(%). Score 1 and score 2 are calculated by the same rule in [7](data A learning network). The weighted average of score 1 and score 2 for the 3 levels of difficulty are 31.1 and 26.8, respectively.

	basic	for 3-dan	for 5-dan
problems	61	43	47
1st	41.0	27.9	21.3
Cumulative	41.0	27.9	21.3
2nd	19.7	27.9	21.3
Cumulative	60.7	55.8	42.6
3rd	3.3	23.3	12.8
Cumulative	63.9	79.1	55.3
4th	4.9	2.3	4.3
Cumulative	68.8	81.4	59.6
5th	6.6	2.3	2.1
total	75.4	83.7	61.7
score 1	41.0	27.9	21.3
score 2	30.3	27.9	21.3

4.2 Future Improvements of a Neural Network for Go

The previous simulation shows that the best three moves contain 55-65% correct answers while the best five moves contain 65-75% correct answers for unknown problems. To check whether these network answers are better than the random choices among most frequent answers in tsume-go problems (e.g., stone positions 2-1 and 2-2), we calculated the percentage occurrence of three most frequent answers among all 3388 problems. The three most frequent answers are 7-8 (equivalent to 2-1 from the current positioning), 5-8 (4-1), and 4-8 (5-1). The total percentage of the three answers is only about 25% that is inferior to the current network performances (55-65%). This indicates that the neural network has at least some ability to learn the stone patterns and to produce the correct moves. It shows clearly that the neural network acquires the classification of the stone patterns that can be used for unknown problems. Thus our results imply that the neural networks can provide the alternative approach to the traditional search algorithms with tsume-go programming.

We still need to improve the rate of correct answers for unknown problems. In tsume-go problems, all the move must be correct to reach the correct solution (move sequences). Here we studied a simple typical neural network to examine the possibility of neural networks for Go programming. As discussed below, there are a variety of ways to modify the current networks to improve the network performance.

The learning target was set as 0.9 for the (one) correct answer and 0.1 for the others (wrong answers), respectively. These values are chosen to attain the clear distinction between the two, while achieving the best convergence (Figure 4). When we set 1.0 and 0.0 for the correct and wrong answers, respectively, the rate of correct answers does not converge to 1 even after learning 10000 rounds; it reaches about 0.85. This is probably because 1.0 and 0.0 are bounded by one side that makes convergence extremely difficult. There should be the optimal combination of the two target values.

The back-propagation method can be also replaced with other network learning methods. We can also classify the types of moves specifically and use different networks for different types of moves. For example, the "Kurosen-Shiroshi" category used in this paper includes many types of moves, such as "hane," "tsuke" and "hourikomi." (Figure 7). If a network learns one type of move, the rate of correct answers is known to rise for the same-type problems[11]. Furthermore, the categories can be classified by a different network.

Thus neural networks can be more complex and different from the current networks and there are some optimal conditions for an arbitrary network.

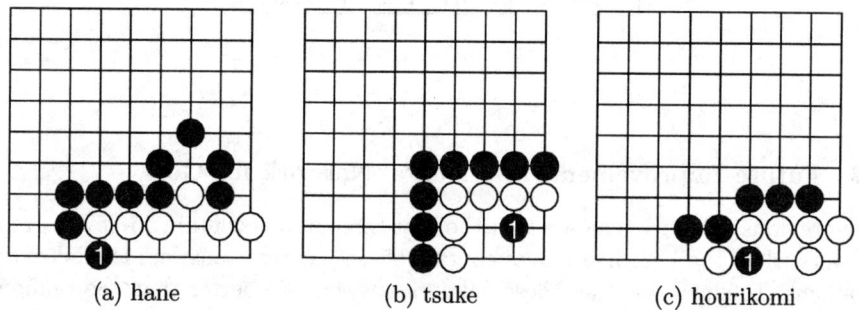

(a) hane   (b) tsuke   (c) hourikomi

Fig. 7. Examples of the types of moves.

We can also build a hierarchical system of networks to produce good candidates. For example, we can build a two-layer network systems: a top network is used as a classifier of the type of moves, and many bottom networks work for a specific type of moves to produce an individual move (Figure 7). Another classifier can be added for the categories of problems on the top of the hierarchy. A classifier can be a conventional search algorithm instead of a neural network.

Furthermore, the network jobs can be divided into many different ways: classification by the size of patterns on the board, classification by the connectivity of stone groups, and division by more simple patterns.

In tsume-go problems where the attacker is to kill the defender's territory/group, even a single wrong choice is not allowed. Therefore, to build a full tsume-go processor program, every single move must be correct even for unknown problems. The performance of the current networks is relatively poor for this respect. However, as discussed above, by modifying the networks and combining several methods, we should be able to improve the performance of the networks.

4.3 Building a Tsume-Go Processor Program

Once we achieve a higher rate of correct answers for each move, we can build a whole tsume-go program. The neural networks described in this paper answers next move only. Tsume-go solver should answer sequential moves to "solve" a given problem. There would be at least the following two types of algorithm for building a tsume-go solver.

One type may consist of only neural networks. In the current simulation, we applied neural networks for only generating a candidate(s) of the attacker in tsume-go problems, and the networks learned the attacker's moves only. But neural networks can also learn the moves of the defender. We can build a pair of special networks to generate sequential moves: a network for moves of the attacker and the defender.

Another type may be a combination of neural networks and conventional search algorithms. At first, the neural networks pick up some candidates, and next a search algorithm preferentially searches the game trees of the candidates. The combined program of neural networks and search algorithms is likely to reach the correct answers faster than the search algorithm only.

5 Conclusions

In the endgames of Go and tsume-go problems, the position that needs a unique solution appears very often. We applied neural networks for solving tsume-go problems on a 9×9 board. After learning a few hundred times, a neural network can answer all known problems perfectly. The network also produces correct answers for unknown problems to some extent. Our result indicates that a neural network has an ability to produce next moves in tsume-go problems. The current neural network program may be used as a component of strong tsume-go and Go programs.

Acknowledgements

The authors would like to thank Prof. Shojiro Kawakami for his useful information on Go programming in the early stage of the work and to thank Dr. Hiroyuki Iida, Dr. Atsushi Yoshikawa and Dr. Takuya Kojima for their helpful comments.

References

1. D.E. Rumelhart, G.E. Hinton, and R.J. Williams. Learning representations by backpropagating errors. *Nature*, vol. 323, pp. 533-536, 1986.
2. M. Enzenberger. The Integration of A Priori Knowledge into a Go Playing Neural Network. available from http://www.cip.physik.uni-muenchen.de/ enz/NeuroGo/NeuroGo.html, 1996.
3. D. Fotland. Knowledge Representation in The Many Faces of Go. available from Go Archive Site as Go/comp/mfg.Z, 1993.
4. N. Richards, D. Moriarty, and R. Miikkulainen. Evolving Neural Networks to Play Go. *Proceedings of the 7th International Conference on Genetic Algorithms*, East Lansing, MI, 1997.
5. T. Wolf. About problems in generalizing a tsumego program to open positions. *Proceedings of the 3rd Game Programming Workshop*, Hakone, pp.20-26, 1996.
6. A. Tozawa. *Tsume-go for 9-kyu to 1-kyu*, (in Japanese). Seibidou Shuppan, ISBN4-415-04423-9, 1992.
7. T. Kojima, A. Yoshikawa, K. Ueda, and S. Nagano. Toward Building a Model of the Go Champion: Fusion between Cognitive Science and Artificial Intelligence, (in Japanese). *Proceedings of 1997 Japanese Cognitive Science Society Winter Symposium "Game and Cognitive Science"*, pp.25-31, 1997.
8. Y. Ishida. *Fundamental tsume-go 100 Problems*, (in Japanese). Nihon Bungeisha, ISBN4-537-01223-4, 1997.
9. Y. Ishida. *Tsume-go 100 Problems to challenge 3-dan*, (in Japanese). Tsuchiya Shoten, ISBN4-8069-1413-4, 1989.
10. Y. Ishida. *Tsume-go 100 Problems to attain 5-dan*, (in Japanese). Tsuchiya Shoten, ISBN4-8069-1414-4, 1989.
11. N. Sasaki. The Neural Network Programs for Games, (in Japanese). Ph.D. thesis, Graduate School of Information Science, Tohoku University, 1998.

Distributed Decision Making in Checkers

J. Ignacio Giráldez and Daniel Borrajo

Universidad Carlos III de Madrid
c/ Butarque, 15
28911 Leganés, Madrid, Spain
{giraldez,dborrajo}@ia.uc3m.es

Abstract. The game of checkers can be played by machines running either heuristic search algorithms or complex decision making programs trained using machine learning techniques. The first approach has been used with remarkable success. The latter approach yielded encouraging results in the past, but later results were not so useful, partly because of the limitations of current machine learning algorithms. The focus of this work is the study of techniques for distributed decision making and learning by Multi-Agent DEcision Systems (MADES), by means of their application to the development of a checkers playing program. In this paper, we propose a new architecture for knowledge based systems dedicated to checkers playing. Our aim is to show how the combination of several known models for checkers playing can be integrated into a MADES, that learns how to combine individual decisions, so that the MADES plays better than any of them, without "a priori" knowledge of the quality or area of expertise of each model. In our MADES, we integrate well known search algorithms along standard machine learning algorithms. We present results that clearly show that the team as a single entity plays better than any of its components working in isolation.

1 Introduction

Computer programs for checkers play have been traditionally built using quite different approaches and paradigms. Heuristic search combined with database lookups has yielded impressive results [13], while machine learning algorithms have fared poorly. As it is stated in [13], some methods like genetic algorithms, neural nets and function optimization have been tried for the task of learning to classify checkers situations (as either win, loss, or draw for a given color), but were discarded because of unacceptably high error rates. The authors of the present paper have experimented with several machine learning paradigms (ID3 [7], C4.5 [8], bayesian learning [14], and backpropagation [10]) and have obtained similar results. In section 2 we discuss this issue and report on our own experience.

On the one hand, heuristic search combined with database lookup requires the use of huge resources and takes a limited advantage of available knowledge, but performs satisfactorily. On the other hand, machine learning programs should provide a satisfactory solution to a problem that is full of learning opportunities, but they fail to do so in the general setting given the huge hypothesis spaces. Since we believe that you can take advantage of both approaches, we propose to integrate them in a distributed checkers playing system. With this aim, we built autonomous decision making and learning systems for playing checkers, based on heuristic search and different machine learning paradigms. Each one of these systems is built as an autonomous agent using a single paradigm, and is able to play checkers. We have organized them as a Multi Agent DEcision System (MADES) [4].

The MADES decides which move to make on the checkers board by means of a distributed decision making procedure as explained in section 3. The idea of building a distributed decision making and learning system to play checkers is based on the following belief: given appropriate conditions, a group of agents forming a MADES is expected to play better than any of them playing in isolation. The aforementioned conditions were discussed at length in [3] by the authors, in a general context. The purpose of integrating individual monolithic systems into a MADES is to obtain a team performance unattainable when the individual systems work in isolation. We believe this to be an issue of paramount importance since it provides a performance enhancement mechanism (again, only when certain conditions are met).

A similar approach was used by Epstein [2]. In that case, she built a set of game-independent *Advisors*, some of them could also learn using different learning techniques. Her system had a meta-theory on how to play independent of the actual game that was been played. In our case, we differentiate between the agents that propose a decision (move in the case of game playing) and the advisors that decide on which agent is more appropriate for that decision. This allows us to learn two different concepts: how to make a decision, and how to give credit to someone making a decision. Also, her meta-theory depends very much on the game playing paradigm, while our architecture is domain-independent.[1] Finally, her advisors did not collaborate for making decisions, while our agents are allowed to ask for advice to the rest.

In section 4 we explain the experiments we have carried out and give the results obtained. We evaluate these results and draw some conclusions in section 5.

2 Machine Learning and the Game of Checkers

The game of checkers, like most games, is full of learning opportunities for machine learning systems. The pioneering work of Arthur Samuel [11], demonstrated the use of two learning mechanisms which noticeably improved the behaviour of his checkers program. The learning mechanisms he used in his program were very primitive, compared to the range of machine learning formalisms available nowadays. Nonetheless, these formalisms have not yet provided a satisfactory

---

[1] We are currently applying it to a hard induction problem, with very encouraging results.

machine learning solution to checkers. The focus of Samuel's work was the study of machine learning techniques in the context of checkers, so he resisted the temptation of hardwiring expert knowledge into his program, because he insisted in letting the program discover that knowledge by itself.

Supervised machine learning paradigms can be used to build game playing programs [6]. These learning systems use past play experience, and create a summarised representation of it, that forms the basis for decision making systems, that can make the decision of which move to make. Past play experience is contained in a training set, usually as a series of board descriptions, each followed by the final outcome (class). Supervised machine learning paradigms try to find common patterns in the boards that belong to the same class, and the collection of patterns encountered is used to build a class membership criterium that is used to classify (possibly) unseen checkers situations.

The authors have built for their experiments four different supervised learning systems, based on the following paradigms: ID3, C4.5, bayesian learning and backpropagation. The training set used was a subset of 29000 randomly chosen elements from the DB5 database [12]. The target concepts were **win**, **draw** and **loss** for white (white to move in all the situations). For selecting which move to make next, the successors of the current situation are presented to these programs. The program classifies every successor and gives it a score; the successor that scores highest is the preferred one, and indicates which move to make next. The situations of DB5 are endgames of 5 pieces at most, so the four systems were trained with endgames of 5 pieces or less.[2] Nonetheless, since we expected to obtain a powerful generalization as a result of the inductive nature of the algorithms involved, we tested the four systems with endgames of 8 pieces or less.

The four programs showed a very selective performance: a given program of the four may play certain endgames very well, but it plays others poorly (the set of checkers problems that an agent plays satisfactorily is known as its *competence region*). Moreover, the endgames that were played well by one program did not coincide with those played well by another[3] except for, as one would expect, in the case of ID3 and C4.5, given that C4.5 uses the same basic techniques as ID3. Since C4.5 handled well most of the situations that ID3 handled well, plus many others, we stopped using ID3 and used C4.5 instead.

With the aim of improving the overall performance of the programs, a second training set with 100,000 situations, randomly taken from DB5, was built. The four systems were trained with this new training set. Surprisingly, the backpropagation system performed worse with this second training set. We modified the topology of the neural network with the purpose of making more expressive power available for internal representation, but none of the enlarged nets per-

---

[2] Examples were randomly selected from the database. In case the database examples have any kind of bias towards a specific type of position, this does not affect to our goal; our aim is not to build the best machine learning system out of that data, but to learn how to better combine it with other systems.
[3] The reasons of this behaviour are explained in [3].

formed any better than the one trained with the first training set; the neural network did not scale up well. On the other hand, the Bayes and C4.5 systems improved remarkably, but still showed the same highly selective behaviour.

None of the four systems that were tested performed satisfactorily. We believe that this is because of the effect of the representation formalism used to represent the checkers situations in the training set. The use of an inadequate representation formalism can cause an unacceptable error rate. Sometimes this is due to the difficulty of expressing the target concept in terms of adequate input attributes or combinations of them. The use of meaningful intermediate concepts, with a more direct relation to the target concept, can alleviate this problem. We will illustrate this point with an example. Suppose that having one more king is to some extent determinant in some situations. Using the raw board description, the machine learning program will only notice that having some men in certain locations leads to some advantage (because of the existence of a crowning chance). To identify the boards that lead to that advantage, raw features will have to be combined, and many such combinations will have to be remembered, and in some way related to that advantage. The feature combinations that are thus grouped will be quite dissimilar. Now, let us imagine a higher level description for the checkers situations, using intermediate concepts (e.g. crowning chance in next move, crowning chance in n moves, capture chance in n moves, dominance of the center, victory chance in n moves, and so on), besides a raw description. The prior series of combinations of raw features is expressed now more easily, because more descriptive features are being used. This means that we are making the work easier for the machine learning program, because the common pattern is now expressed in a simpler way (that involves less features of the checkers situation description, combined more simply). This could be achieved by careful hand writing of the input features, or by use of automatic methods, such as *constructive induction* [9].

Other learning approaches applied to game playing have ranged from chunking in chess [1], temporal differences in backgammon [15], or bayesian learning of evaluation functions in Othello [5]. A similar multi-agent approach applied to game playing was followed by Wiering [17] by learning game evaluation functions using hierarchical neural networks architectures. In his case, all the agents implement the same paradigm (they are all neural networks). All the expert networks used by Wiering are equally suitable for being specialized to deal with any subset of the domain, as opposed to MADES hybrid approaches like ours, where some agents are more likely to correctly classify some situations, given that the paradigm they implement is better suited for that task.

3 Multi Agent Decision Systems for Checkers Play

In this section we describe how the overall architecture works as a problem solving (decision making) and learning model.

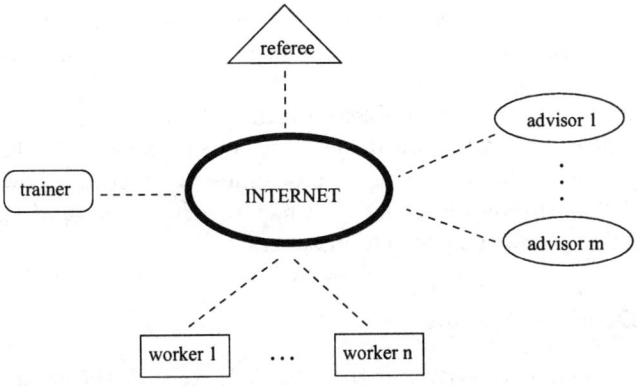

Fig. 1. The Intelligent Agents Organisation.

3.1 The Composition of a MADES

The Intelligent Agents Organization is a model composed by multiple intelligent heterogeneous agents that cooperate to attain a common overall goal. The IAO structure, is shown in Figure 1.

- One agent, known as the **referee**, is in charge of the overall system control. It broadcasts problem instance descriptions (in our application they are checkers situations), and control signals to the rest of the team. It then receives the respective replies from the rest of the agents. These replies may be either advice, or problem solving proposals (move proposals in our application). The services the referee may request to an agent are: solution proposal synthesis (only to worker agents), execution of a learning session (if the agent has learning capabilities), and advice request (only to selected agents). These service requests are scheduled in a way that maximizes parallellism (every agent runs on a different machine), so the MADES response time is minimized.
- The **worker agents** receive problem descriptions (checkers situations) from either the referee, or another worker, and reply with solution proposals. They work in parallel on a solution proposal to the same problem instance, are capable of autonomous decision making, and, some of them, have learning capabilities. Any of them could be the basis of a monolithic system aimed to solve each problem. The MADES should learn how to organize these worker agents to obtain a joint performance superior to the one that would be obtained in case we built a monolithic system with just one of the worker agents. The learning mechanism that accomplishes this task is distributed reinforcement learning of workers competencies [4].
- Several agents may play the role of **advisors**. They are contacted by other agents that wish to know who is the worker that is expected to handle best a given problem instance. The advisor replies with the identification of the worker that is expected to solve best the problem instance. The advice is used

by the referee as an aid for conflict resolution,[4] and it is also used by workers who wish to know which worker is the most appropriate for collaborating in the solution of a problem instance.
- A **trainer** agent produces problem instances that are used for training and testing. The criteria for problem synthesis affects the success of the learning effort. We are currently working on procedures to determine how to produce problems that speedup learning, and to force the learning of knowledge to handle the worst solved problem instances.

3.2 IAO Decision Making

When a problem instance arrives at the referee, it consults the advisors to determine whether any worker agent is expected to solve that instance of the problem satisfactorily. In that case, the proposal that this agent provides will be given a privileged status when it has to compete with the proposals of its fellow workers. The problem instance description is broadcast to the workers, so that they can work on it, and reply with a solution proposal. One advantage IAO presents, is that the advisors and the workers work in parallel, so the IAO response time is a very small overhead longer than the one that would be obtained from a monolithic system built from the most time consuming IAO worker.

When the referee receives the proposals of all the worker agents, and the advice from its advisors, it has to decide which proposal to use (most of the times this proposals will be incompatible and contradictory). The referee uses a poll mechanism for conflict resolution: the proposal that gets the greatest support is the one the referee will follow. The advisors' candidates receive extra votes in this poll, so they have some advantage over less credited workers.

One of the problems we perceived in previous experiments was that when the number of classifier workers was greater than the number of searcher workers, the system biased towards decisions made by classifiers, producing undesired results. So, we defined an automatic weighting mechanism that equals the maximum number of votes attainable by classifiers, and the maximum number of votes attainable by searchers.

3.3 Learning in IAO

Two different kinds of learning take place in IAO. First, the autonomous worker agents with learning capabilities can learn on their own about how to do their respective work. This is usually called *centralized learning* [16]. And, second, the advisors learn the workers competencies. This is a form of *distributed learning*. Centralized learning deals with knowledge about solutions, that will permit a worker agent to solve the problems it is presented, so it can be carried out locally by the agent, isolated from the rest of the team. Conversely, distributed learning of agents competencies requires the use of global information, because it is based

---

[4] For instance, when the agents disagree about which move should be made, the referee uses this advice to decide about which alternative to take.

on distributed credit assignment, that analyzes the performance of the MADES as a whole and of the workers individually, with the goal to learn competencies. This kind of learning will be used to make the synergetic effect possible.

The centralized learning algorithms are the same ones used in monolithic systems. Distributed learning is actually what will allow the team of agents to perform better than the individuals on their own. In this process, the advisors analyze how satisfactory the solution the MAS produced is.

We have designed an algorithm to learn workers competencies under the following hypothesis: if a worker is the most competent in the solution of a certain problem, it is also expected to be the most competent in the solution of another problem of the same difficulty and *appearance*. If the problem space is partitioned in subsets that contain similar problem instances, the competence data known for a certain problem instance, is expected to be also valid for the rest of the problem instances lying in the same subset of the partition. This is a generalization mechanism whose success depends on the similarity measure used.

How many subsets are used, and what the similarity metric is, depend on the kind of problem. The goal of this partition is to enclose, in a single subset, problem instances for which any worker agent deserves the same credit. The intended learning will be as reliable as the degree of fulfillment to this requirement. A reinforcement table is associated to every subset. In such a table, a reinforcement is associated to every worker agent, whose meaning is how adequate is the worker agent for the solution of problem instances lying in the subset. This tables are used by a reinforcement advisor agent: once a problem instance arrives to it, it locates the subset of the partition the problem belongs to, and the associated reinforcement table. Then, it determines who is the most adequate worker according to the table. If such a worker exists, the advisor replies to the referee with the worker's identification. Otherwise, it informs the referee about the lack of discerning data. As a result of "a posteriori" problem solving episode analysis, the participation of workers in the solution is determined, and they are consequently reinforced. The analysis and the reinforcement learning effort are carried out by the reinforcement advisor (the referee has been collecting and preparing data for this process during the problem solving episode).

4 Implementation and Experiments

To evaluate the IAO model, we have built a MADES composed of 8 agents (see Figure 2):

- The **referee** agent.
- An advisor agent, known as **reinfAG**, that builds and consults the appropriate reinforcement table, in order to advise other agents about the agent that is expected to solve most satisfactorily a given problem instance.
- The **C4.5AG** agent, based on Quinlan's C4.5 running in C trained on 100,000 instances randomly taken from the Schaeffer's DB5 database of checkers endgames [12].

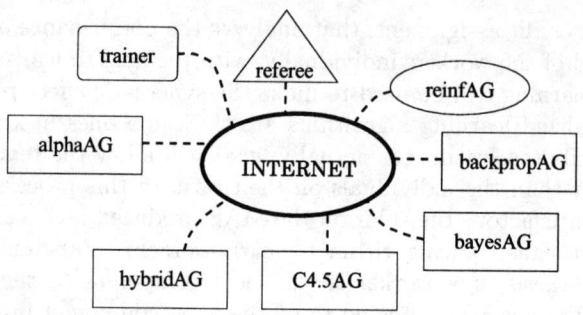

Fig. 2. Architecture of the checkers Multi-Agent System.

- **backpropAG**, a connectionist worker agent. We have built a neural network that learns by means of the backpropagation with momentum learning algorithm [10]. It has been trained with 29,000 examples taken from DB5.
- **bayesAG**, a bayesian classifier trained on the same 100,000 instances used by C4.5AG, running in C.
- **alphaAG**, an alpha-beta based worker agent with decision making capabilities only. Search has been constrained, so that the maximum search depth is limited to 5, and the maximum number of moves that the move generator outputs is 12. The purpose of this severe search constraint is to impose a time limit per move, brief enough to make possible the execution of many experiments,[5] and long enough to yield interesting play. Again, the main goal of this research (for now) is to learn how to combine several agents (strong or not), but not to build the best ever player.

  The knowledge this agent uses is hardwired into its evaluation function and into its move generator (the moves believed to be most interesting are generated first). A simple evaluation function has been used, so that the time it takes to compute it allows to compute it many times. The computation of the evaluation function evaluates the material difference between both sides, weighing the pieces with an amount that reflects the importance of the board area it dominates. This naive evaluation function provides reasonably good play in most common situations.
- **hybridAG**, a heuristic searcher, based on alpha-beta. When a search tree leaf is reached, this worker asks reinfAG who is the worker that is expected to handle best this leaf situation. In case that worker is available, hybridAG requests from it the evaluation of the leaf node. If that worker is not available, hybridAG performs the evaluation of the leaf node locally. Notice that this

---

[5] Currently the average speed of problems played is around 100 per day, when the MADES runs using 3 Linux PCs, 1 SPI 4MP, and 2 HP Apollo workstations, shared with other users and applications. The availability of more powerful hardware would make it possible to reach selective search depths in accordance to competition programs.

is a loosely coupled hybrid system, and that the searcher will be coupled with different classifiers at different moments.
- A **trainer** agent that produces problem instances that are used for training and testing. These problems are produced in a balanced fashion: there are as many situations that are wins (or losses) for white as there are for black. This has been accomplished by: first, a checkers situation is randomly generated according to some restrictions (e.g. the total amount of pieces must be equal or less than a given constant); then, its inverted form is computed. If the first situation was a win for black, the next situation produced will be its inversion, i.e. a win for white. So, none of the sides is favored by the trainer.

The agents communicate using the TCP/IP protocol over the Internet. Since the computers reside in the same network segment, the communication process takes much less time than the time local servicing of requests takes. In case computers in very distant locations were used, the network slowness in the heavy traffic hours would need to be considered.

For training the MADES, 16,000 checkers problems were generated by the trainer agent. The checkers problems were played until either one side wins, or a draw is reached. The draw criterium we used is to test when a series of moves was being cyclicly repeated.

The MADES played these problems against alphaAG, and a learning session was executed after every problem was played. Since our aim is to prove that the MADES can learn to make decisions in such a way that it beats any of its worker agents playing on its own, the test games were played between the MADES and each of its workers in turn. A set of 100 test problems was produced by the trainer, and this set was used in all the matches. The following results were obtained:

opponent	MADES advantage
c4.5AG	42%
backpropAG	35%
bayesAG	34%
hybridAG	4%
alphaAG	4%

The MADES advantage is computed as the difference between the number of games won by the MADES, minus the number of games won by its opponent, divided by the number of games played, and multiplied by 100 to obtain a percentage. This equals to expressing the game equity as a percentage. We express this calculation mathematically in the following formula:

$$adv(G) = \frac{[w(G) - l(G)] \times 100}{|G|}$$

where G is the set of games played, $|G|$ is the number of played games, $w(G)$ is the number of games won by the MADES, and $l(G)$ is the number of games lost by the MADES. Since we played 100 games per match, the formula above is simplified in this case to the evaluation of the difference between the games won by the MADES and the games won by its opponent. Further experiments with matches consisting of 24 and 50 games, closely approximated the results shown here.

The results show that the MADES beats any of its members. We believe that this gain in the quality of play justifies by itself the construction of the MADES from the standalone systems. We are currently in the MADES training stage, where, after every training game, reinforcement tables are updated. Since the reinforcement tables have not yet converged, we expect that the results will improve as the tables get near their convergence values.

The flexibility of the IAO model allows the replacement of a worker by another, and the adaption of the rest of the system to the new MADES composition, thanks to the adaptive behaviour of the advisor. We replaced two workers during the MADES's lifetime; the first replacement improved the MADES' score in 6.8%, and the second in 4.68%.

5 Discussion

What the results show might seem obvious to anyone: an isolated system should be beaten (or drawn) by a team formed by itself plus other agents. What is not obvious at all is to determine under which conditions this is feasible, and to learn to control the system in a way that assures inter-agent cooperation and prevents inter-agent hampering. It should be noted that there is no "a priori" clue about how the individual systems should be combined, and that the MADES learns how to perform this combination on its own. This is the main contribution of the present work.

Initially, the competence and preference regions of the worker agents are unknown. They are learnt by the system, and this information is used to influence the way the MADES makes decisions. This is quite dissimilar to putting together a program with good openings, a program with good middlegame play and a program with good endgame play. Because, in the latter case, one has "a priori" knowledge of the individual systems capabilities, while, in our model, the individual capabilities are not known "a priori", so the MADES has to learn them. This is very common in machine learning: when one builds a decision system based on the output of a machine learning paradigm, one usually cannot foresee which instances will be either correctly or incorrectly classified (i.e. one lacks "a priori" knowledge about the competence region of the system, except for some very coarse grain guesses based on the composition of the training set). So if computer checkers is to take advantage of decision systems based on multiple autonomous learning (or not) agents, a method for learning dynamic competencies like the one we propose here is needed, because there will not be available "a priori" knowledge in the general case for determining which agents should be heeded at a given time.[6]

---

[6] There will be no informed nor reliable way to combine individual decisions to reach a common overall decision.

The focus of this work is the study of techniques for distributed decision making and learning in MADES, and their application to the construction of a checkers playing program. We are aware that the checkers system presented here does not take advantage of the latest heuristic search enhancements, as championship level programs do. Moreover, we restricted ourselves to checkers problems of 8 pieces at most in our experiments.[7] The assesment of the results reported here provides an experimental background to support the adequacy of the theoretical IAO model. Now that we have tested the adequacy of the techniques involved, we wish to start working on producing a championship level program, enlarging our scope to the whole game of checkers, and refining and improving the existing agents.

The results improve as learning progresses due to two effects. On the one hand, the more extensively the reinforcement tables cover the domain, the more information that is available for **reinfAG**. We are in this first stage currently, and the domain is not completely covered yet. This means that **reinfAG** can not give advice in some situations because there is no reinforcement table corresponding to the subsets those situations belong to. For the MADES to be mature, the whole domain must be covered, and the reinforcement tables must get near its convergence values.

On the other hand, the convergence of the reinforcement tables is a second learning stage that will primarily take place after the covering is done. We believe the MADES has not even passed the covering stage, because new reinforcement tables are often created during learning sessions. We expect the results to improve when the problem domain be totally covered with the union of the competence regions of the worker agents, and the reinforcement tables have approached their convergence values.

6 Acknowledgements

We would like to thank Jonathan Schaeffer for making the DB5 and DB6 databases available, as well as for his assistance with the code for its access.

References

[1] Hans Berliner and Murray Campbell. Using chunking to solve chess pawn endgames. *Artificial Intelligence Journal*, 23:97–120, 1984.

[2] Susan L. Epstein. *Heuristic Programming in Artificial Intelligence*, chapter The Intelligent Novice. Learning to Play Better, pages 273–284. Ellis Horwood, 1989.

[3] Jos I. Girldez and Daniel Borrajo. The intelligent agents organization. In Eugnio Oliveira and Nick Jennings, editors, *Proceedings of the Workshop on Multi-Agent Systems: Theory and Applications (MASTA'97)*, pages 43–56, Coimbra, Portugal, October 1997.

---

[7] Due to an initial lack of computing power, and to our desire to avoid that the intricacies of the problem domain would shift our attention from our primary goal.

[4] Jos I. Girldez and Daniel Borrajo. Distributed reinforcement learning in multi-agent decision systems. In Helder Coelho, editor, *Progress in Artificial Intelligence, Iberamia 98*, number LNAI 1484 in Lecture Notes in Artificial Intelligence, pages 148–159, Lisboa, Portugal, October 1998. Springer-Verlag.

[5] Kai-Fu Lee and Sanjoy Mahajan. A pattern classification approach to evaluation function learning. *Artificial Intelligence Journal*, 36:1–25, 1988.

[6] J. Ross Quinlan. Learning efficient classifiation procedures and their application to chess end games. In R. S. Michalski, J. G. Carbonell, and T. M. Mitchell, editors, *Machine Learning, An Artificial Intelligence Approach, Volume I*. Morgan Kaufman, 1983.

[7] J. Ross Quinlan. Induction of decision trees. *Machine Learning*, 1(1):81–106, 1986.

[8] J. Ross Quinlan. *C4.5: Programs for Machine Learning*. Morgan Kaufmann, San Mateo, CA, 1993.

[9] L. De Raedt and M. Bruynooghe. Interactive concept-learning and constructive induction by analogy. *Machine Learning*, 8(2):107–150, March 1992.

[10] D.E. Rummelhart, J.L. McClelland, and the PDP Research Group. *Parallel Distributed Processing Foundations*. The MIT Press, Cambridge, MA, 1986.

[11] Arthur Samuel. Some studies in machine learning using the game of checkers. In E. Feigenbaum and J. Feldman, editors, *Computers and Thought*. McGraw-Hill, New York, NY, 1963.

[12] Jonathan Schaeffer. DB5: Checkers endgames database. ftp cs.ualberta.ca:/pub/chinook/DB5/, 1994.

[13] Jonathan Schaeffer, Robert Lake, Paul Lu, and Martin Bryant. Chinook, the world man-machine checkers champion. *AI Magazine*, 17(1):21–29, Spring 1996.

[14] J.Q. Smith. *Decision Analysis, a Bayesian Approach*. Chapman & Hall, 1988.

[15] Gerald Tesauro. Practical issues in temporal difference learning. *Machine Learning*, 8(3/4):257–277, May 1992.

[16] G. Weiss. Some studies in distributed machine learning and organizational design. Technical Report FKI-189-94, Institut fur Informatik, Technische Universität München, 1994.

[17] M. A. Wiering. TD learning of game evaluation functions with hierarchical neural architectures. Master's thesis, University of Amsterdam, 1995.

Game Tree Algorithms and Solution Trees

Wim Pijls and Arie de Bruin

Department of Computer Science, Erasmus University,
P.O.Box 1738, 3000 DR Rotterdam, The Netherlands.
{pijls, adebruin}@few.eur.nl

Abstract. In this paper a theory of game tree algorithms is presented, entirely based upon the concept of a solution tree. Two types of solution trees are distinguished: max and min trees. Every game tree algorithm tries to prune as many nodes as possible from the game tree. A cut-off criterion in terms of solution trees will be formulated, which can be used to eliminate nodes from the search without affecting the result. Further, we show that any algorithm actually constructs a superposition of a max and a min solution tree. Finally, we will see how solution trees and the related cutoff criterion are applied in major game tree algorithms like alphabeta and MTD.
Keywords: Game tree search, Minimax search, Solution trees, Alphabeta, SSS*, MTD.

1 Introduction

A game tree models the behavior of a two-player game. Each node n in such a tree represents a position in a game. An example of a game tree with game values is found in Figure 1. The players are called Max and Min. Max is moving from the square nodes, Min from the circle nodes. The game value $f(p)$ for a position p may be defined as the guaranteed pay-off for Max. This function obeys the minimax property. An algorithm computing the guaranteed pay-off in a node n is called a game tree algorithm. Over the years many algorithms have been designed. Every algorithm tries to eliminate as many nodes as possible from the game tree search. So every algorithm has its own cut-off criterion. We will design a cut-off criterion derived from a theory of game trees. In this theory the notion of a solution tree is the key notion, which turns out to be a powerful tool for establishing such a criterion. We show that, in the family of search algorithms obeying the cut-off criterion, alphabeta is the depth-first instance, whereas MT-SSS is the best-first instance. Besides, we show in an obvious way that every algorithm necessarily builds a critical tree.

This paper is organized as follows. In Section 2 we recall some facts on solution trees mentioned earlier by Stockman[9]. In Section 3 the notion of a search tree is recalled. This notion has been introduced by Ibaraki [4]. Next, minimax functions on a search tree are defined, and the role of solution trees in a search tree is discussed. Section 4 presents a general theory on game tree algorithms

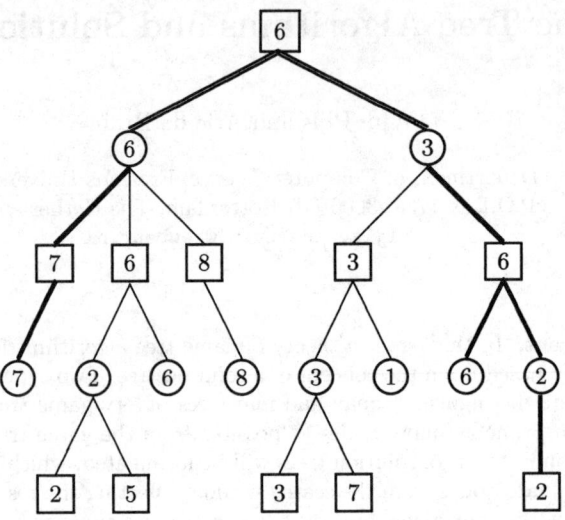

Fig. 1. A game tree with f-values.

based upon solution trees. A general cut-off criterion is the most important result. Section 5 connects two well-known game tree algorithms to the cut-off criterion. The reader is referred to [7] for details on this connection.

To conclude the opening section some preliminaries are given. A game tree is denoted by G and its root is denoted by r throughout this paper. Given a statement related to a game tree, replacing the terms max/min by min/max yields the so-called *dual* statement.

2 Solution Trees

A *strategy* of Max in a tree G is defined as a subtree, including in each max node exactly one continuation and in each min node all continuations (all countermoves to Max). Since the choice of Max in each position is known in such a subtree, Max is able to calculate the outcome for each series of choices that his opponent can make. In this paper a subtree with exactly one child in an internal max node and all children in a min node, which we have called a strategy for Max, will also be referred to as a *min solution tree*, or briefly a *min tree*. Dually a strategy for Min is defined, also called a *max solution tree* or a *max tree*. In Figure 1 the bold edges generate a max tree. A max tree is denoted by T^+ and a min tree by T^- in this paper.

Given a min solution tree, the most beneficial choice for Min in each min node is a move towards a terminal with minimal value. Consequently, in a given min tree (Max strategy) T^-, the profit for Max under optimal play of Min is equal to the minimum of all pay-off values in the terminals of T^-. Therefore we introduce the following function g for a max tree T^+ and a min tree T^-:

$$g(T^+) = \max\{f(p) \mid p \text{ is a terminal in } T^+\} \qquad (2.1)$$
$$g(T^-) = \min\{f(p) \mid p \text{ is a terminal in } T^-\} \qquad (2.2)$$

The intersection of a max tree T^+ and a min tree T^- consists of exactly one path. The g-definition implies that $g(T^-) \leq f(p_0) \leq g(T^+)$, where p_0 denotes the terminal at the end of the intersection path. It follows that $g(T^-) \leq g(T^+)$ for any two solution trees T^+ and T^- in a game tree.

Suppose that the Max player confines himself to a certain tree T^-. Then Max achieves a pay-off of $g(T^-)$, if Min replies consistently towards a terminal with value equal to $g(T^-)$. If Min deviates from a path towards a terminal equal to $g(T^-)$, Max gets a higher pay-off. Hence, $g(T^-)$ is the guaranteed pay-off for Max playing in T^-. It follows that the highest attainable pay-off for Max is equal to the maximum of the values $g(T^-)$ in the set of all min trees T^-. Dually, the most beneficial pay-off from the viewpoint of Min is equal to the minimum of the values $g(T^+)$ in the set of all max trees T^+. Since the guaranteed pay-off is equal to the game value by definition, we come to the following equality holding in each node n of a game tree:

$$f(n) = \max\{g(T^-) \mid T^- \text{ a min tree rooted in } n\}$$
$$= \min\{g(T^+) \mid T^+ \text{ a max tree rooted in } n\}$$

This equality can be proved formally by means of induction on the height of n. Since this equality is due to Stockman [9], it will be referred to as *Stockman's theorem* in this paper.

3 The Search Tree

So far, we were dealing with complete game trees. However, in every game tree algorithm the tree is built up step by step. At any time during execution a subtree of the game tree has been generated. Such a subtree is called a *search tree*. We assume that, as soon as at least one child of a node n is generated, all other children of n are also added to the search tree. If the children of a node n have been generated, n is called *expanded* or *closed*. If a non-terminal n has no children in a search tree (and hence n is a leaf in this search tree), then n is called *open*. A terminal n is called *closed* or *open* respectively, according to whether its pay-off value has been computed or not. The foregoing definitions of *open* and *closed* imply, that an open leaf in a search tree either is a non-terminal, whose children have not been generated yet, or is a terminal, whose game value has not been computed yet. Obviously, every closed leaf in a search tree is a terminal in the game tree.

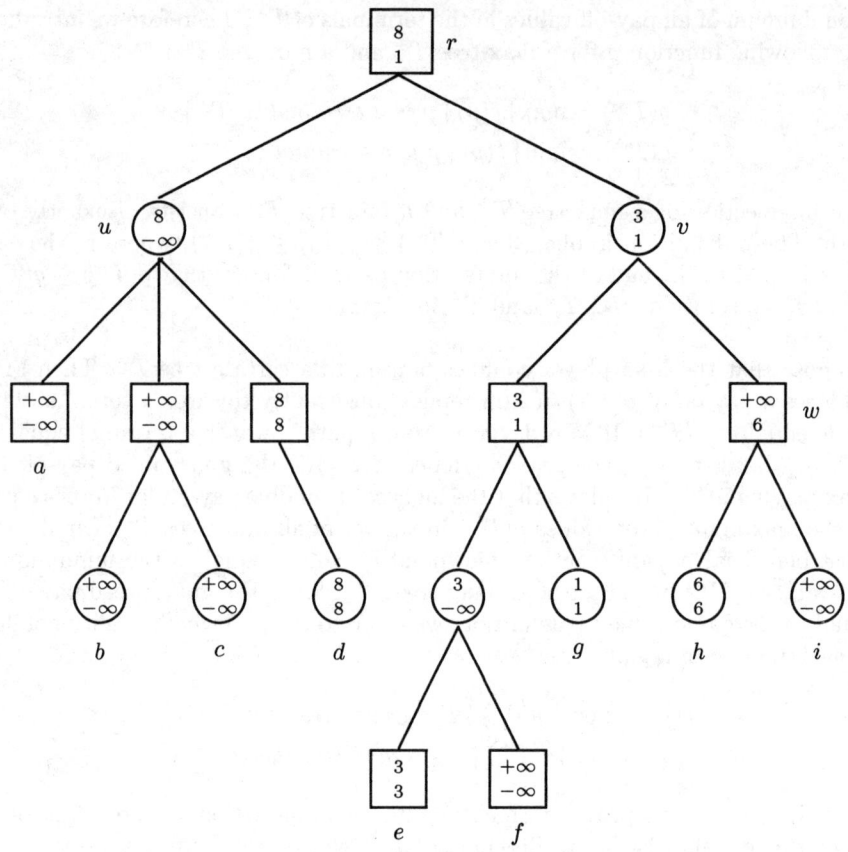

Fig. 2. A search tree derived from Figure 1.

Since $f(p)$ is not known yet in an open node p of a search tree S, the minimax function cannot be applied in S. To get an idea of the game values we assign two preliminary values to each open leaf. First, we assign $+\infty$ as a preliminary value. This gives rise to a function f^+ in a search tree S, defined as the minimax function in S assuming $f^+(p) = +\infty$ as game value in each open leaf p and $f^+(p) = f(p)$ in each closed leaf. Second, we assume $-\infty$ as game value in the open leaves. The related minimax function is called f^-. In every node n the inequality $f^-(n) \leq f(n) \leq f^+(n)$ holds, which can be shown by induction on the height of n. See Figure 2 for an instance of a search tree derived from Figure 1. The nodes a, b, c, f and i are open leaves, whereas d, e, g and h are closed leaves (terminals that have their game values evaluated). In each node n of Figure 2 the top value denotes $f^+(n)$ and the bottom value denotes $f^-(n)$.

In a search tree with minimax function f^+ Stockman's theorem can be applied. Likewise, this theorem can be applied to the f^--function. To rule out the annoying nodes with infinite values, we introduce a new definition. For a max and a min tree in a search tree this new g-definition will be given below (similar to the c-function in [5]). This definition is a generalization of definitions (2.1) and (2.2), which only hold for solution trees in a complete game tree, i.e., a tree with solely closed nodes.

$$g(T^+) = \max\{f(p) \mid p \text{ is a closed } terminal \text{ in } T^+\} \tag{3.1}$$
$$g(T^-) = \min\{f(p) \mid p \text{ is a closed } terminal \text{ in } T^-\} \tag{3.2}$$

Applying Stockman's theorem to f^+ and f^- respectively leads to the equalities below. Although Stockman's theorem deals with the old g-definition, these equalities are also valid for the new g-definition. By a *closed solution tree* we mean a solution tree in a search tree with solely closed leaves.

$$f^+(n) = \min\{g(T^+) \mid T^+ \text{ is a closed } max \text{ tree with root } n\} \tag{3.3}$$
$$= \max\{g(T^-) \mid T^- \text{ is a } min \text{ tree with root } n\} \tag{3.4}$$
$$f^-(n) = \max\{g(T^-) \mid T^- \text{ is a closed } min \text{ tree with root } n\} \tag{3.5}$$
$$= \min\{g(T^+) \mid T^+ \text{ is a } max \text{ tree with root } n\} \tag{3.6}$$

Here we assume that the minimum/maximum of the empty set is $+\infty/-\infty$.

We will comment on the formulas for f^+. (The formulas for f^- are dual.) For a min tree T^- with $+\infty$ as the game value in the open nodes and for a closed max tree T^+, the old and the new g-definition yield the same value. For a non-closed max tree T^+, the g-value in old sense equals $+\infty$. Consequently, the above equalities for $f^+(n)$ should be regarded as an application of Stockman's theorem, where non-closed max trees (with infinite g-value) are left out of consideration in the right-hand side of (3.3).

4 A General Theory

In this section a general theory on game tree algorithms is developed. A key role is played by the notions *alive* and *dead*.

4.1 Alive Nodes

In this subsection the definition and the significance of the notion *alive* is discussed. The definition of an *alive* node is as follows. A node n in a search tree S is called alive if n is on the intersection path of a max tree T^+ and a min tree T^- (either rooted in r) with $g(T^+) < g(T^-)$.

Given an alive node n in a search tree S, we can construct a game tree $G_n \supset S$, whose game value can only be obtained if one particular open descendant[1] of

---
[1] In this paper each node n is assumed to be its own descendant.

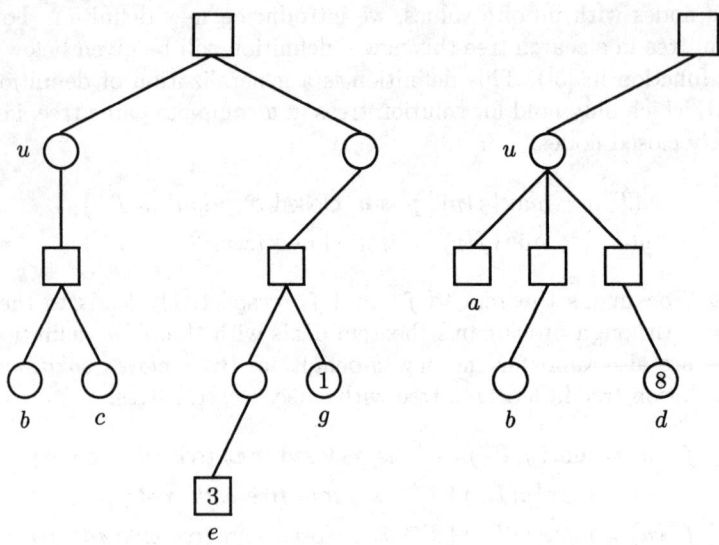

Fig. 3. A max and a min tree, derived from Figure 2.

n is expanded. The construction of G_n proceeds as follows. Denote the actual values $g(T^+)$ and $g(T^-)$ by g_1 and g_2 respectively. The leaf p_0 at the end of the intersecting path must be open, since, if it was not, we would have $g(T^-) \leq f(p_0) \leq g(T^+)$. Choose a value f_0 with $g_1 \leq f_0 \leq g_2$. Define $f(p_0) = f_0$ and $f(p) \leq g_1$ for any open node $p \neq p_0$ in T^+ and $f(q) \geq g_2$ for any open node $q \neq p_0$ in T^-. To complete G_n, the other open nodes in S (if any) are closed arbitrarily. After being extended, both T^+ and T^- have g-values equal to f_0. Stockman's theorem entails, that the game value of G_n equals f_0. As long as p_0 is not closed in G_n, T^+ and T^- satisfy $g(T^+) = g_1 < g_2 = g(T^-)$ and every value in the range $[g_1, g_2]$ is still achievable as game value for r.

The above construction is illustrated using the Figures 2 and 3. Figure 3 shows solution trees T^+ and T^- with $g(T^+) = 3$ and $g(T^-) = 8$. Node u is on the intersecting path and is therefore alive. The game tree G_u is constructed by defining $f(b) = f_0$ with $f_0 \in [3, 8]$, $f(c) \leq 3$ and $f(a) \geq 8$. The nodes f and i in Figure 2 may be closed arbitrarily.

4.2 Definition of the h-Functions

In this subsection, we give an alternative definition for the notion *alive*. This definition implies a practical method to establish whether a node is alive, using the so-called h-functions. These h-functions in a search tree S are defined as:

$$h^-(n) = \min\{g(T^+) \mid T^+ \text{ a max tree in } S \text{ through } r \text{ and } n\} \qquad (4.1)$$
$$h^+(n) = \max\{g(T^-) \mid T^- \text{ a min tree in } S \text{ through } r \text{ and } n\} \qquad (4.2)$$

It is easily seen that a node n is *alive* iff $h^-(n) < h^+(n)$.
As a result of (3.4) and (3.6) respectively, the definition of the h-functions reduces in the root to $h^+(r) = f^+(r)$ and $h^-(r) = f^-(r)$. Since every solution tree considered in the above definition goes through r, r has a maximal h^+-value and a minimal h^--value in any given search tree S.
Extending the equality $f^+(r) = h^+(r)$, we will give formulas for the h-functions in any other node. Those formulas are of highly practical significance. To this end we need a new notion. Denote by $AMAX(n)/AMIN(n)$ the set of max/min nodes, that are proper ancestors of n. The following interesting formulas hold for the h-values in a node n of a search tree S.

$$h^-(n) = \max\{f^-(m) \mid m \in \{n\} \cup AMAX(n)\} \qquad (4.3)$$
$$h^+(n) = \min\{f^+(m) \mid m \in \{n\} \cup AMIN(n)\} \qquad (4.4)$$

We give a sketch of the proof for (4.3). As a result of (3.6), every node $m \in \{n\} \cup AMAX(n)$ is the root of a max tree T_m with $g(T_m) = f^-(m)$. In the superposition of all those trees T_m, we choose arbitrarily a max tree T^+ through r and n. This tree T^+ satisfies the following equality: $g(T^+) = h^-(n) = \max\{f^-(m) \mid m \in \{n\} \cup AMAX(n)\}$. See [7] for a full proof.

4.3 Dead Nodes

A node that is not alive is called *dead*. It is easily shown that every ancestor of an alive node is alive as well. As a result, a descendant of a dead node is dead. In terms of the h-functions, we may state that n is dead iff $h^-(n) \geq h^+(n)$. The h-functions will be utilized in this subsection to derive some properties of dead nodes.
Consider a given max tree T^+ including an open dead node p. By the definition of h^- we have $g(T^+) \geq h^-(p)$. There is a node $m \in AMIN(p)$ with $f^+(m) = h^+(p)$ due to (4.4), and m is the root of a closed max tree T' with $g(T') = f^+(m)$ due to (3.3). Hence, $h^+(p)$ is associated not only with a min tree through r and p (by definition), but also with a max tree rooted in a node $m \in AMIN(p)$. We perform the following transformation to the given tree T^+. Remove the subtree below m from T^+ and append T' to T^+ in m. Since $g(T^+) \geq h^-(p) \geq h^+(p) = f^+(m) = g(T')$, the g-value does not increase by this transformation. Since T' is a closed solution tree, the transformed tree has solely closed leaves below m. In a similar way, any other open dead node can be eliminated from T^+. The resulting tree does not include any open dead node and its g-value does not exceed the original value $g(T^+)$.

To illustrate the above transformation, see Figure 2. It is easily seen using (4.3) and (4.4) that $h^-(i) = 6$ and $h^+(i) = 3$, meaning that i is dead. Any max tree T^+ through i has g-value ≥ 6. The subtree below v in such a tree T^+ may be replaced by the max tree rooted in v and ending up in the terminals e and g.

Given an alive node n, the above transformation can be applied to a max tree associated with $h^-(n)$, i.e., a max tree T^+ such that $g(T^+) = h^-(n)$. This results into a new tree avoiding open dead nodes and not exceeding $h^-(n)$ by its g-value. Since $f^+(m') \geq h^+(n) > h^-(n) = g(T^+)$ for every node $m' \in \{n\} \cup AMIN(n)$, replacing a subtree of T^+ rooted in m' with a closed subtree would raise the g-value. We conclude that no node from $\{n\} \cup AMIN(n)$ is involved in the above transformation of T^+ eliminating open dead nodes. It follows that, given an alive node n, a solution tree T^+ through n associated with $h^-(n)$ can be found avoiding any open dead node. As long as the algorithm does not expand an open node of this tree T^+, the value $h^-(n)$ is unaffected. Therefore, while expanding a dead node, the h^--value of any alive node in a search tree is not affected. For reasons of duality, any alive h^+-value isn't affected either.

4.4 Main Theory

We now come to our theory consisting of four observations.

a) We have shown in subsection 4.1 that, if n is alive, a game tree G_n can be constructed, in which $f(r)$ is unknown as long as one particular open descendant p_0 of n is not expanded. The conclusion is that any alive node n cannot be discarded.

b) As a result of a), an algorithm must continue as long as the search tree contains any alive nodes. Therefore, the algorithm may only stop when all nodes in the search tree are dead. We might say therefore, that any game tree algorithm actually aims at killing the entire search tree.

c) All nodes in a search tree S are dead iff $g(T^+) \geq g(T^-)$ for any two solution trees T^+ and T^- in S. As a result of (3.4) and (3.6), this condition is equivalent to the equality $f^-(r) = f^+(r)$, which is the stop criterion of every game tree algorithm therefore. When the condition $f^-(r) = f^+(r)$ is achieved, both a closed max tree and a closed min tree with g-value equal to $f(r)$ are present in the search tree. The superposition of these two trees is called a *critical tree*. This notion has been introduced in [6], with a totally different definition however. Since the algorithm must continue until $f^-(r) = f^+(r)$ holds, we conclude that every game tree algorithm needs to build a critical tree.

The intersection of the max and the min tree in a critical tree is a path with constant f-value, as can easily be shown using Stockman's theorem.

d) Expanding descendants of a dead node does not affect the h-values of any alive node, as we have shown in subsection 4.3. Consequently, an alive node can only be killed by expanding an alive node. For a game tree algorithm to achieve its goal, every node needs to be killed. Therefore, expanding a dead node is useless. Since every dead node has solely dead descendants, a dead node along with the subtree underneath may be neglected during the search.

Notice that the notes a) and d) constitute a general cut-off criterion for game tree algorithms: alive nodes must be respected, dead nodes may be neglected. Note c) describes the situation on termination of a game tree algorithm.

5 Two Well-Known Game Tree Algorithms

In this section we will discuss, how some well-known game tree algorithms fit into our theory. The results are not proved. See [7] for details.

5.1 The Alphabeta Algorithm

An extensive treatment of the alphabeta procedure can be found in [6]. A procedure call $alphabeta(n, \alpha, \beta)$ has three parameters: a node n in a game tree, and two numbers α and β. The precondition is $\alpha < \beta$. We present a postcondition of alphabeta, which extends the postconditions in [6] and [3], in that it relates the new functions f^+ and f^- to the return value of an alphabeta call. For an alphabeta call with return value v, the following postcondition applies:

$$v \leq \alpha \Rightarrow v = f^+(n),$$
$$\alpha < v < \beta \Rightarrow v = f(n)$$
$$v \geq \beta \Rightarrow v = f^-(n).$$

In case of $v = f^+(n)$, the search tree contains a closed max tree T^+ with $g(T^+) = f^+(n)$, cf.(3.3). An extra feature is, that this max tree is unique in the search tree. The case $v = f^-(n)$ is dual.

A call $alphabeta(r, -\infty, +\infty)$ computes the exact value of a game tree and is referred to as the alphabeta algorithm. Using the new postcondition, the following property can be proved. When a node n is parameter in a nested call $alphabeta(n, \alpha, \beta)$ of the alphabeta algorithm, this node is open and alive and satisfies the relation $h^-(n) = \alpha < \beta = h^+(n)$. Moreover, every node to the left of n is dead. Consequently, the alphabeta algorithm may be characterized as the algorithm *expanding the leftmost open alive node* in each step.

5.2 SSS* and MT-SSS

The SSS*-algorithm was published in 1979 by Stockman [9]. Before 1994, it was never used in actual applications. Nevertheless, SSS* has drawn considerable attention in literature. The originating paper is one of the top 50 referenced in the AI Journal[1]. A suitable explanation of the algorithm can be found in [5]. It was shown recently[8] that SSS* is equivalent to a series of alphabeta calls with a null-window, i.e., an α-β-window with $\beta = \alpha + 1$. This new formulation is called MT-SSS and has turned out to be a very convenient and efficient algorithm. The above characterization of alphabeta has its counterpart for SSS* and MT-SSS. Without going into details, we mention the following properties. When a node n is expanded by a null-window call in MT-SSS, then $\beta = h^+(n) = h^+(r) = f^+(r)$ being a maximal h^+-value in the actual search tree according to subsection 4.2. In addition, every node to the left of n with the same h^+-value is dead. Similarly, the merit for a node in the list maintained during SSS* is equivalent to the h^+-function and, whenever a node n is selected from the list, every node to the left of n is dead. However, since the h^--value is not taken into account in MT-SSS or

SSS*, aliveness is not guaranteed for a node n when being selected or expanded during SSS* or MT-SSS. Fortunately, only in very rare situations n is dead. See [7] for an example. The costs of the search overhead caused by expanding a dead node in rare cases does not outweigh the costs of maintaining the h^--function in the search tree. We are allowed to state, that in virtually each step during SSS* or MT-SSS, the selected or expanded node n is *the leftmost open alive node with maximal h^+-value*. So a characterization of MT-SSS or SSS* is obtained.

There exist also dual versions of SSS* and MT-SSS, named Dual and MT-DUAL, where the h^--function is considered instead of the h^+-function.

6 Concluding Remarks

In this paper we developed a full theory on game tree algorithms, entirely based upon the notion solution tree. Stockman's theorem on solution trees was the basic principle underlying the theory. The main points were presented in Section 4. As a result we obtained a pruning criterion. In Section 5, we showed how two major algorithms fit into the theory. We may say, that MT-SSS and MT-Dual are best-first instances, whereas alphabeta is a depth-first instance.

Two fairly important algorithms are not discussed yet, viz. proof number search and Negascout. The role of solution trees in those algorithms was described in [2].

References

1. Bobrow, D.: Artificial Intelligence in perspective, a retrospective on fifty volumes of the Artificial Intelligence Journal. *Artificial Intelligence*, 59:5-20, ISSN 0004-3702, 1993.
2. de Bruin, A., Pijls, W., Plaat, A.: Solution Trees as a Basis for Game Tree Search. *ICCA Journal*, 17(4): 207-219, ISSN 0920-234X, 1994.
3. Finkel, R.A., Fishburn, J.P. Parallelism in alpha-beta search. *Artificial Intelligence*, 19:89-106, ISSN 0004-3702, 1982.
4. Ibaraki, T.: Generalization of alpha-beta and SSS* search procedures. *Artificial Intelligence*, 29:73-117, ISSN 0004-3702, 1986.
5. Kumar, V., Kanal, L.N.: A General Branch and Bound Formulation for Understanding and Synthesizing And/Or Tree Search Procedures. *Artificial Intelligence*, 21:179-198, ISSN 0004-3702, 1983.
6. Knuth D.E., Moore, R.W.: An analysis of alpha-beta pruning. *Artificial Intelligence*, 6:293-326, ISSN 0004-3702, 1975.
7. Pijls, W., de Bruin, A.: Game tree algorithms and solution trees. Technical Report EUR-CS-98-02, Erasmus University Rotterdam, 1998, available as: http://www.cs.few.eur.nl/few/inf/publicaties/rapporten.eur-few-cs-98-02.ps
8. Plaat, A., Schaeffer, J., Pijls, W., de Bruin, A.: A Minimax Algorithm Better than SSS*. *Artificial Intelligence*, 84:299-337, ISSN 0004-3702, 1996.
9. Stockman, G. A minimax algorithm better than alpha-beta? *Artificial Intelligence*, 12:179-196, ISSN 0004-3702, 1979.

A New Heap Game

Aviezri S. Fraenkel and Dmitri Zusman

Department of Applied Mathematics and Computer Science
The Weizmann Institute of Science
Rehovot 76100, Israel
fraenkel@wisdom.weizmann.ac.il
http://www.wisdom.weizmann.ac.il/~fraenkel
dimaz@wisdom.weizmann.ac.il

Abstract. Given $k \geq 3$ heaps of tokens. The moves of the 2-player game introduced here are to either take a positive number of tokens from at most $k-1$ heaps, or to remove the *same* positive number of tokens from all the k heaps. We analyse this extension of Wythoff's game and provide a polynomial-time strategy for it.

Keywords: multi-heap games, efficient strategy, Wythoff game

1 Introduction

We propose the following two-player game on k heaps with finitely many tokens, where $k \geq 3$. There are two types of moves: (i) remove a positive number of tokens from up to $k-1$ heaps, possibly $k-1$ entire heaps, or, (ii) remove the *same* positive number of tokens from all the k heaps. The player making the last move wins.

Any position in this game can be described in the following standard form: $(m_0, \ldots, m_{k-1})$ with $0 \leq m_0 \leq \ldots \leq m_{k-1}$, where m_i is the number of tokens in the i-th heap. Given any game Γ, we say informally that a *P-position* is any position u of Γ from which the *P*revious player can force a win, that is, the opponent of the player moving from u. An *N-position* is any position v of Γ from which the *N*ext player can force a win, that is, the player who moves from v. The set of all *P*-positions of Γ is denoted by $\mathcal{P}$, and the set of all *N*-positions by $\mathcal{N}$. Denote by $F(u)$ all the followers of u, i.e., the set of all positions that can be reached in one move from the position u. It is then easy to see that:

$$\text{For every position } u \text{ of } \Gamma \text{ we have } u \in \mathcal{P} \text{ if and only if } F(u) \subseteq \mathcal{N} ;$$
$$\text{and } u \in \mathcal{N} \text{ if and only if } F(u) \cap \mathcal{P} \neq \emptyset . \quad (1)$$

For $n \in \mathbb{Z}^0$, denote the n-th *triangular number* by $T_n = \frac{1}{2}n(n+1)$. We prove,

Theorem 1. *Every P-position of the game can be written in the form $(T_n, m_1, \ldots, m_{k-1})$, where the $(k-1)$-tuples $(m_1, \ldots, m_{k-1})$ range over all the (unordered) partitions of $(k-1)T_n + n$ with parts of size $\geq T_n$. In other words, $\mathcal{P} = \bigcup_{n=0}^{\infty} \mathcal{P}_n$, where*

$$P_n = \{(T_n, m_1, \ldots, m_{k-1}) : \sum_{i=1}^{k-1} m_i = (k-1)T_n + n,$$
$$T_n \leq m_1 \leq \ldots \leq m_{k-1},\ n \in \mathbb{Z}^0\}. \qquad (2)$$

Example. For $k = 4$,

$$P_n = \{(T_n, m_1, m_2, m_3) : m_1 + m_2 + m_3 = 3T_n + n,\ n \in \mathbb{Z}^0\}.$$

The first few P-positions are:

$P_0 = \{(0,0,0,0)\}$
$P_1 = \{(1,1,1,2)\}$
$P_2 = \{(3,3,3,5),\ (3,3,4,4)\}$
$P_3 = \{(6,6,6,9),\ (6,6,7,8),\ (6,7,7,7)\}$
$P_4 = \{(10,10,10,14),\ (10,10,11,13),\ (10,10,12,12),\ (10,11,11,12)\}$
$P_5 = \{(15,15,15,20),\ (15,15,16,19),\ (15,15,17,18),$
$\qquad\qquad\qquad\qquad\qquad\qquad (15,16,16,18),\ (15,16,17,17)\}.$

2 The Proof

Throughout, as in (2), every k-tuple $(T_n, m_1, \ldots, m_{k-1})$, $(m_0, \ldots, m_{k-1})$ or $(k-1)$-tuple $(m_1, \ldots, m_{k-1})$ is arranged in nondecreasing order. Any of the first two tuples is also called a position (of the game) or partition (of $kT_n + n$); and the third is also a partition (of $(k-1)T_n + n$). The terms m_i are called components (of the tuple) or parts (of the partition).

Lemma 1. *Given any partition $(m_1, \ldots, m_{k-1})$ of $(k-1)T_n + n$, where each part has size $\geq T_n$. Then each part has size $< T_{n+1}$.*

Proof. We have,

$$(k-1)T_n + n - m_{k-1} = \sum_{i=1}^{k-2} m_i \geq (k-2)T_n.$$

Hence for all $i \in \{1, \ldots, k-1\}$, $m_i \leq m_{k-1} \leq T_n + n = T_{n+1} - 1$. ∎

Lemma 2. *Let $k \geq 3$ and $n \in \mathbb{Z}^0$. Every integer in the semi-closed interval $t \in [T_n, T_{n+1})$ appears as a component in some position of P_n. It appears in P_m for no $m \neq n$.*

Proof. The smallest component in P_n is T_n, and by Lemma 1, the largest part cannot exceed $T_n + n = T_{n+1} - 1$. Hence $t \in [T_n, T_{n+1})$ appears as a component in P_m for no $m \neq n$. Let $t \in [T_n, T_{n+1})$, say $t = T_n + j$, $0 \leq j \leq n$. Then for $k \geq 3$, $T_n + j$ appears in the partition $\{m_1, \ldots, m_{k-1}\} = \{T_n^{k-3}, T_n + n - j, T_n + j\}$ of $(k-1)T_n + n$, where T_n^{k-3} denotes $k-3$ copies of T_n, and so $T_n + j$ appears in some position of P_n. ∎

Proof of Theorem 1. It follows from (1) that it suffices to show two things: (I) A player moving from any position in P_n lands in a position which is in P_m for no m. (II) From any position which is in P_m for no m, there is a move to some P_n, $n \in \mathbb{Z}^0$. The fact that (I) and (II) suffice in general for characterizing $\mathcal{P}$ and $\mathcal{N}$, is shown in [6] for the case of games without cycles, based on a formal definition of the P- and N-positions, and a proof of (1). (It is not true for cyclic games: given a digraph consisting of two vertices u and v, and an edge from u to v, and an edge from v to u. Place a token on u. The two players alternate in pushing the token to a follower. The outcome is clearly a *draw*, since there is no last move. However, putting $\mathcal{P} = \{u\}$, $\mathcal{N} = \{v\}$, satisfies (1).)

(I) Let P_n be any k-tuple of the form (2). Removing tokens from up to $k-1$ heaps, including the first heap, results in a position Q such that the first element is in P_j for some $j < n$, yet there is a heap whose size is a component in P_n. Thus $Q \in P_m$ for no m by Lemma 2. Removing tokens from up to $k-1$ heaps, excluding the first heap, results in a position Q whose last $k-1$ components sum to a number $< (k-1)T_n + n$. Since, however, the first component is in T_n, Q is not of the form (2). Hence $Q \in P_m$ for no m.

So consider the move from P_n which results in $Q = (T_n - t, m_1 - t, \ldots, m_{k-1} - t)$ for some $t \in \mathbb{Z}^+$. If $Q \in P_m$ for some $m < n$, then $T_n - t = T_m$. Then $(T_n - t) + (m_1 - t) + \ldots + (m_{k-1} - t) = kT_n + n - kt = kT_m + m$. Thus, $0 = k(T_n - T_m - t) = m - n < 0$, a contradiction. Hence $Q \in P_m$ for no m.

(II) Let $(m_0, \ldots, m_{k-1})$ be any position which is in P_m for no m. Since $\bigcup_{n=0}^{\infty}[T_n, T_{n+1})$ is a partition of $\mathbb{Z}^0$, we have $m_0 \in [T_n, T_{n+1})$ for precisely one $n \in \mathbb{Z}^0$. Put $L = \sum_{i=1}^{k-1} m_i$.

CASE (i). $m_0 = T_n$. If $L > (k-1)T_n + n$, then removing $L - (k-1)T_n - n$ from a suitable subset of $\{m_1, \ldots, m_{k-1}\}$, results in a position in P_n. So suppose that $L < (k-1)T_n + n$. Then $L = (k-1)T_n + j$ for some $j \in \{0, \ldots, n-1\}$. Subtracting $T_n - T_j$ from all components then leads to a position in P_j. Indeed, $m_0 - (T_n - T_j) = T_j$, and $\sum_{i=1}^{k-1}(m_i - (T_n - T_j)) = (k-1)T_j + j$.

CASE (ii). $T_n < m_0 < T_{n+1}$, say $m_0 = T_n + j$, $j \in \{1, \ldots, n\}$. Suppose first that $L \geq (k-1)T_n + n + j$. If $m_1 < T_{n+1}$, subtract j from m_0 to get to T_n. By the first part of Lemma 2, m_1 is a part in some partition of $(k-1)T_n + n$. Then reduce, if necessary, a subset of the m_i for $i > 1$, so that $m_1 + \sum_{i=2}^{k-1} m'_i = (k-1)T_n + n$. Here and below, m'_i denotes m_i after a suitable positive integer may have been subtracted from it. If $m_1 \geq T_{n+1}$, then decrease m_1 to T_n. Then $T_n + \sum_{i \neq 1} m_i \geq T_n + j + T_n + (k-2)T_{n+1} \geq kT_n + (k-2)(n+1) + 1 \geq kT_n + n + 2 > kT_n + n$, since $k \geq 3$. Again by Lemma 2, m_0 is a part in some partition of $(k-1)T_n + n$. So reducing, if necessary, a subset of the m_i for $i \geq 2$, we get $m_0 + \sum_{i=2}^{k-1} m'_i = (k-1)T_n + n$.

So consider the case $L \leq (k-1)T_n + n + j$. We claim that subtracting $m_0 - T_m$ from all components of $(m_0, \ldots, m_{k-1})$ leads to a position in T_m, where $m = L - (k-1)m_0$. First note that $m = \sum_{i=1}^{k-1} m_i - (k-1)m_0 \geq 0$, and $m = L - (k-1)m_0 \leq (k-1)T_n + n + j - (k-1)m_0 = n - (k-2)j \leq n - j < n$ (since $k \geq 3$), so $0 \leq m < n$, as required. Secondly, $m_0 - (m_0 - T_m) = T_m$, and

$\sum_{i=1}^{k-1}(m_i - (m_0 - T_m)) = L - (k-1)(m_0 - T_m) = (k-1)T_m + m$. (Note that for $L = (k-1)T_n + n + j$ we provided two winning moves. The second leads to a win faster than the first.)

In conclusion, we see that $\bigcup_{i=0}^{\infty} P_i = \mathcal{P}$. ∎

3 Aspects of the Strategy

We observe that the *statement* of Theorem 1 tells a player whether or not it is possible to win by moving from any given position. The *proof* of the theorem shows how to compute a winning move, if it exists. Together they form a *strategy* for the game.

The strategy can, in fact, be computed in polynomial time. Given any position $Q = (m_0, \ldots, m_{k-1})$ of the game. Its input size is $\Theta\left(\sum_{i=0}^{k-1}(\log m_i)\right)$. Solving $m_0 = n(n+1)/2$ leads to $n = \lfloor(\sqrt{1+8m_0} - 1)/2\rfloor$. By Theorem 1, $Q \in \mathcal{P}$ if and only if $m_0 = T_n$, where $n = (\sqrt{1+8m_0} - 1)/2$ is an integer, and $\sum_{i=1}^{k-1} m_i = (k-1)T_n + n$. Otherwise $Q \in \mathcal{N}$, and the proof of Theorem 1 indicates how to compute a winning move to a $\mathcal{P}_n$-position. All of this can be done in time which is polynomial in the input size.

It is also of interest to estimate the density of the $\mathcal{P}$-positions in the set of all game positions. Subtracting $T_n - 1$ from each m_i in the sum of (2), we get partitions of the form

$$x_1 + \ldots + x_{k-1} = n + k - 1, \quad 1 \leq x_1 \leq \ldots \leq x_{k-1} \leq n + 1,$$

where $x_i = m_i - (T_n - 1)$. The number $p_{k-1}(n+k-1)$ of partitions of $n+k-1$ into $k-1$ positive integer parts is estimated in [7, Ch. 4]. It is a polynomial of degree $k-1$ in $n+k-1$, whose leading term is $(n+k-1)^{k-2}/(k-2)!$. Thus the number of positions $\mathcal{P}_n$ for $n \leq N$ is estimated by $\pi(N) = \sum_{n=0}^{N}(n+k-1)^{k-2}/(k-2)!$. It is easy to see that

$$\int_{-1}^{N}(x+k-1)^{k-2}/(k-2)!\,dx \leq \pi(N) \leq \int_{0}^{N+1}(x+k-1)^{k-2}/(k-2)!\,dx,$$

leading to

$$\frac{(N+k-1)^{k-1} - (k-2)^{k-1}}{(k-1)!} \leq \pi(N) \leq \frac{(N+k)^{k-1} - (k-1)^{k-1}}{(k-1)!}.$$

The total number of positions up to $\mathcal{P}_N$ is the number of partitions of the form $m_0 + \ldots + m_{k-1} = n$, $0 \leq m_0 \leq \ldots \leq m_{k-1}$, where n ranges from 0 to $kT_N + N$. Adding 1 to all the parts, we get partitions of the form $x_0 + \ldots + x_{k-1} = n + k$, $1 \leq x_0 \leq \ldots \leq x_{k-1} \leq n+k$, whose number is $p_k(n+k)$. As above, the total number of positions is thus estimated by $\nu(N) = \sum_{n=0}^{kT_N + N}(n+k)^{k-1}/(k-1)!$. Using integration as above, we get

$$\frac{(kT_N + N + k)^k - (k-1)^k}{k!} \leq \nu(N) \leq \frac{(kT_N + N + k + 1)^k - k^k}{k!}.$$

For large N, the ratio is thus about

$$\frac{\pi(N)}{\nu(N)} \approx \frac{k}{kT_N + N + k} \left(\frac{N+k}{kT_N + N + k}\right)^{k-1}.$$

Dividing the numerator and denominator of the second fraction by N^{k-1} results in $\pi(N)/\nu(N) = O(1/N^{k+1})$. We see that the P-positions are rather rare, so our game sticks to the majority of games in the sense of [9] and [10]. The rareness of P-positions in general, is, in fact, consistent with the intuition suggested by (1): a position is in $\mathcal{P}$ if and only if all of its followers are in $\mathcal{N}$, whereas for a position to be in $\mathcal{N}$ it suffices that one of its followers is in $\mathcal{P}$. The scarcity of the P-positions is the reason why game strategies are usually specified in terms of their P-positions, rather than in terms of their N-positions.

4 Epilogue

In the heap games known to us, such as those discussed in [1], the moves are restricted to a single heap (which might, in special cases, be split into several subheaps). We know of two exceptions. One is Moore's Nim$_k$, [8], where up to k heaps can be reduced in a single move (so Nim$_1$ is ordinary Nim). The other is Wythoff's game, Wyt, [11], [2], [3], [12], where a move may affect up to two heaps. The motivation for the present note was to extend Wythoff's game to more than two heaps.

Wyt is played on two heaps. The moves are to either remove any positive number of tokens from a single heap, or to remove the same positive number of tokens from both heaps. Denoting by (x, y) the positions of Wyt, where x and y denote the number of tokens in the two heaps with $x \leq y$, the first eleven P-positions are listed in Table 1. The reader may wish to guess the next few entries of the table before reading on.

For any finite subset $S \subset \mathbb{Z}^0$, define the Minimum EXcluded value of S as follows: $\text{mex } S = \min \mathbb{Z}^0 \setminus S$ = least nonnegative integer not in S [1]. Note that if $S = \emptyset$, then $\text{mex } S = 0$. The general structure of Table 1 is given by:

$$A_n = \text{mex}\{A_i, B_i : 0 \leq i < n\}, \quad B_n = A_n + n \quad (n \in \mathbb{Z}^0).$$

Since the input size of Wyt is succinct, namely $\Theta(\log(x+y))$, one can see that the above characterization of the P-positions implies a strategy which is exponential. A polynomial strategy for Wyt can be based on the observation that $A_n = \lfloor n\alpha \rfloor$, $B_n = \lfloor n\beta \rfloor$, where $\alpha = (1+\sqrt{5})/2$ is the golden section, $\beta = (3+\sqrt{5})/2$. Another polynomial strategy depends on a special numeration system whose basis elements are the numerators of the simple continued fraction expansion of α. These three strategies can be generalized to Wyt$_a$, proposed and analysed in [4], where $a \in \mathbb{Z}^+$ is a parameter of the game. The moves are as in Wyt, except that the second type of move is to remove say $k > 0$ and $l > 0$ from the two heaps subject to $|k - l| < a$. Clearly Wyt$_1$ is Wyt.

Table 1. The first few P-positions of Wyt.

n	A_n	B_n
0	0	0
1	1	2
2	3	5
3	4	7
4	6	10
5	8	13
6	9	15
7	11	18
8	12	20
9	14	23
10	16	26

The generalization of Wyt to more than two heaps was a long sought-after problem. In [5] it is shown that the natural generalization to the case of $k \geq 2$ heaps is to either remove any positive number of tokens from a single heap, or say $l_1, \ldots, l_k$ from all of them simultaneously, where the l_i are nonnegative integers with $\sum_{i=1}^{k} l_i > 0$ and $l_1 \oplus \ldots \oplus l_k = 0$, and where $\oplus$ denotes Nim-sum (also known as addition over GF(2), or XOR). In particular, the case $k = 2$ is Wyt. But the actual computation of the P-values seems to be difficult.

The heap-game considered here is a generalization of the *moves* of Wyt, but not of its strategy. In fact, it doesn't specialize to the case $k = 2$; we used the fact that $k \geq 3$ in several places of the proof. However, the P-positions of the present game have a compact form, the exhibition of which was the purpose of this note.

We remark finally that the *Sprague-Grundy* function g of a game provides a strategy for the *sum* of several games. The computation of g for Nim_k, $k \geq 2$, and Wyt_a, $a \geq 1$ seems to be difficult. It would be of interest to compute the g-function for the present game. Perhaps this is also difficult.

Acknowledgment

We thank Uriel Feige for a simplification in the statement and proof of Theorem 1.

References

1. E. R. Berlekamp, J. H. Conway and R. K. Guy 1982, *Winning Ways* (two volumes), Academic Press, London.
2. H. S. M. Coxeter 1953, The golden section, phyllotaxis and Wythoff's game, *Scripta Math.* **19**, 135–143.

3. A. P. Domoryad 1964, *Mathematical Games and Pastimes* (translated by H. Moss), Pergamon Press, Oxford.
4. A. S. Fraenkel 1982, How to beat your Wythoff games' opponent on three fronts, *Amer. Math. Monthly* **89**, 353–361.
5. A. S. Fraenkel 1996, Scenic trails ascending from sea-level Nim to alpine chess, in: *Games of No Chance*, Proc. MSRI Workshop on Combinatorial Games, July, 1994, Berkeley, CA (R. J. Nowakowski, ed.), MSRI Publ. Vol. 29, Cambridge University Press, Cambridge, pp. 13–42.
6. A. S. Fraenkel [≥1999], *Adventures in Games and Computational Complexity*, to appear in Graduate Studies in Mathematics, Amer. Math. Soc., Providence, RI.
7. M. Hall 1986, *Combinatorial Theory*, 2nd edition, Wiley, New York.
8. E. H. Moore 1909 – 1910, A generalization of the game called nim, *Ann. of Math.* **11** (Ser. 2), 93–94.
9. D. Singmaster 1981, Almost all games are first person games, *Eureka* **41**, 33–37.
10. D. Singmaster 1982, Almost all partizan games are first person and almost all impartial games are maximal, *J. Combin. Inform. System Sci.* **7**, 270–274.
11. W. A. Wythoff 1907, A modification of the game of Nim, *Nieuw Arch. Wisk.* **7**, 199–202.
12. A. M. Yaglom and I. M. Yaglom 1967, *Challenging Mathematical Problems with Elementary Solutions* (translated by J. McCawley, Jr., revised and edited by B. Gordon), Vol. II, Holden-Day, San Francisco.

Infinite Cyclic Impartial Games

Aviezri S. Fraenkel and Ofer Rahat

Department of Applied Mathematics and Computer Science
The Weizmann Institute of Science
Rehovot 76100, Israel
fraenkel@wisdom.weizmann.ac.il
http://www.wisdom.weizmann.ac.il/~fraenkel
ofer@wisdom.weizmann.ac.il

Abstract. We define the family of *locally path-bounded* digraphs, which is a class of infinite digraphs, and show that on this class it is relatively easy to compute an optimal strategy (winning or nonlosing); and realize a win, when possible, in a finite number of moves. This is done by proving that the Generalized Sprague-Grundy function exists uniquely and has finite values on this class.

Keywords: infinite cyclic games, locally path-bounded digraphs, generalized Sprague-Grundy function

1 Introduction

We are concerned with *combinatorial games*, which, for our purposes here, comprise 2-player games with perfect information, no chance moves and outcome restricted to (lose, win), (draw, draw) for the two players. A draw position is a position in the game such that no win is possible from it, but there exists a next move which guarantees, for the player making it, not to lose. You *win* a game by making a last move in it. A game is *impartial* if for every position in it, both players have the same set of next moves; otherwise it's *partizan*. Nim is impartial, chess partizan. A game is *cyclic* if it contains cycles (the possibility of returning to the same position), or loops (pass-positions). These notions, slightly changed here, can be found in [1]. It is clear that a necessary (yet not sufficient) condition for the existence of draw positions is that the game be cyclic.

For Partizan cyclic games, see [3], [12], [4], [10], [5]; finite impartial cyclic games are discussed only briefly in [1], [2]. Particular finite impartial cyclic games are analyzed in [9], [7]. Infinite impartial games are treated briefly at the end of [14], where both the "generalized Sprague-Grundy function" γ, defined below, and its associated "counter function" were permitted to be transfinite ordinals.

Our purpose here is to define a certain class of infinite digraphs on which γ assumes only finite values, but the counter function may contain transfinite ordinal values. The motivation for doing this is based, in part, on the following considerations. It is easier to compute with finite than with transfinite ordinals. For just determining who of the two players can win (or that both can at most draw) in a *sum* of games, the "generalized Nim-sum" of a finite set of γ-values seems to be required (§4). The generalized Nim-sum is based on the binary expansion of ordinals. It's easy to see that every ordinal, finite or transfinite, has a unique expansion as a *finite* sum of powers of ordinals (based on the greedy algorithm and the fact that the ordinals are well-ordered — see [13, XIV, §19]). For example, $\omega = 2^\omega$. We do not wish to enter here into the question of the computational complexity of computing with transfinite ordinals. But it seems possible that it's easier to compare the size of ordinals with each other, which suffices for counter function values, than to compute and work with their binary expansions, as needed for the γ-values. The counter function helps for consummating a win, and this can be done, in a finite number of moves even if the counter function value is a transfinite ordinal, since the ordinals are well-ordered. Moreover, often the structure of the digraph is such that the γ-function itself suffices to provide a winning strategy, without the need of an additional counter function (§4).

The connection between games and digraphs is simple: with any impartial game Γ we associate a digraph $G = (V, E)$ where V is the set of positions of Γ and $(a, b) \in E$ if and only if there is a move from position a to position b. It is called the *game-graph* of Γ. We identify games with their corresponding game-graphs, game positions with digraph vertices and game moves with digraph edges, using them interchangeably. It is thus natural to define a *cyclic* digraph as a digraph, finite or infinite, which may contain cycles or loops.

In §2 we provide basic tools needed for the statement and proof of the result (Theorem 1), and §3 contains the proof. An example demonstrating Theorem 1 is given in the final §4.

2 Preliminaries

The subset of nonnegative integers is denoted by $\mathbb{Z}^0$, and the subset of positive integers by $\mathbb{Z}^+$.

Given a digraph $G = (V, E)$. For any vertex $u \in V$, the set of *followers* of u is $F(u) = \{v \in V : (u, v) \in E\}$. A vertex u with $F(u) = \emptyset$ is a *leaf*. The set of *predecessors* of u is $F^{-1}(u) = \{w \in V : (w, u) \in E\}$. A *walk* in G is any sequence of vertices $u_1, u_2, \ldots$, not necessarily distinct, such that $(u_i, u_{i+1}) \in E$, i.e., $u_{i+1} \in F(u_i)$ ($i \in \mathbb{Z}^+$). Edges may be repeated. A *path* is a walk with all vertices distinct. In particular, there's no repeated edge in a path. The *length* of a path is the number of its edges. If every path in G has finite length, then G is called *path-finite*. If there exists $b \in \mathbb{Z}^0$ such that every path in G has length $\leq b$, then G is *path-bounded*.

Definition 1. A cyclic digraph is *locally path-bounded* if for every vertex u_i there is a bound $b_i(u_i) = b_i \in \mathbb{Z}^0$ such that the length of every (directed) path emanating from u_i doesn't exceed b_i. The integer b_i is the *local path bound* of u_i.

Note that every path-bounded digraph is locally path-bounded, and every locally path-bounded digraph is path-finite. But neither of the two inverse relationships needs to hold. Our main result is concerned with locally path-bounded digraphs.

Given a digraph $G = (V, E)$. The *Generalized Sprague-Grundy function*, also called γ-*function*, is a mapping $\gamma \colon V \to \mathbb{Z}^0 \cup \{\infty\}$, where the symbol ∞ indicates a value larger than any natural number. If $\gamma(u) = \infty$, we say that $\gamma(u)$ is infinite. We wish to define γ also on certain subsets of vertices. Specifically: $\gamma(F(u)) = \{\gamma(v) < \infty \colon v \in F(u)\}$. If $\gamma(u) = \infty$ and $\gamma(F(u)) = K$, we also write $\gamma(u) = \infty(K)$. Next we define equality of $\gamma(u)$ and $\gamma(v)$: if $\gamma(u) = k$ and $\gamma(v) = \ell$ then $\gamma(u) = \gamma(v)$ if one of the following holds: (a) $k = \ell < \infty$; (b) $k = \infty(K)$, $\ell = \infty(L)$ and $K = L$. We also use the notations

$$V^f = \{u \in V \colon \gamma(u) < \infty\}, \qquad V^\infty = V \setminus V^f,$$

$$\gamma'(u) = \operatorname{mex} \gamma(F(u)) = \operatorname{mex}\{\gamma(v) < \infty \colon v \in F(u)\}, \qquad (1)$$

where for any finite subset $S \subset \mathbb{Z}^0$, the *Minimum EXcluded* value mex is defined by

$$\operatorname{mex} S = \min(\mathbb{Z}^0 \setminus S) = \text{minimum term in } \mathbb{Z}^0 \text{ not in } S.$$

We need some device to tell the winner where to go when we use the γ-function. For example, suppose that there is a token on vertex u (Fig. 1). It turns out that it's best for the player moving now to go to a position with γ-value 0. There are two such values: one (the leaf) is an immediate win, and the other (v) is only a nonlosing move. This digraph may be embedded in a large digraph where it's not clear which option leads to a win. The device which overcomes this problem is a counter function, as used in the following definition. For realizing an optimal strategy, we will normally select a follower of least counter function value with specified γ-value. The counter function also enables us to prove assertions by induction.

Definition 2. Given a cyclic digraph $G = (V, E)$. A function $\gamma \colon V \to \mathbb{Z}^0 \cup \{\infty\}$ is a γ-*function* with *counter function* $c \colon V^f \to J$, where J is any infinite well-ordered set, if the following three conditions hold:

A. If $\gamma(u) < \infty$, then $\gamma(u) = \gamma'(u)$.

B. If there exists $v \in F(u)$ with $\gamma(v) > \gamma(u)$, then there exists $w \in F(v)$ satisfying $\gamma(w) = \gamma(u)$ and $c(w) < c(u)$.

C. If $\gamma(u) = \infty$, then there is $v \in F(u)$ with $\gamma(v) = \infty(K)$ such that $\gamma'(u) \notin K$.

Remarks.

- In **B** we have necessarily $u \in V^f$; and we may have $\gamma(v) = \infty$ as in **C**.
- To make condition **C** more accessible, we state it also in the following equivalent form:

    C'. If for every $v \in F(u)$ with $\gamma(v) = \infty$ there is $w \in F(v)$ with $\gamma(w) = \gamma'(u)$, then $\gamma(u) < \infty$.

- If condition **C'** is satisfied, then $\gamma(u) < \infty$, and so by **A**, $\gamma(w) = \gamma'(u) = \gamma(u)$.

- To keep the notation simple, we write $\infty(0)$, $\infty(1)$, $\infty(0,1)$ etc., for $\infty(\{0\})$, $\infty(\{1\})$, $\infty(\{0,1\})$, etc.

The γ-function was first defined in [14]. It was found independently in [8]. The simplified version given above, and two other versions, appear in [11]. Since this function is not well-known, we repeated its definition above. The γ-function exists uniquely on any finite cyclic digraph, but its associated counter function exists nonuniquely.

We say that a digraph G has a γ-function, if γ exists on G with values restricted to $\mathbb{Z}^0 \cup \{\infty\}$, i.e., no transfinite ordinal values.

We are now ready to state our main result.

Theorem 1. *Every locally path-bounded digraph $G = (V, E)$ has a unique γ-function with an associated counter function; and for every $u \in V^f$, $\gamma(u)$ doesn't exceed the length of a longest path emanating from u. The statement doesn't hold in general for path-finite digraphs.*

3 The Proof

We wish to examine some properties of path-finite and locally path-bounded digraphs. To begin with, is it clear that for a path-finite digraph, if $v \in F(u)$, then every path emanating from v is not longer than any path emanating from u?

Perhaps it is clear, but it's also wrong: in a path-finite graph, every path originating at some vertex u and continuing to its ultimate end, terminates at a vertex v, where v is either a leaf or a predecessor of some w on the path. Thus a path of minimum length emanating from u in Fig. 1 terminates at the leaf, whereas a path of maximum length beginning at u terminates at y. It has length 3. But a maximal-length path emanating from $v \in F(u)$ clearly has length 4.

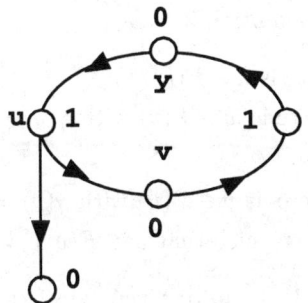

Fig. 1. The numbers are γ-values.

If a digraph $G = (V, E)$, possibly with infinite paths, has no leaf, then the label ∞ on all the vertices is evidently a γ-function: **A** and **B** are satisfied vacuously, and **C** is satisfied with $\gamma'(u) = 0$ for all $u \in V$. If G has a leaf, then some of the vertices have a γ-function, such as the leaf and its predecessors, but possibly γ doesn't exist on some of the vertices. For the case where γ exists on a subset $V' \subseteq V$, we define $\gamma'(u) = \text{mex}\{\gamma(F(u)) : F(u) \subseteq V'\}$. Since $F(u)$ may, nevertheless, be infinite for any vertex u in a locally path-bounded digraph, it is not clear a priori that $\gamma'(u)$ exists. The following lemma takes care of this point.

Lemma 1. *Let u be any vertex with local path bound b in a locally path-bounded digraph $G = (V, E)$. Then $\gamma'(u)$ exists (i.e., it is a nonnegative integer), and in fact, $\gamma'(u) \leq b$.*

Proof. We consider two cases.

(i) Suppose that u has finite γ-value m. Then $\gamma'(u)$ exists, and in fact, $\gamma'(u) = \gamma(u) = m$ by **A**. Moreover, there exists $u_1 \in F(u)$ with $\gamma(u_1) = m-1$, there exists $u_2 \in F(u_1)$ with $\gamma(u_2) = m-2, \ldots$, there exists $u_m \in F(u_{m-1})$ with $\gamma(u_m) = 0$. Then $u, u_1, \ldots, u_m$ is a path of length m, so $m \leq b$. (The path may continue beyond u_m, but in any case $\gamma'(u) = m \leq b$.)

(ii) Suppose that u has either no γ-value or value ∞. It suffices to show that if $v \in F(u) \cap V^f$, then $\gamma(v) < b$. Indeed, $|F(u)|$ may be infinite, and $F(u)$ may contain vertices with no γ-value. But if $\gamma(v) < b$ for all $v \in F(u) \cap V^f$, then clearly $\gamma'(u)$ exists and $\gamma'(u) \leq b$. Note that we cannot use the argument of case (i) directly on v, since a path from v may be longer than a path from u, as we just saw. So suppose there is $v_0 \in F(u) \cap V^f$ with $\gamma(v_0) = n \geq b$. As in case (i), there is a path $v_0, v_1, \ldots, v_n$ of length n with $\gamma(v_i) = n - i$ ($i \in \{0, \ldots, n\}$). Then $u, v_0, v_1, \ldots, v_n$ is a walk of length $n+1 > b$ emanating from u. Hence it cannot be a path. But $v_i \neq v_j$ for all $i \neq j$, since $\gamma(v_i) \neq \gamma(v_j)$. Hence $v_j = u$ for some $j \in \{0, \ldots, n\}$. The contradiction is that v_j does and u doesn't have a finite γ-value. Thus $\gamma(v) < b$ for all $v \in F(u) \cap V^f$, hence $\gamma'(u)$ exists, and in fact, $\gamma'(u) \leq b$. ∎

Proof of Theorem 1. Let $V' \subseteq V$ be a maximal subset of vertices on which γ exists, together with an associated counter function c, subject to the following additions to **B** and **C** of Definition 2:

If $\gamma(u) < \infty$ and there is $v \in F(u) \cap V_\nu$,

$$\text{then there is } w \in F(v) \text{ with } \gamma(w) = \gamma(u),\ c(w) < c(u)\ , \quad (2)$$

if $\gamma(u) = \infty$, then there is $v \in F(u)$ with $\gamma(v) = \infty$

$$\text{such that } w \in F(v) \cap V_\nu \Longrightarrow \gamma'(w) \neq \gamma'(u)\ , \quad (3)$$

where $V_\nu = V \setminus V'$. (In (3) we have $\gamma'(w) \neq \gamma'(u)$, instead of $w \in F(v)$ and $\gamma(w) \neq \gamma'(u)$ in **C**.) In addition we require:

$$\text{If } \gamma(u) = \infty \text{ with } \gamma'(u) = l, \text{ then } \gamma'(v) \geq l \text{ for all } v \in V_\nu\ . \quad (4)$$

The subset V' is maximal in the sense that adjoining any $u \in V_\nu$ into V' violates either Definition 2, or (2) or (3) or (4). If $V_\nu \neq \emptyset$, let $u \in V_\nu$. By Lemma 1,

$\gamma'(u) = k$ exists for some $k \in \mathbb{Z}^0$. It follows that there is a minimum value $m = \min\{k \in \mathbb{Z}^0 : u \in V_\nu, \gamma'(u) = k\}$. Let $K = \{u \in V_\nu : \gamma'(u) = m\}$. Then $V_\nu \neq \emptyset \implies K \neq \emptyset$. We consider four cases.

Case 1. For every $u \in K$ we have $m \in \gamma'(F(u))$, where, consistent with (1),

$$\gamma'(F(u)) = \{\gamma'(v) : v \in F(u)\} = \{\max \gamma(F(v)) : v \in F(u)\}$$
$$= \{\max\{\gamma(w) < \infty : w \in F(v), v \in F(u)\}\}.$$

Note that $u \in K$, $v \in F(u) \implies \gamma(v) \neq m$ by the definition of mex, so $v \in F(u)$, $\gamma'(v) = m \implies v \in V_\nu$, in fact, $v \in K$. Thus putting $\gamma(u) = \infty$ for all $u \in K$ satisfies **C**, and is also consistent with (3); and with (4) by the minimality of m. Furthermore, it doesn't violate **A**, and is consistent with **B** by (2). This contradicts the maximality of V'.

We may thus assume henceforth that there exists $u \in K$ such that

$$m \notin \gamma'(F(u)). \tag{5}$$

Case 2. There exist $u \in K$ and $v \in F(u)$ with $\gamma(v) = \infty$, such that for every $w \in F(v)$, either $\gamma(w) \neq m$, or $w \in V_\nu$ with $\gamma'(w) \neq m$. Putting $\gamma(u) = \infty$ is clearly consistent with **C**, and (3); and it doesn't violate **A**. In view of (2), also **B** is satisfied. This contradicts the maximality of V'. So we may assume that

$$\forall u \in K \text{ and } \forall v \in F(u) \text{ with } \gamma(v) = \infty,$$
$$\exists w \in F(v) \text{ (with } \gamma(w) = m \text{ or } w \in V_\nu \text{ with } \gamma'(w) = m).$$

We subdivide this into the following two cases:

$$\exists u \in K \text{ such that } \forall v \in F(u) \text{ with } \gamma(v) = \infty,$$
$$\exists w \in F(v) \text{ with } \gamma(w) = m, \tag{6}$$

or

$$\forall u \in K \text{ and } \forall v \in F(u) \text{ with } \gamma(v) = \infty,$$
$$\exists \text{ no } w \in \cap V', \text{ but } \exists w \in F(v) \cap V_\nu \text{ with } \gamma'(w) = m. \tag{7}$$

Case 3. (6) holds. We repeat that for any $u \in K$, since $\gamma'(u) = m$, u has no follower with γ-value m. Suppose that there exists $y \in F^{-1}(u)$ with $\gamma(y) = m$. Then by (2), there exists $v \in F(u)$ with $\gamma(v) = m$, contradicting $\gamma'(u) = m$. Thus putting $\gamma(u) = m$ is consistent with **A**. It is also consistent with (3): putting $\gamma(u) = m$ could presumably increase $\gamma'(y)$ for some $y \in F^{-1}(u) \cap V_\nu$, and thus upset (3) for the value $\gamma(z) = \infty$ of some grandparent $z = F^{-1}(F^{-1}(y))$ of y. Now by (4), $\gamma'(u) \geq \gamma'(z)$. If indeed $\gamma'(y)$ increased, then for the new value we have $\gamma'(y) > \gamma'(u)$, so $\gamma'(y) > \gamma'(z)$ and $\gamma(z) = \infty$ remains unaffected. Consistency with **C** thus follows from (3) which becomes **C** when u is labeled m. Since $y \in F^{-1}(u) \implies \gamma(y) \neq m$, as we saw at the beginning of this case, the potential adverse effect on any grandparent z of y considered above, cannot happen.

We now show that also **B** holds. Suppose first that $F(u) \subseteq V'$. For every $v \in F(u)$ for which $\gamma(v) > m$, there exists $w \in F(v)$ with $\gamma(w) = m$. This follows from **A** if $\gamma(v) < \infty$, and from (6) if $\gamma(v) = \infty$. It remains to define $c(u)$ sufficiently large so that $c(w) < c(u)$. This will be done below.

In view of the minimality of m and by (5), the second possibility is that for every $v \in F(u)$ for which $v \in V_\nu$, we have $\gamma'(v) > m$. For every such v there exists $w \in F(v)$ with $\gamma(w) = m$ by the definition of mex. Again we have to define $c(u)$ sufficiently large to satisfy $c(w) < c(u)$.

Let $S = \{v \in F(u) : \gamma(v) > m\} \cup \{v \in F(u) : v \in V_\nu, \gamma'(v) > m\}$. We have just seen that for every $v \in S$ there is $w \in F(v)$ with $\gamma(w) = m$. Put $T = \{w \in F(v) : v \in S, \gamma(w) = m\}$. Let $c(u)$ be the smallest ordinal $> c(w)$ for all $w \in T$. Then also **B** is satisfied. This contradicts the presumed maximality of V'.

Note that the case $F(u) \subseteq V \setminus V^\infty$ satisfies (6) vacuously, and so is also included in the present case.

Case 4. (7) holds. If (7) holds nonvacuously, then as in Case 1, putting $\gamma(u) = \infty$ for all $u \in K$ is consistent with **C**, (3) and the other conditions. This contradicts again the maximality of V'. Hence $K = \emptyset$, and so also $V_\nu = \emptyset$.

Whenever γ exists on a digraph, finite or infinite, it exists there uniquely. See [6]. Finally, if b is the local bound of $u \in V$, then $\gamma'(u) \leq b$ by Lemma 2. Hence if $u \in V^f$, then $\gamma(u) \leq b$ by **A**.

For proving the last statement of the theorem, consider the digraph G which consists of a vertex u, and $F(u) = \{u_0, u_1, \ldots\}$, where, for all $i \in \mathbb{Z}^0$, u_i is the top vertex of a Nim-heap of size i, so $\gamma(u_i) = i$ ($i \geq 0$). It is clear that G is path-finite but not locally path-bounded. Also $\gamma(F(u)) = \{0, 1, \ldots\}$, so $\gamma(u)$ cannot assume any finite value. In fact, $\gamma(u) = \omega$, where ω is the smallest transfinite ordinal bigger than all the natural numbers. If, however, $F(u_i) = u_{i-1}$, and $F(u) = \{u_0, u_1, \ldots\}$ as above, then again G is path-finite but not locally path-bounded, yet $\gamma(u) = 2$. ■

4 An Example

We specify below a locally path-bounded digraph $G = (V, E)$ on some of whose vertices we place a finite number of tokens. A move consists of selecting a token and moving it to a follower. Multiple occupancy of vertices is permitted. The player first unable to move loses, and the opponent wins. If there is no last move, the outcome is a draw.

For any $r \in \mathbb{Z}^0$, a *Nim-heap* of size r is a digraph with vertices $u_0, \ldots, u_r$ and edges (u_j, u_i) for all $0 \leq i < j \leq r$. In depicting G (Fig. 2), we use the convention that bold lines and the vertices they connect constitute a Nim-heap, of which only adjacent (bold) edges are shown, to avoid cluttering the drawing. Thin lines denote ordinary edges.

All the horizontal lines are thin, and each vertex u_i on this horizontal line connects via a vertical thin edge to a Nim-heap G_i pointing downwards, of size $\lfloor (4i+8)/3 \rfloor$, $i \in \mathbb{Z}^0$. From u_i there also emanates a Nim-heap H_j of size $j = \lfloor (i+2)/3 \rfloor$ pointing upward. Each G_i has a back edge to its top vertex forming a cycle of length $i+1$. Thus G_0 has a loop at the top of its Nim-heap (of size 2). There is an additional back edge to the vertex u_i on the horizontal line, forming a cycle of length $i+3$.

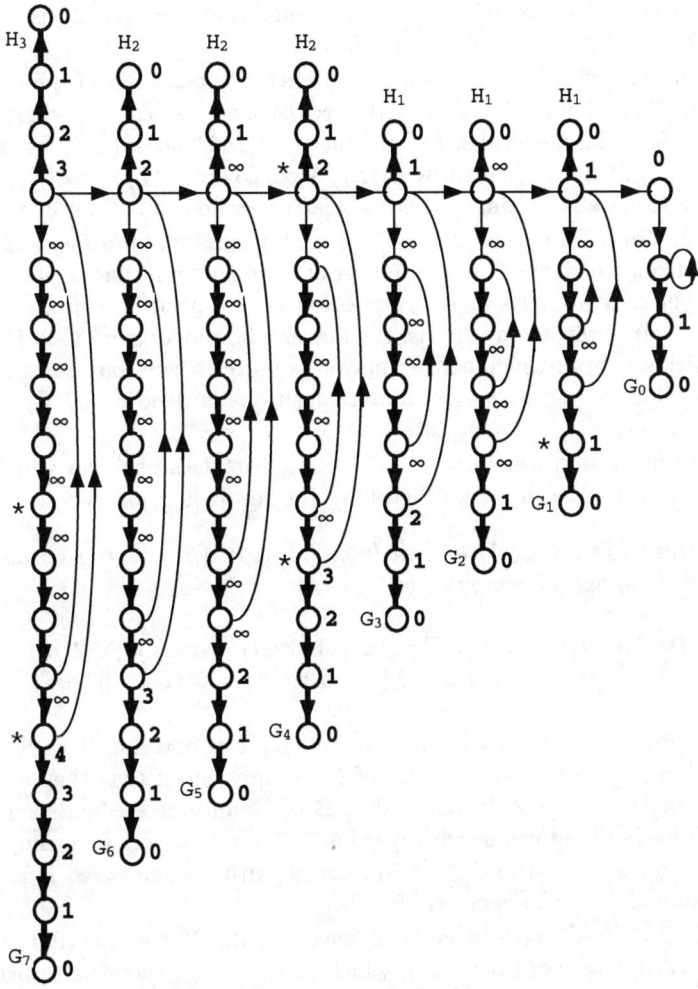

Fig. 2. The tail-end of a locally path-bounded digraph.

From any vertex u on the horizontal line there is a longest path, via G_i, of length $\lfloor (4i+11)/3 \rfloor$, and the other vertices have shorter maximal length. Thus

G is locally path-bounded. But it is not path-bounded, since i can be arbitrarily large.

Sample Problem. Compute an optimal strategy for the 5-token game placed on the 5 starred vertices of G.

To solve this problem, we introduce the generalized Nim-sum ([14], [11], [6]). For any nonnegative integer h we write $h = \sum_{i \geq 0} h^i 2^i$ for the binary encoding of h ($h^i \in \{0,1\}$). If a and b are nonnegative integers, then their *Nim-sum* $a \oplus b = c$, also called *exclusive or*, *XOR*, or *addition over* $GF(2)$, is defined by $c^i \equiv a^i + b^i \pmod 2$, $c^i \in \{0,1\}$ ($i \geq 0$).

The Generalized Nim-sum of a nonnegative integer a and $\infty(L)$, for any finite subset $L \subset \mathbb{Z}^0$, is defined by $a \oplus \infty(L) = \infty(L) \oplus a = \infty(L \oplus a)$, where $L \oplus a = \{\ell \oplus a : \ell \in L\}$. The Generalized Nim-sum of $\infty(L_1)$ and $\infty(L_2)$, for any finite subsets L_1, L_2 of $\mathbb{Z}^0$, is defined by $\infty(L_1) \oplus \infty(L_2) = \infty(L_2) \oplus \infty(L_1) = \infty(\emptyset)$. Clearly the Generalized Nim-sum is associative and $a \oplus a = 0$ for every a.

Given any finite or infinite game Γ, we say informally that a *P-position* is any position u from which the *P*revious player can force a win, that is, the opponent of the player moving from u. An *N-position* is any position v from which the *N*ext player can force a win, that is, the player who moves from v. A *D*-position is any position u from which neither player can force a win, but has a nonlosing next move. The set of all P-, N- and D-positions is denoted by $\mathcal{P}$, $\mathcal{N}$ and $\mathcal{D}$ respectively.

For any finite multiset $\boldsymbol{u} = (u_1, \ldots, u_n)$ of vertices of G on which tokens reside, one token on each u_i, we then have the result [6]:

Proposition. *The P-, N- and D-labels of $\boldsymbol{u} = (u_1, \ldots, u_n)$ in any locally path-bounded digraph G are given by*

$$\mathcal{P} = \{\boldsymbol{u} \in V : \sigma(\boldsymbol{u}) = 0\}, \quad \mathcal{D} = \{\boldsymbol{u} \in V : \sigma(\boldsymbol{u}) = \infty(K),\ 0 \notin K\}$$
$$\mathcal{N} = \{\boldsymbol{u} \in V : 0 < \sigma(\boldsymbol{u}) < \infty\} \cup \{\boldsymbol{u} \in V : \sigma(\boldsymbol{u}) = \infty(K),\ 0 \in K\}. \quad \blacksquare$$

We are now ready to solve the above problem, by observing that the symbols appearing on Fig. 2 are the γ-values of G. Simply check that they satisfy the conditions of Definition 2. In particular, **B** of Definition 2 is satisfied if every vertex on the horizontal line with γ-value $< \infty$ gets a counter-value between ω and $\omega 2$, and every vertex in the Nim-heaps with γ-value $< \infty$ is assigned a counter value $< \omega$, which is clearly feasible.

For the 5 starred vertices we then have $1 \oplus 3 \oplus 2 \oplus 4 \oplus \infty(0,1,2,3,4) = 4 \oplus \infty(0,1,2,3,4) = \infty(4,5,6,7,0)$, which contains 0, hence the position is in $\mathcal{N}$. Thus the player moving from this position can win by going to a position of Nim-sum 0, namely, pushing the token on the infinity label to 4. Indeed the resulting Nim-sum is $1 \oplus 3 \oplus 2 \oplus 4 \oplus 4 = 0$.

We remark that any tokens on two vertices with γ-value ∞ is a draw position, no matter where the other tokens are, if any. Also note that for realizing a win in this game we do not really need a counter function.

Epilogue

We have defined locally path-bounded digraphs, and shown that the generalized Sprague-Grundy function γ exists on such digraphs with finite, though not necessarily bounded, values. Of course local path-boundedness is only a sufficient condition for the existence of γ. Any finite or infinite digraph without a leaf, satisfies trivially $\gamma(u) = \infty$ for all its vertices u.

A large part of combinatorial game theory is concerned, however, with digraphs which do have leaves. If we exclude digraphs without leaves, then the condition of local boundedness is, in a sense, best possible, as stated in the last part of Theorem 1.

References

1. E. R. Berlekamp, J. H. Conway and R. K. Guy 1982, *Winning Ways* (two volumes), Academic Press, London.
2. J. H. Conway 1976, *On Numbers and Games*, Academic Press, London.
3. J. H. Conway 1978, Loopy Games, *Ann. Discrete Math.* **3**, Proc. Symp. Advances in Graph Theory, Cambridge Combinatorial Conf. (B. Bollobás, ed.), Cambridge, May 1977, pp. 55–74.
4. J. A. Flanigan 1981, Selective sums of loopy partizan graph games, *Internat. J. Game Theory* **10**, 1–10.
5. J. A. Flanigan 1983, Slow joins of loopy games, *J. Combin. Theory* (Ser. A) **34**, 46–59.
6. A. S. Fraenkel ≥1999, *Adventures in Games and Computational Complexity*, to appear in Graduate Studies in Mathematics, Amer. Math. Soc., Providence, RI.
7. A. S. Fraenkel and A. Kotzig 1987, Partizan octal games: partizan subtraction games, *Internat. J. Game Theory* **16**, 145–154.
8. A. S. Fraenkel and Y. Perl 1975, Constructions in combinatorial games with cycles, *Coll. Math. Soc. János Bolyai*, **10**, Proc. Internat. Colloq. on Infinite and Finite Sets, Vol. 2 (A. Hajnal, R. Rado and V. T. Sós, eds.) Keszthely, Hungary, 1973, North-Holland, pp. 667–699.
9. A. S. Fraenkel and U. Tassa 1975, Strategy for a class of games with dynamic ties, *Comput. Math. Appl.* **1**, 237–254.
10. A. S. Fraenkel and U. Tassa 1982, Strategies for compounds of partizan games, *Math. Proc. Camb. Phil. Soc.* **92**, 193–204.
11. A. S. Fraenkel and Y. Yesha 1986, The generalized Sprague-Grundy function and its invariance under certain mappings, *J. Combin. Theory* (Ser. A) **43**, 165–177.
12. A. S. Shaki 1979, Algebraic solutions of partizan games with cycles, *Math. Proc. Camb. Phil. Soc.* **85**, 227–246.
13. W. Sierpiński 1958, *Cardinal and Ordinal Numbers*, Hafner, New York.
14. C. A. B. Smith 1966, Graphs and composite games, *J. Combin. Theory* **1**, 51–81. Reprinted in slightly modified form in: *A Seminar on Graph Theory* (F. Harary, ed.), Holt, Rinehart and Winston, New York, NY, 1967.

On the Complexity of Tsume-Go

Marcel Crâşmaru

SCVR Vatra Dornei 5975 SV, ROMANIA,
mi@assist.cccis.ro

Abstract. In this paper, we explain why Go is hard to be programmed. Since the strategy of the game is closely related to the concept of *alive-dead* group, it is plainly necessary to analyze this concept. For this a mathematical model is proposed. Then we turn our research to Tsume-Go problems in which one of the players has always a unique good move and the other has always only two good moves available to choose from. We show that this kind of problems are NP-*complete*.

1 Introduction

The creation of a program that plays the game of Go, even at a medium level, is known to be difficult. Our aim in this paper is to give a reasonable explanation to this problem by analyzing the fundamentals of Go, more precisely, the concept of *living group*. What territories are, is a problem that can be reduced eventually to the life and death problem (tsume-Go) of certain groups placed on the board. Strong players always look to groups to find out if they are thick or weak, that is if they can live easily or not. In fact, the core of the strategy of the game is the understanding of what is a living or a dead group.

The Go-programmer dream is to find a good pruning function that would reduce a lot the searching tree in the emerging problems during the game, problems that can be reduced, as we said before, to life and death problems. What we will show is that even if one has an exceptionally good pruning function, there are positions in which the complexity of the search is NP-complete.

In the past, results on the complexity of Go were obtained by Lichtenstein and Sipser [1], who have constructed Go positions without kos which are PSPACE-hard. Using kos, Robson [3] has shown Go is EXPTIME-complete. On the other hand, Morris [2] showed that playing sums of relatively simple combinatorial games is NP-hard and then, Yedwab [4] and Moews [5] showed that even a sum of games of the form $a||b|c$ is NP-hard. For further references see Berlekamp [6].

Let $\mathcal{G}$ be a group and $\mathcal{T}(\mathcal{G})$ a tree constructed recursively as follows: the root is the (initial) position to which $\mathcal{G}$ belongs, and the children of any node v are all the positions obtained throw a legal move from v. We also denote by $\Psi(v)$ a pruning function for $\mathcal{T}(\mathcal{G})$, that is, a set containing some descendents of the node v. Let $\mathcal{T}_\Psi(G)$ be the subtree generated by $\Psi(v)$ with the same root as $\mathcal{T}(\mathcal{G})$ and then, recursively, $\Psi(v)$ is the set of children of any node v.

Let G_{NP} be the set of all groups $\mathcal{G}$ for which there is a pruning function $\Psi(v)$, computable in polynomial time, that satisfies $|\Psi(v)| = 1$ always for one of the players and $|\Psi(v)| = 2$ for the other for all nodes v of $\mathcal{T}_\Psi(G)$ and one can determine if $\mathcal{G}$ can be removed or not from the board only by scanning in $\mathcal{T}_\Psi(G)$ at a depth which is polynomial of the size of the board.

Let us consider the problem:

(GNP)   Given $\mathcal{G} \in G_{\mathrm{NP}}$, decide whether $\mathcal{G}$ is a dead group.

Our main result is:

Theorem 1. *The problem (GNP) is NP−complete.*

In Sect. 1 we present a mathematical model of the game of Go, in the second we give an algorithm that checks if a group is alive or not, while in the last we prove Theorem 1.

2 A Mathematical Model of Go without the Rule of Ko

Let $\mathcal{T}$, be a nonempty finite set, which we call the *board*, and $\mathcal{N} \subset \mathcal{T} \times \mathcal{T}$ a symmetric binary relation on $\mathcal{T}$, named the *neighboring relation*. If $(x, y) \in \mathcal{N}$ we say that x and y are *neighbors* and write $x \mathcal{N} y$. Let $\mathcal{S} = \{B, W, 0\}$ the set of *colors* (B stands for Black, W for White and 0 for an empty) and $\mathcal{S}^* = \{B, W\}$, the set of *players*. If $x \in \mathcal{S}^*$ we denote by $\bar{x}$ the other element of $\mathcal{S}^*$.

Let $\mathcal{F}$ be the set of *configurations* which are defined to be the set of all functions from $\mathcal{T}$ to $\mathcal{S}$. (For example, the fact that $f(x) = W$ means that in x a white stone is placed). For each configuration f, it is natural to associate a binary relation $\mathcal{N}_f \subset \mathcal{T} \times \mathcal{T}$. Thus, we write $x \mathcal{N}_f y$ iff $x \mathcal{N} y$ and $f(x) = f(y)$. It is easy to see that the reflexive and transitive closure of $\mathcal{N}_f$ is an equivalence relation and its classes of equivalence will be called *groups*. We write by $\mathcal{G}_f^x$ the group that contains x. The color of the group $\mathcal{G}_f^x$ is $Color(\mathcal{G}_f^x) = f(x)$.

For any Black or White group we define the set of *liberties* by

$$\mathcal{L}_f^x = \{y \in \mathcal{T} : f(y) = 0 \text{ and } y \mathcal{N} z \text{ for some } z \in \mathcal{G}_f^x\},$$

that is the empty neighbors of $\mathcal{G}_f^x$. The number of liberties of $\mathcal{G}_f^x$ is the cardinal of $\mathcal{L}_f^x$ and will be denoted by $\|\mathcal{G}_f^x\|$.

During the game, any configuration is paired with a player whose turn is make the next move, thus we need to introduce the set of *positions* which is defined to be $\mathcal{P} = \mathcal{F} \times \mathcal{S}^*$. The fact that $p = (f, a) \in \mathcal{P}$ means that in the configuration f is a's turn to move and for a group that contains x in position p, the notation $\mathcal{G}_p^x$ will be used rather than $\mathcal{G}_f^x$.

Now we are ready to define the *moving function*, which we denote by $\phi \colon \mathcal{P} \times \mathcal{T}^* \to \mathcal{P}$. Let $\mathcal{T}^* = \mathcal{T} \cup \{pass\}$. If $p, p' \in \mathcal{P}$ and $x \in \mathcal{T}^*$ then $p' = \phi(p, x) = \phi((f, a), x)$ will mean that in the position p the player a has placed his stone in x and the outcome is the position p'.

There are two cases.

1. $x = pass$ or $f(x) \neq 0$. We define
$$\phi((f,a),x) = (f,\bar{a}) ,$$
which means that if one of the players moves *pass* then we get the same configuration with the other player's turn to move and playing on an occupied place is the same as playing *pass*.

Let us note that this also says that to play *pass* is allowed in the initial position $p_0 = (f_0, B)$, where $f_0(x) = 0$ for any $x \in \mathcal{T}$.

2. $f(x) = 0$. In this case, we may obtain or not a different configuration if the move captures or not some stones.

Let
$$g(z) = \begin{cases} f(z) & \text{if } z \neq x \\ a & \text{if } z = x \end{cases}$$
be the configuration obtained by placing a stone of color a on x. Let
$$Capture(p,x) = \{\mathcal{G}_g^z \colon g(z) = \bar{a},\ z \mathcal{N} x \text{ and } \|\mathcal{G}_g^z\| = 0\}$$
be the groups captured by the move, that is the opposite color groups which are neighbors to x and have no liberties.

If $Capture(p,x) \neq \emptyset$, then let
$$h(z) = \begin{cases} g(z) & \text{if } z \notin Capture(p,x) \\ 0 & \text{if } z \in Capture(p,x) \end{cases}$$
be the configuration obtained from g after removing the groups with no liberties.

Then, since the suicide is forbidden, we define:
$$\phi((f,a),x) = \begin{cases} (f,\bar{a}) & \text{if } Capture(p,x) = \emptyset \text{ and } \|\mathcal{G}_g^x\| = 0 \\ (h,\bar{a}) & \text{if } Capture(p,x) \neq \emptyset \\ (g,\bar{a}) & \text{if } Capture(p,x) = \emptyset \text{ and } \|\mathcal{G}_g^x\| \neq 0 \end{cases}.$$

Given $p_1, \ldots, p_n \in \mathcal{P}$, let $\Psi(\{p_1, \ldots, p_n\})$ be the set of all positions that can be obtained from $p_1, \ldots, p_n$ by playing any possible move. Thus, $\Psi \colon 2^{\mathcal{P}} \to 2^{\mathcal{P}}$ and
$$\Psi(\{p_1, \ldots, p_n\}) = \bigcup_{i=1}^{n} \bigcup_{x \in \mathcal{T}^*} \phi(p_i, x) ,$$
where $2^{\mathcal{P}}$ denotes the set of all subsets of $\mathcal{P}$. In particular, $\Psi^{(k)}(\{p_0\})$ is the set of all positions that can be obtained from the initial position p_0 in exactly k moves. If the position p' is obtained from p in k moves we write shortly $p \stackrel{k}{\Longrightarrow} p'$.

The following theorem shows that the "history" that created a position plays no role in the study of a legal position of the game.

Theorem 2. *Let $G(p_0)$ be the set of all positions generated by p_0. Then, $p = (f, a) \in G(p_0)$ iff $\|\mathcal{G}_p^x\| \neq 0$ for any $x \in \mathcal{T}$ with $f(x) \neq 0$.*

Proof. Clearly p_0 has the property that $\|\mathcal{G}_{p_0}^x\| \neq 0$ for any $x \in \mathcal{T}$ with $f_0(x) \neq 0$. Suppose now that for the position $p = (f, a)$ we have that $\|\mathcal{G}_p^x\| \neq 0$ for any $x \in \mathcal{T}$ with $f(x) \neq 0$. Then, for any position p' with $p \stackrel{1}{\Longrightarrow} p'$, we have that $\|\mathcal{G}_{p'}^x\| \neq 0$ because Ψ removes the groups without liberties. By induction, we obtain that for any $p \in G(p_0)$ we have that $\|\mathcal{G}_p^x\| \neq 0$ for any $x \in \mathcal{T}$ with $f(x) \neq 0$.

To prove the other implication, let $p = (f, a)$ be a position and let $x_1, \ldots, x_{|\mathcal{T}|}$ be an ordering of $\mathcal{T}$ such that the sequence $f(x_1), \ldots, f(x_{|\mathcal{T}|})$ has the form

$$\underbrace{B, W, B, W, \ldots, B, W}_{l \text{ colors}}, \underbrace{X, \ldots, X}_{m \text{ colors}}, \underbrace{0, \ldots, 0}_{n \text{ colors}},$$

for some $l, m, n \geq 0$, $l + m + n = |\mathcal{T}|$, and $X \in \mathcal{S}^*$. Then $p_0 \stackrel{k}{\Longrightarrow} p$ through $x_1, \ldots, x_{|\mathcal{T}|}$ in $k \leq l + 2m + 1$ steps (the step from W to X may need a *pass* move, from X to X requires a *pass* move and from X to a may need a *pass* move). Since $\|\mathcal{G}_p^x\| \neq 0$ for any $x \in \mathcal{T}$ with $f(x) \neq 0$ it follows that the same is true for all the intermediary positions between p_0 and p, therefore all these positions are correctly constructed (without suicidal moves). Although is not necessary, note that the way in which p is obtained from p_0 takes a minimum number of steps. This completes the proof of the theorem.

3 The Life and Death Algorithm

By the definition of the function move, one can see that if the group $\mathcal{G}_p^x$ has only one liberty (i.e. $\|\mathcal{G}_p^x\| = 1$) where $p = (f, a)$ and $Color(\mathcal{G}_p^x) = \bar{a}$ then the group $\mathcal{G}_p^x$ can be captured by taking that liberty. Let then

$$\mathcal{D}_0 = \{\mathcal{G}_p^x : p \in G(p_0), \ p = (f, a), \ Color(\mathcal{G}_p^x) = \bar{a} \text{ and } \|\mathcal{G}_p^x\| = 1\}.$$

be the set of groups which can be captured in one move. For any $i \geq 1$, we define recursively the sets $\mathcal{D}_i$ of groups that can be captured in at most $i+1$ moves. For this, let us note that Go players have different points of view on their groups. Thus, the player a who has a group $\mathcal{G}_p^x$, will convince himself that the group is dead if for all his possible moves he will get a dead group, and the opponent $\bar{a}$ needs to find only one move to kill it. This leads to the following definition

$$\mathcal{D}_i = \mathcal{D}_{i-1} \cup \{\mathcal{G}_p^x : p \in G(p_0), \ p = (f, a), \ Color(\mathcal{G}_p^x) = \bar{a}$$
$$\text{and } \mathcal{G}_{p'}^x \in \mathcal{D}_{i-1} \text{ for some } p' \in \Psi(\{p\})\}$$
$$\cup \{\mathcal{G}_p^x : p \in G(p_0), \ p = (f, a), \ Color(\mathcal{G}_p^x) = a$$
$$\text{and } \mathcal{G}_{p'}^x \in \mathcal{D}_{i-1} \text{ for every } p' \in \Psi(\{p\})\}.$$

Now, we are ready to define the set of all *dead groups* by

$$\mathcal{D} = \bigcup_{i \in \mathbb{N}} \mathcal{D}_i$$

and the set of *living groups* by

$$\mathcal{V} = \{\mathcal{G}_p^x : p \in G(p_0),\ \mathcal{G}_p^x \notin \mathcal{D}\}\ .$$

Since $\mathcal{P}$ is finite and the set of groups in a certain position is also finite, it follows that $\mathcal{D}$ is finite. This implies that there is an integer, call it α, for which $\mathcal{D}_\alpha = \mathcal{D}_{\alpha+1}$, therefore, by the definition of $\mathcal{D}$ it follows that $\mathcal{D} = \mathcal{D}_\alpha$. For α we may take the trivial upper bound $2|\mathcal{T}|3^{|\mathcal{T}|}$.

The following is an algorithm that analyses if a group can be killed or not.

```
procedure Can_kill(p : G(p₀), x : T, depth : N) : boolean
var w:boolean; Y ∈ 2^G(p₀);
begin
  if Gₚˣ ∈ D₀ then return TRUE;
  if depth > α then return FALSE;
  if (p = (f, a) AND Color(Gₚˣ) ≠ a)
    w =FALSE; Y = Ψ(p);
    while (Y ≠ ∅ AND NOT w)
      p' ∈ Y; Y = Y \ {p'};
      w = w OR Can_kill(p', x, depth + 1);
    endwhile
    return w;
  else
    w =TRUE; Y = Ψ(p);
    while (Y ≠ ∅ AND w)
      p' ∈ Y; Y = Y \ {p'};
      w = w AND Can_kill(p', x, depth + 1);
    endwhile
    return w;
  endif
end.
```

The next theorem proves that this algorithm is correct.

Theorem 3. $\mathcal{G}_p^x \in \mathcal{D}$ iff $Can\_kill(p, x, 0) = \text{TRUE}$.

Proof. Since trivially $|G(p_0)| < 2 \cdot 3^{|\mathcal{T}|}$, the algorithm stops eventually. By induction on i we immediately see that $\mathcal{G}_p^x \in \mathcal{D}_i$ implies that $Can\_kill(p, x, 0) = \text{TRUE}$, and by the definition of $\mathcal{D}$ it follows that the same is true for $\mathcal{D}$.

Conversely, let $Can\_kill_n(p, x, 0)$ be the above algorithm with α replaced by n. It is not difficult to prove by induction on n that $Can\_kill_n(p, x, 0) = \text{TRUE}$ implies $\mathcal{G}_p^x \in \mathcal{D}_n$ for any $n \geq 0$. This completes the proof of the theorem.

Let us remark that to check that $\mathcal{G}_p^x \in \mathcal{D}_0$ takes $O(|\mathcal{T}|)$ steps. The calculation of Ψ needs $O(|\mathcal{T}|)$ steps, too. We also mention that if the depth of the calculation is d then $O(d \cdot |\mathcal{T}|)$ is the memory space needed.

We conclude this section with some observations on kos. So far no restrictions on the cycling of positions was imposed in our mathematical model. Thus, since repetitions were allowed the set $\mathcal{V}$ of living groups is larger than in an usual game with the rule of ko.

For example, in Fig. 1 and 2, if White moves then his bigger groups are alive, although in a normal game the first is dead while the second is in an undetermined position and special extra rules apply. Since our main object in this work is to prove the complexity of a certain set of groups in which, as we will next see, no kos are involved, we don't go in further detail with the ko rule.

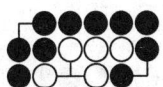

Fig. 1. A 2-kos position in which White lives.

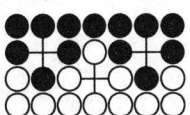

Fig. 2. In a 3-kos position both Black and White live.

4 Proof of Theorem 1

The proof has two parts. Firstly we show that GNP is a NP−problem and secondly we reduce the 3 SAT problem to the life and death problem of certain groups from G_{NP}. The theorem will then follow, since 3 SAT is NP complete, result proved by Cook in 1971.

We begin by showing that GNP belongs to NP. For this, let us consider a nondeterministic Turing machine which performs the following algorithm:

```
input  𝒢ₚˣ ∈ G_NP
guess  p₁,...,p_|T|  with p = p₁
check that p_{i+1} ∈ Ψ(pᵢ) for 2 ≤ i ≤ |T| − 1
if (Color(𝒢ₚˣ) = a AND |Ψ({f, a})| = 2)
        check that 𝒢ₚˣ ∉ 𝒟₀ for every pᵢ
        if this holds then return FALSE
else        // Color(𝒢ₚˣ) = a AND |Ψ({f, ā})| = 2
        check that 𝒢ₚˣ ∈ 𝒟₀ for some pᵢ
        if this holds then return TRUE
endif
end
```

The computation of this algorithm takes at most $O(|\mathcal{T}|^2)$ steps.

For the second part of the proof, we show how the 3 SAT problem can be reduced to the life and death problem of certain groups. We first state the 3 SAT problem. Let us consider the formula $F = c_1 \wedge \cdots \wedge c_n$ where $c_j = \tilde{u}_{i_1} \vee \tilde{u}_{i_2} \vee \tilde{u}_{i_3}$, $\tilde{u}_{i_k} \in \{u_{i_k}, \overline{u_{i_k}}\}$ for $1 \leq i \leq n$ and $1 \leq j \leq n$ and $\mathcal{U} = \{u_1, \ldots, u_m\}$ is the set of logical variables that appear in F. Then the 3 SAT problem is:

(3 SAT) *Decide whether F evaluates to true.*

Let $p(i)$ and $q(i)$ be the number of appearances in F of u_i and $\overline{u_i}$ respectively. For each i, $(1 \leq i \leq m)$ we construct the diagrams A(i) (see Fig. 3).

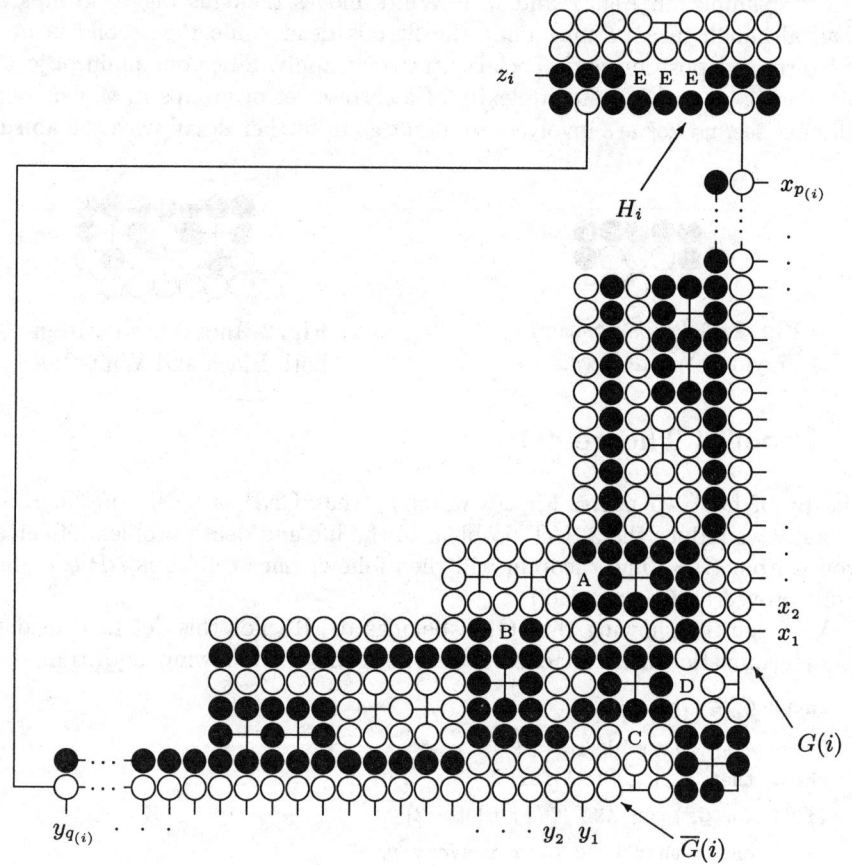

Fig. 3. Diagram A(i).

Remark 1. Suppose that in diagram A(i), White moves. Then:
1. $\|G(i)\| = 2$, $\|\overline{G}(i)\| = 2$ and $\|H_i\| = 3$.
2. If White plays A and B then H_i is dead.

3. If White plays A and Black plays C then $G(i)$ and H_i are alive and $\overline{G}(i)$ is dead.

4. If White plays B and Black plays D then $\overline{G}(i)$ and H_i are alive and $G(i)$ is dead.

For any $c_j = \tilde{u}_{i_1} \vee \tilde{u}_{i_2} \vee \tilde{u}_{i_3}$, $1 \leq j \leq n$, we associate a diagram B(j) (see Fig. 4). For any k, $1 \leq k \leq 3$, this diagram is connected with that from diagram A(i_k) as follows. If $\tilde{u}_{i_k} = u_{i_k}$ then $\tilde{G}(i_k)$ is $G(i_k)$, the connection being through x_l for some l, $1 \leq l \leq p(i_k)$. If $\tilde{u}_{i_k} = \overline{u_{i_k}}$ then $\tilde{G}(i_k)$ is $\overline{G}(i_k)$, the connection being through y_l for some l, $1 \leq l \leq q(i_k)$.

Remark 2. In diagram B(j) the following hold true:
1. $\|\tilde{G}(i_l)\| = 2$ and $\|L(i_l)\| = 3$ for $1 \leq l \leq 3$.
2. The group $K(j)$ is alive iff the group $\tilde{G}(i_l)$ alive for some l, $1 \leq l \leq 3$.

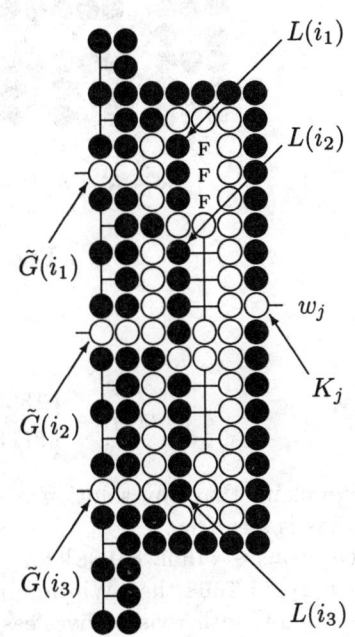

Fig. 4. Diagram B(j).

Finally, all the previous diagrams are interconnected also with diagram C (see Fig. 5). Thus, all the groups H_i, are connected to those from diagram A(i) through z_i for $1 \leq i \leq m$. Similarly, all the groups K_j, are connected to those from diagram B(j) through w_j for $1 \leq j \leq n$.

Remark 3. Suppose that in diagram C, White moves. Then we have:
1. $\|\Lambda\| = \|\Omega\| = 2$.
2. The group Λ is alive iff either H_i is dead for some i, $1 \leq i \leq m$ or K_j is alive for all j, $1 \leq j \leq n$.

The construction of all these diagrams needs $O(n)$ steps, so the reduction is made in a polynomial number of steps.

Remark 4. (Easy life). For White, an easy way to live with Λ is by moving A (or B) in diagram A(i) and Black plays poorly by failing to respond with C (or D) in the same diagram. This allows White to occupy both A and B and then, by 2 of Remark 1 H_i is dead, so Λ is alive by 2 of Remark 3.

Let us now see that the 3 SAT problem is equivalent to the life and death problem of the group Λ from diagram C. In the position given by the above construction suppose that White moves. We denote by $v(u_i) \in \{\text{TRUE}, \text{FALSE}\}$ any assignation of the logical variables u_i, $1 \leq i \leq m$. The correspondence between our construction and $v(u_i)$ is given by the convention:

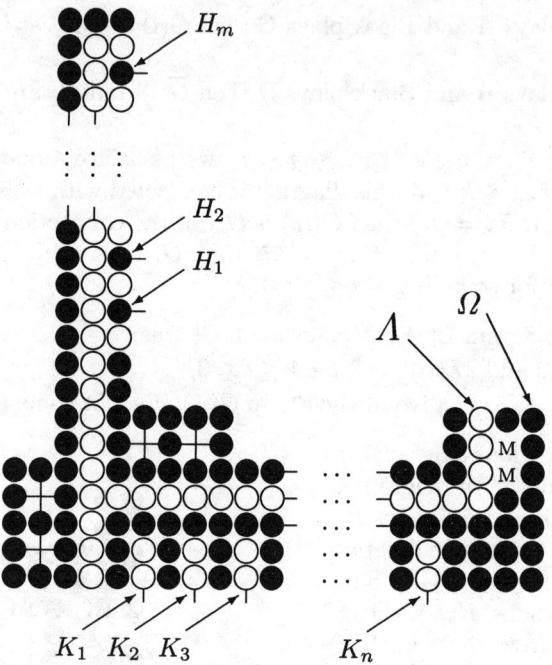

Fig. 5. Diagram C.

$$v(u_i) = \begin{cases} \text{TRUE} & \text{if } G(i) \text{ is alive} \\ \text{FALSE} & \text{if } \overline{G}(i) \text{ is alive} \end{cases} \quad (1)$$

We claim that any solution of the 3 SAT problem makes Λ alive and the converse is also true.

Let $v(u_i) \in \{\text{TRUE}, \text{FALSE}\}$ for $1 \leq i \leq m$ be an assignation which validates F. If $v(u_1) = \text{TRUE}$ then White moves in A in diagram $A(1)$ so Black is forced to respond C (otherwise Λ lives easily, confer Remark 4). If $v(u_1) = \text{FALSE}$ then White moves in B in diagram $A(1)$ so Black is forced to respond D (otherwise Λ lives easily, confer Remark 4). White will then turn to the subsequent diagrams for $i = 2, \ldots, m$ and play in the same way, giving Black only unique-choice moves. Playing in this manner yields

$$v(\tilde{u}_i) = \text{TRUE} \quad \text{iff} \quad \tilde{G} \text{ is alive for } i = 1, \ldots, m \ . \quad (2)$$

Let us also note that White moves make K_j alive for $j = 1, \ldots, n$. For if K_j is dead for some j, then by 2 of Remark 2 the groups $\tilde{G}(i_l)$ from diagram $B(j)$ are all dead. By (2) it follows that $v(\tilde{u}_{i_l}) = \text{FALSE}$ for $l = 1, 2, 3$, so c_j is FALSE which implies that F is FALSE, contradicting our assumption. Since K_j are alive for $j = 1, \ldots, n$, by 2 of Remark 3 it follows that Λ is alive.

Conversely, suppose that Λ is alive. As we observed before in Remark 4, Λ can live easily if Black plays poorly. Suppose that Black plays well. Then by 2 of Remark 3, the only way for White to live with Λ is to live with all K_j for $j = 1, \ldots, n$. By 2 of Remark 2, K_j lives iff $\tilde{G}(i_l)$ lives for some $l = 1, 2, 3$. But $\tilde{G}(i_l)$ lives iff White plays in diagram $A(i_l)$ in A if $\tilde{G}(i_l) = G(i_l)$ and B if $\tilde{G}(i_l) = \overline{G}(i_l)$. Therefore, by the construction of $B(j)$ and 3, 4 of Remark 1, the only moves for White which make Λ alive are A or B in diagram $A(i)$ for $i = 1, \ldots, m$ and since Black plays well, he is forced to respond with C or D. Let us also note that the order of diagrams $A(i)$ in which White makes the moves is not important, since the result (after his m moves) is the same.

With the convention (1) White moves give values to the variables u_i for $i = 1, \ldots, m$. Now, because K_j lives for all $j = 1, \ldots, n$, it means that c_j is true for $j = 1, \ldots, n$, which implies that F is TRUE. This concludes the proof of our claim.

It remains to show that Λ is in G_{NP}. In the above discussion, we saw that if White wants to live with Λ he has to move A or B in diagram $A(i)$ for $i = 1, \ldots, m$ and we observed that the order of i's is not important. Moreover, trying to kill Λ, Black had unique choice-responses. Thus, a pruning function with the required properties exists. Note also that to check that Λ lives after White has played his first m moves takes only $O(n)$ steps (in every diagram $B(j)$ it is enough to check if one of the groups $\tilde{G}(i_l)$ for $l = 1, 2, 3$ has 2 liberties), and the theorem is proved.

References

1. Lichtenstein, D. and Sipser, M., GO is Polynomial-Space Hard, Journal ACM, Vol. **27**, No. 2, (April 1980) 393-401.
2. Morris, F.L., Playing Disjunctive Sums is Polynomial Space Complete, Int. Journal Game Theory, Vol. **10**, No.3-4, (1981) 195-205.
3. Robson, J., The Complexity of Go, Proc. IFIP (International Federation of Information Processing), (1983) 413-417.
4. Yedwab, L., On Playing Well in a Sum of Games, Master Thesis, MIT/LCS/TR-348, 1985.
5. Moews, D.J., On Some Combinatorial Games Connected with Go, Ph.D. Thesis, University of California at Berkeley, 1993.
6. Berlekamp, E., Wolfe, D. Mathematical Go: Chilling Gets the Last Point, A. K. Peters, 1994.

Extended Thermography for Multiple Kos in Go

William L. Spight

Oakland, CA 94609, USA
billspight@aol.com

Abstract. In evaluating a local position in go, players want to know its current territorial count and the value of a local play. Many go positions are combinatorial games; the mean value of the game corresponds to the count and its temperature corresponds to the value of the play. Thermography finds the mean value and temperature of a combinatorial game. However, go positions often include kos, repetitive positions which are not classical combinatorial games. Thermography has been generalized to include positions containing a single ko. This paper extends thermography further to include positions with multiple kos. It also introduces a method for pruning redundant branches of the game tree.

Keywords: go, ko, multiple kos, thermography

1 Introduction

Thermography finds the mean values and temperatures of combinatorial games. Classical thermography fails for kos, which have cyclical game graphs. Berlekamp [1] breaks the cycle by allowing one player, the komaster, to win the ko. But when more than one ko is active at one time, no player can win them all, as a rule.

The method presented here converts the game graph to a pair of and/or trees. As long as the ko rules prevent the ko from continuing interminably, the trees are finite. Then we may apply thermography to the trees.

1.1 Definition of the Thermograph

The thermograph of a combinatorial game, G, shows, for each temperature, t, the Left and Right scores of G cooled by t [2]. (By convention, in go Black is Left and White is Right.) The Left (Right) score of a game is the result of alternating minmax play if player Left (Right) plays first. The set of all (Left score (Right score), temperature) pairs is the Left (Right) wall of the thermograph. A game is cooled by t by imposing a tax equal to t on each play. Instead of cooling we shall make use of the relation between the thermograph and minmax play in a universal enriched environment (UEE) [1].

A UEE is a sum of simple games (called *switches*) of the form $\{v|-v\}$, where Black can play to a position worth v points and White can play to a position worth $-v$ points (for Black). (Switches {0|0} are familiar to go players. They call them *dame*.) The temperature of {v|–v} is v. The temperature, t, of the UEE is the temperature of its hottest component. To construct a UEE, $U_{n,t}$, let

$$\delta_n = 1/\text{lcm}\{1, 2, ..., n\} . \quad (1)$$

Let $U_{n,t}$ consist of $(2n+1)$ switches at each temperature 0, δ_n, $2.\delta_n$, ..., t. Let n be large enough that the Left score of $U_{n,t}$ is

$$\text{LS}(U_{n,t}) = t/2 , \quad (2)$$

the Right score is

$$\text{RS}(U_{n,t}) = -t/2 , \quad (3)$$

the Left value of the thermograph of game G at temperature t is

$$L(G_t) = \text{LS}(G + U_{n,t}) - \text{LS}(U_{n,t}) , \quad (4)$$

and correspondingly the Right value is

$$R(G_t) = \text{RS}(G + U_{n,t}) - \text{RS}(U_{n,t}) . \quad (5)$$

Defining the walls of the thermograph by minmax considerations allows us to generalize thermography to include multiple kos.

1.2 Kos and Thermographs

Kos are go positions which potentially repeat. Since repetition of board positions in go may lead to hung games, different versions of ko rules prevent some or all repetitions. For any version of ko rules we may draw a thermograph for a position containing one or more kos as long as the rules prohibit a hung game.

A board position in which at least one of the options is prohibited by the ko rules is called ko-banned [1]. In a ko-banned position minmax play may be in the environment. The lines of a classical thermograph reflect only alternating local plays, but the lines for ko thermographs must be able to reflect plays in the environment as well. Let us call a thermograph which reflects plays in the environment an extended thermograph (cf. [2]).

2 Deriving Thermographs

By convention the axes of a thermograph are rotated counterclockwise 90°. The temperature, t, is plotted on the vertical axis with positive values above the origin. The score for Left (Black), v, is plotted on the horizontal axis with positive values to the left of the origin. Each thermographic line represents a mast. If a go position is terminal, its thermograph is a vertical mast with the equation, $v = m$, the local score. In the extended thermograph, if a sequence of play from the original position reaches a terminal position with a score of m, the equation of the corresponding thermographic line is $v = m + (w-b)\, t$, where w is the number of local plays in the sequence by White, and b is the number of local plays in the sequence by Black.

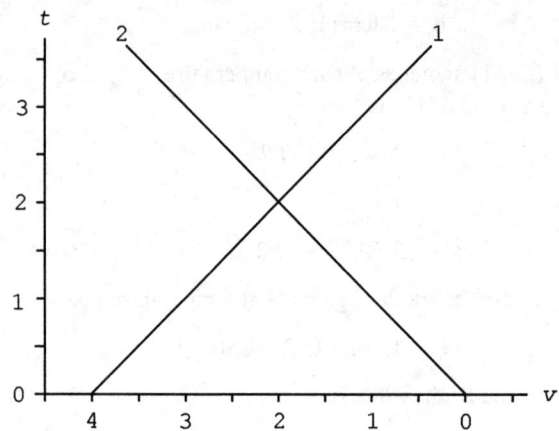

Fig. 1. Scaffolds for game $\{4\mid 0\}$

In the combinatorial game $\{4\mid 0\}$, Black can play to a position worth 4 points and White can play to a position worth 0. Figure 1 shows the thermographic lines which represent plays in that game, line (1) when Black plays first and line (2) when White plays first. Line (1) is called the Left scaffold; line (2) is the Right scaffold [1]. Their equations are

$$v = 4 - t \tag{1}$$

$$v = t. \tag{2}$$

The scaffolds intersect at temperature $t = 2$. When $t < 2$, each player will prefer to play in the game rather than in the environment, and the scaffolds represent the Right and Left walls at t.

But when $t > 2$, minmax play for each player is in the environment. Minmax play at t is not sufficient to determine the Right and Left walls. When $t = 3$, for instance, it only tells us that the Left wall is greater than 1 and the Right wall is less than 3.

However, minmax play through a range of temperatures, in this case a drop from temperature 3 to 2, will establish the Right and Left walls at temperature 3. Since each player plays in the environment, the gain of the first player between those temperatures is the same amount as it would be in the environment alone, without the game. The Left or Right wall at temperatures greater than 2 is the same as it is at temperature 2.

When $t \geq 2$ the Left and Right walls coincide at $v = 2$. The coincident walls form a mast with equation

$$v = 2. \tag{3}$$

The temperature of the game is 2 and its mean value is 2. Figure 2 shows the thermograph of $\{4 \mid 0\}$.

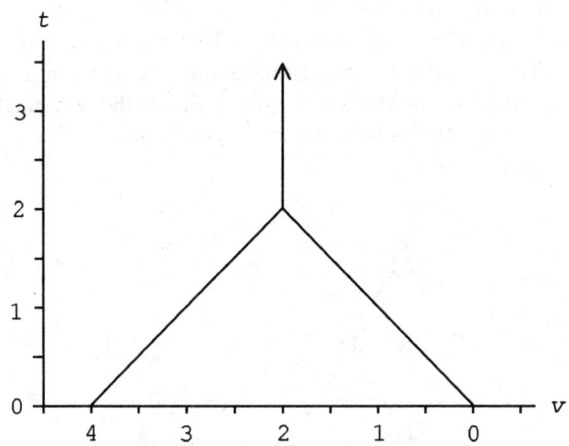

Fig. 2. Thermograph of game $\{4 \mid 0\}$

Note:
- Plays in the environment are not sufficient to determine thermographic values.
- The value of a simple mast, where both players prefer to play in the environment, is determined at its base.

(Some kos have more complicated masts. Berlekamp [1] covers this important topic.)

In this case, which is typical, the intersection of the Left and Right scaffolds determines the mast. It may also happen that the walls do not intersect, even at the lowest temperature. An example is the game $\{-6 \mid 8\}$. Then the game is terminal, and its value may be different under different rules. Its mast value is just its terminal value under the rules in use.

(As a combinatorial game, the value of $\{-6 \mid 8\}$ is 0. Go players recognize this game as a *seki*, and give it a territorial value of 0. Territory counting, as under Japanese rules, typically coincides with the value of such a game in combinatorial game theory.)

In theory we can determine any thermograph by finding the Left and Right values at each temperature for each line of play and eliminating the suboptimal ones. When minmax play for each player is in the environment, however, we need not consider all possible line of play. We may simply assume that each player plays in the environment until it is wrong to do so.

Since simple masts are determined at their bases, a typical method of drawing thermographs is to derive them bottom up, from the thermographs of their followers. That approach will not work with kos, however, unless modified, since ko positions ultimately follow themselves.

2.1 Thermographs for And/Or Trees

In combinatorial game trees Left branches represent options for Left and Right branches represent options for Right. By contrast, in and/or game trees the branches represent options for the player with the move. This extension of thermography utilizes and/or trees. To convert a combinatorial game tree to an and/or game tree, we make plays in the environment explicit. Figure 3 shows the combinatorial game tree for the game {2 || 0 | –6} on the left, and its corresponding and/or game trees on the right.

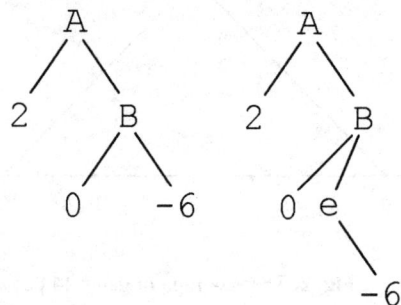

Fig. 3. Two kinds of game tree

There are actually two trees on the right, depending on who plays first from the root. The Left tree represents Black's play to a position worth 2 points. In the Right tree White can play to B. From B Black can play to 0 or play in the environment. The 'e' indicates a play in the environment, which leaves the game node the same, but changes who has the move. From B, White can play to –6.

This and/or tree representation leaves out many lines of play. It ignores inferior plays in the environment, but they are dominated. It also ignores sequences of plays in the environment. After White plays to B and Black plays in the environment, it might be correct for White to play in the environment, too. In fact, if Black's play was correct, White should almost certainly play in the environment.

But in that case, there is a mast at B at the current temperature. Such lines of play give no thermographic information. That is why we can prune them.

Black's environmental option establishes the fact that there is a mast at B. The Left and Right options at B carry all the information needed to find that mast. That fact leads to our basic principle of pruning:

Ignore environmental plays at established masts.

The first 'e' establishes the mast at B. No further 'e's are allowed from B. We also disallow environmental plays from the root. They would also be meaningless, since either player can play from the root.

How can we construct the thermograph of the and/or tree? The process is essentially the same as with combinatorial game trees. We work bottom up from the followers.

To reach −6 from the root there are 2 plays in the game (local plays) by White, none by Black. That gives us the Right scaffold

$$v = -6 + 2t. \qquad (1)$$

To reach 0 there is one local Black play and one local White play, which gives us the Left scaffold

$$v = 0. \qquad (2)$$

The 'e' tells us that there is a mast at B. Since it and 0 are both Left options from B, we know that we use the Left wall of the thermograph of B in deriving the final thermograph. The mast of B has the equation

$$v = -3 + t. \qquad (3)$$

Of course, the mast of B would be vertical in its own thermograph, but not in the thermograph of A. Figure 4 shows the construction of the Left wall of B, which is indicated by thickness.

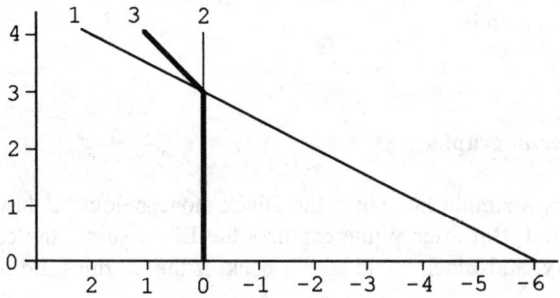

Fig. 4. Left wall of B

Now we can construct the thermograph for A. Its Right scaffold is the Left wall of B, and its Left scaffold is the line

$$v = 2 - t. \qquad (4)$$

Figure 5 shows the construction of the thermograph of A. The thick lines form the thermograph.

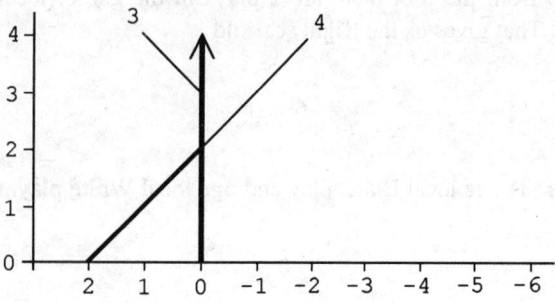

Fig. 5. Thermograph of $\{2 \;\|\; 0 \;|\; -6\}$

The scaffolds intersect at a temperature of 2 and a score of 0, the temperature and mean value of the game. The mast is the line

$$v = 0 \,. \tag{5}$$

Playing through a Node. If White plays first from A at temperature 2, Black's correct response is to play to 0 immediately. White plays through B. In go parlance, White's play is *sente*. If we knew that White played through B, we could dispense with the calculation of the mast of B. The thermograph does not depend on it.

2.2 Ko Thermographs

Figure 6 shows a simple ko. Since the Black stone below 'a' has only one liberty, White can take it. But after White captures the Black stone, the capturing stone has only one liberty, and Black could take it back, if the ko rules did not prohibit the re-capture.

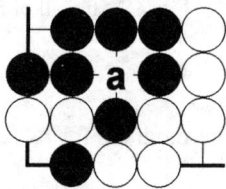

Fig. 6. Ko

If Black captures the two White stones, the score is 5; if White captures the Black stone and later fills, the score is –4. Figure 7 shows the game graph of this ko, on the left. The original position is labeled 'A' and the position after White takes is labeled 'B'. The U-curve between A and B indicates the ko.

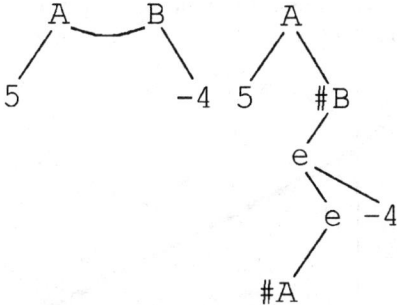

Fig. 7. A simple ko graph and its trees

To draw the thermograph of A we derive Left and Right and/or game trees from the game graph and the ko rules. In Figure 7 the trees for this graph are on the right. The Prolog program in the appendix converts a game graph to its and/or trees.

In the Right tree White plays from A to #B. The '#' indicates that there is a ko ban at B. A ko-banned node does not form a mast (unless it is terminal), because a play in the environment removes the ko ban, and that changes the game. (If a position is terminal, there is no environment left, not even a *dame*.)

Next Black plays in the environment. Now White has two options, one to −4, and another environmental play. Since the first environmental play removed the ko ban, the second one establishes a mast at B. Next Black plays to #A, where the tree stops.

That seems unusual, to say the least. But after Black plays to #A, an environmental play would simply return to a position equivalent to one after an initial environmental play from the root, and yield no thermographic information. (Note that, because of the intervening environmental plays, it would play to a position equivalent to that after 3 alternating environmental plays from the root.)

Working bottom up, we first evaluate #A. It has a value only if it is terminal, so we assume that it is. Since White has no play, we might evaluate it the same as the combinatorial game $\{5 \mid \} = 6$. Different go rules evaluate terminal ko positions differently. The value of 6 is consistent with territorial counting under American Go Association rules. Having no useful play, not even a *dame*, White would have to fill in a point of his own territory, sacrifice a stone, or pass and surrender a stone; each choice would cost 1 point. In this paper I shall use AGA territorial rules.

The scaffolds for B have the equations

$$v = 6 \tag{1}$$

$$v = -4 + 2t. \tag{2}$$

They intersect at temperature 5 to produce the mast

$$v = 1 + t. \tag{3}$$

Since White plays from B, we use the Right wall of the thermograph for deriving further thermographs. Figure 8 shows the graphical derivation of the Right wall of B.

Note that B would have a quite different thermograph if it were the root. That is typically the case for subsidiary ko nodes in the and/or tree.

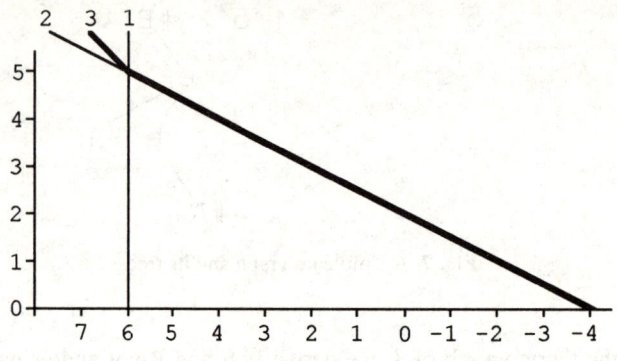

Fig. 8. The Right wall of B

As a ko-banned node, #B does not form a mast. The Right wall of B is its Left wall.

We are now ready to form the thermograph for A. Its Right scaffold is the Left wall of #B and its Left scaffold is the line

$$v = 5 - t. \tag{4}$$

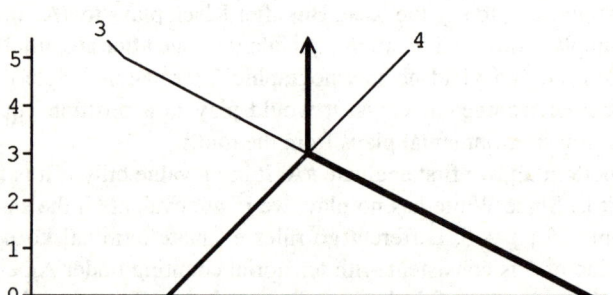

Fig. 9. Simple ko thermograph

Figure 9 shows the derivation of the thermograph. The ko has a temperature of 3 and a mast value of 2. (The mast value and mean value of a ko do not necessarily coincide.)

The value of #A made no difference. That is typical, as it must be at least as good for Black as A. White's environmental play from B is senseless.

2.3 Pruning Redundant Branches

To eliminate redundancies in the and/or game tree, we disallow sequences in which an environmental play follows the root node or an environmental play returns to an established mast.

An environmental play establishes a mast for all previous nodes which are not ko-banned. The play sequence of nodes $A...e...A\ e$ is equivalent to $A\ e\ e$ if the number of local Black and White plays are the same between the two occurrences of A. (It is possible to return to the same board position after an unequal number of Black and White plays. But then the net number of captured stones differs, as well. Such positions should not be considered the same node in the game graph.)

An environmental play cannot establish a mast for a ko-banned node, because it destroys the ko ban. A ko-banned node may have a mast only if it is terminal. Its terminal value depends on the rules.

In the Prolog program the predicate mast checks whether the mast for a node has been established:

```
mast([Root], Root)       :- !.
mast([e | History], Node) :- !, mast1(History, Node).
mast([_ | History], Node) :- mast(History, Node).

mast1([Node | _], Node)   :- !.
mast1([e | History], Node) :- !, mast2(History, Node).
mast1([_ | History], Node) :- mast1(History, Node).

mast2([Node | _], Node)   :- !.
mast2([# Node | _], Node) :- !.
mast2([_ | History], Node) :- mast2(History, Node).
```

History is the ancestor list of Node in reverse order. Searching back, if mast finds no play in the environment ('e'), a mast is established only for the root. If it finds one 'e', a mast is established for any previous occurrence of Node which is not ko-banned. If it finds a second 'e', a mast is established for any previous occurrence of Node.

2.4 Double Ko Life

In Figure 10 Black is alive in double ko. Let us call this position J. White can play at 'a' to position D, threatening to take Black's stones. But Black can then play at 'b' to position K, and White is banned from taking back. Figure 11 shows the double ko graph and tree.

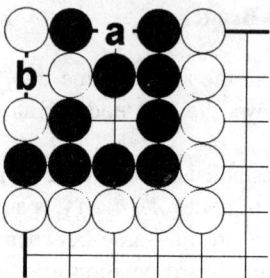

Fig. 10. Double ko life

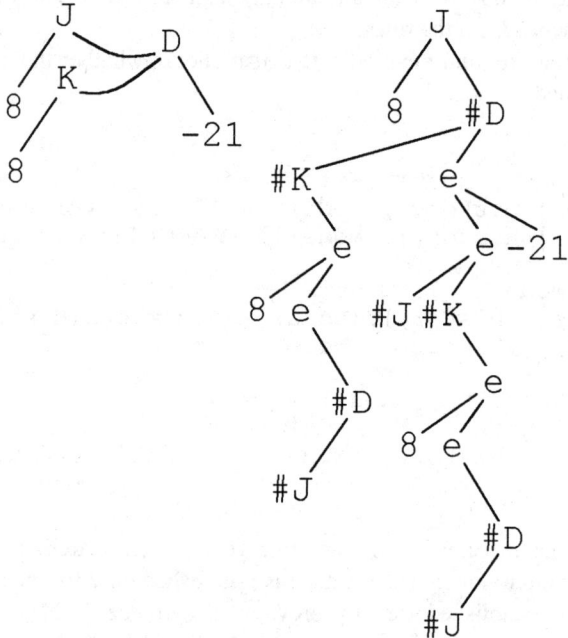

Fig. 11. Double ko graph and tree

To derive the thermograph, let us start in the bottom right of the tree. First we evaluate the ko-banned node, #J, as terminal. $\{8 \mid \} = 9$ in combinatorial game theory, and both AGA and Japanese rules agree. We now find the thermograph for K. The Right scaffold is

$$v = 9. \tag{1}$$

The Left scaffold is

$$v = 8 - t. \tag{2}$$

Figure 12 shows the derivation.

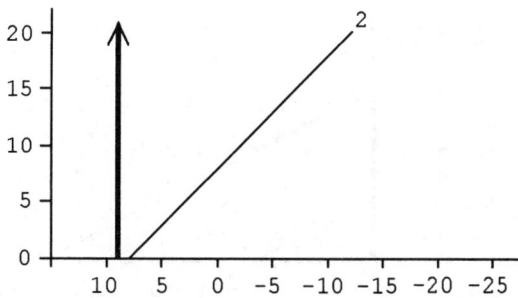

Fig. 12. Thermograph of K

Since the Left scaffold lies to the Right of the Right scaffold, this is a terminal position. As with #J, the terminal value is 9. The mast of K is the Right wall of #K.

From D Black can play to #K or #J. Each one has a value of 9. The Left scaffold of D is equation (1). The Right scaffold is

$$v = -21 + 2t. \qquad (3)$$

In Figure 13 we find the Right wall of D.

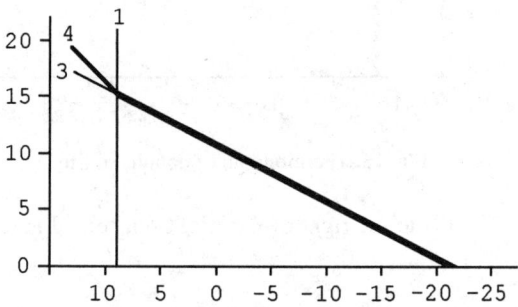

Fig. 13. Right wall of D

The mast of D is

$$v = -6 + t. \qquad (4)$$

The thermograph of #D has no mast. Since Black has the move, it is the maximum of the Right walls of D and #K. The tree below this #K node is the same as below the other #K node, so the thermograph is the same: the vertical line at 9.

In Figure 14 we find the thermograph of #D.

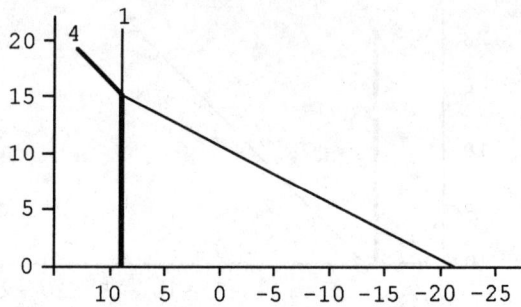

Fig. 14. Thermograph of #D

Below temperature 15 line (1) dominates; above that line (4) does.

Now we can find the thermograph of J. Its Right scaffold is the thermograph of #D. Its Left scaffold is line (2). Figure 15 shows the derivation.

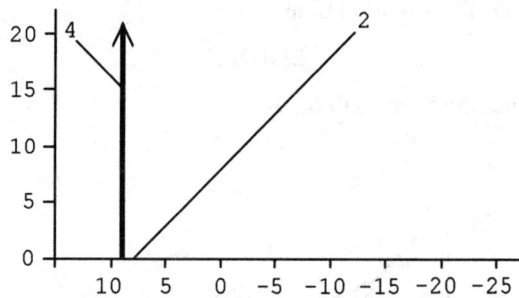

Fig. 15. Thermograph of double ko life

Again the Left scaffold is to the right of the Right scaffold. J is terminal, with a mast value of 9.

2.5 Molasses Ko

Molasses ko (Figure 16) is a curiosity, reported only once in go history [3]. If neither player has a sufficiently large ko threat, a molasses ko may slow the game down, causing it to drag on and on. If White has the move in position A, he plays at W1, Black takes the ko at B2, White takes two stones at W3, and Black takes the ko at B4, reaching position F, and then White plays elsewhere. Then Black continues the ko in similar fashion, and four moves later returns to A. The player who starts the molasses ko by threatening two stones is the one who must find a ko threat to win it. If neither player can win it, the game slows down, with one play elsewhere for every four plays in the ko.

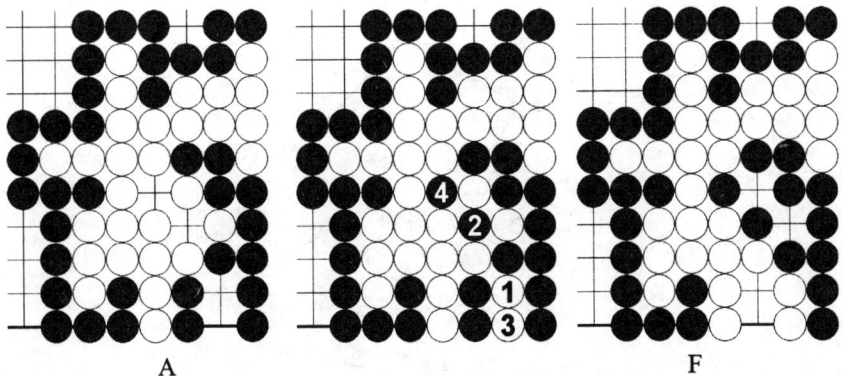

Fig. 16. Molasses ko

What happens when neither player wishes to play elsewhere? That depends on the rules. If the first player to pass is at a disadvantage, the molasses ko is truly interminable. Both players fill in territory until that results in the sacrifice of a group. But under AGA rules both players can safely pass, and the molasses ko remains on the board as a *seki*, worth 0 points of territory.

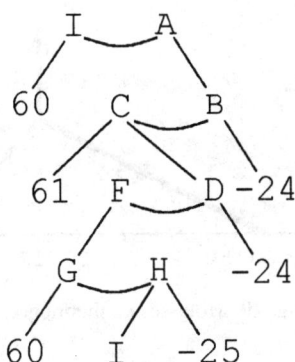

Fig. 17. Molasses ko game graph

In analyzing this ko, it helps to realize that we play through B, C, G and H (Figure 17). From A it takes Black two moves to capture White for 60 points. From C it takes only one move for 61 points. If White does not continue to D, he should not have played to B in the first place. A similar argument applies to F and G. By not finding masts for those nodes we can save ourselves some trouble. Then we get the game tree in Figure 18.

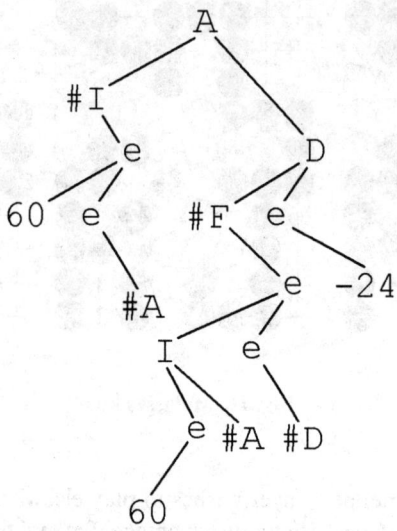

Fig. 18. Molasses ko game tree

This tree yields the thermograph in Figure 19.

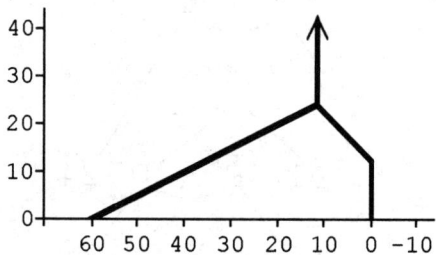

Fig. 19. Molasses ko thermograph

The temperature of the molasses ko is 24 and its mast value is 12. The Right wall shows that when $t < 12$ and White plays first, Black will reply and set the molasses ko in motion. When $t > 12$ Black will respond in the environment.

3   Comparison with Generalized Ko Thermography

Players frequently contest a ko by making threats which the opponent must answer, and then taking the ko back. With correct play, the player who has enough sufficiently large threats can win the ko.

Berlekamp [1] utilizes the concept of komaster to extend thermography to ko positions. The komaster of a ko may win it, but, once she has taken the ko to repeat the global position, she must make another local move immediately. (Berlekamp's economist rules permit two consecutive plays by the same player.)

The komaster rule retains desirable properties of classical thermography:
- For a sum of games which does not include multiple kos which are active at the same time on the same board, the mast value of the sum is the sum of the mast values.
- The temperature of the sum is at most the temperature of its hottest summand.

This extension retains neither of these properties. However, by making ko threats explicit, it allows the exploration of positions which lie between the extremes of komaster, and of intricacies of the relationships among threats and kos.

3.1 Modeling Komaster

Berlekamp [1] discovered that some ko positions are hyperactive; the mast value depends on who is komaster. Whether a position is hyperactive is important information. It is desirable that this extension model komaster conditions.

However, with multiple kos active at the same time, it is impossible for one player to be the komaster for all of them. If she takes one, her opponent is free to take another. We can model only a weaker version of komaster.

For any given ko-banned position, the komaster is allowed to break the ko ban only once, after which she loses komaster status. She pays no price for this privilege, although it is usually correct to continue locally if the opponent responds in the environment.

Since this method makes all ko threats explicit, we use a dummy threat for each position to make a player komaster. We assume that the opponent must answer the threat, and that it evaluates to 0 in a position with no ko.

After the komaster uses the dummy threat, it is impossible to return to any previous node, since each one contained the threat. Thus when the komaster takes the ko back, her opponent may play in the environment.

4 Summary

By determining thermographs through reference to minmax play in a universal enriched environment, we may extend thermography to positions with multiple kos. From the game graphs of ko positions we derive a pair of and/or trees and apply thermography to them.

While this extension sacrifices some valuable properties of classical thermography, it allows us in theory to evaluate any go position which does not lead to a hung game, and to explore relations among kos and threats.

Appendix: Prolog Program for Game Tree Conversion

```
/************************************************************
 * This program converts a combinatorial game graph          *
 * to a pair of and/or trees                                 *
 ************************************************************/

:- op(600, fx, #).
:- op(200, xfx, ::).
:- op(700, xfx, <::).
:- op(700, xfx, ::>).

/************************************************************
 * graph_tree(Node, Graph, LeftTree, RightTree)              *
 * converts a game graph, Graph, to a thermographic tree     *
 * with Node as the root. The tree consists of two left-     *
 * right (and/or) trees, LeftTree and RightTree, which       *
 * indicate alternating play.                                *
 ************************************************************/

graph_tree(Node, Graph, LeftTree, RightTree) :-
   graph_tree(left, Node, Node, [], Graph, LeftTree),
   graph_tree(right, Node, Node, [], Graph, RightTree).
   % The empty list ([]) is the history (ancestor list) of
   % Node.
graph_tree(_, Node, _, _, _, Node) :- number(Node), !.
   % If Node is a number, it is the Tree.
graph_tree(Direction, Node, Root0, History, Graph, Tree)
   :- !, followers(Direction, Node, Graph, Followers),
   % Find the immediate Followers in Direction from Node
   % in Graph.
   kobanned(History, Root0, Root, Followers, Children),
   % The provisional Root (Root0) and Followers map to
   %the Root and its Children.
   other_way(Direction, OppositeDir),
   branches(OppositeDir, Node, [Root | History], Graph,
      Children, Branches),
   % Each Child is the root of a sub-tree, Branch.
   tree(Direction, Root, Branches, Tree).
   % Tree is derived from the Root, the Direction, and the
   % Branches.

/************************************************************
 * followers(Direction, Node, Graph, Followers) finds        *
 * the immediate Followers in Direction from Node in         *
 * Graph. Graph is a list of Nodes and Followers of the      *
 * form Node - LeftFollowers :: RightFollowers.              *
 ************************************************************/

followers(_, _, [], []) :- !.
followers(left, Node, [Node - LeftFollowers :: _ | _],
```

```
        LeftFollowers) :- !.
followers(right, Node, [Node - _ :: RightFollowers | _],
    RightFollowers) :- !.
followers(Direction, Node, [_ | Nodes], Followers) :-
    followers(Direction, Node, Nodes, Followers).

/**************************************************
* kobanned(History, Root0, Root, Followers, Children)  *
* checks History to select Children which are not ko-  *
* banned from among Followers. In case of a ko ban     *
* Root = # Root0. Any ko ban depends on the ko rules   *
* used. This version uses the positional super ko rule,*
* which prohibits any repetition of a whole board      *
* position.                                            *
***************************************************/

kobanned(_, e, e, Followers, Followers) :- !.
    % An environmental play (e) removes any ko ban.
kobanned(_, Root, Root, [], []) :- !.
kobanned(_, _, _, [], []) :- !.
kobanned(History, Root0, Root, [Follower | Followers],
    Children0) :- repeated(History, Follower, YesNo),
  map(YesNo, Root0, Root, Follower, Children0, Children),
  kobanned(History, Root0, Root, Followers, Children).

/**************************************************
* repeated(History, Follower) searches for Follower in *
* History, which is in reverse order. If Follower      *
* occurs before an environmental play (e)is encoun-    *
* tered, there is a ko ban.                            *
***************************************************/

repeated([], _, no) :- !.
repeated([e | History], Follower, YesNo) :- !,
    repeated1(History, Follower, YesNo).
repeated(_,Follower, no) :- number(Follower), !.
repeated([Follower | _], Follower, yes) :- !.
repeated([# Follower | _], Follower, yes) :- !.
repeated([_ | History], Follower, YesNo) :-
    repeated(History, Follower, YesNo).

/**************************************************
* repeated1(History, Follower, YesNo)                  *
* ascertains the node at which an environmental play   *
* (e) was made. If it was Follower, then there is a ko *
* ban; otherwise there is not.                         *
***************************************************/

repeated1([e, # Follower | _], Follower, yes) :- !.
repeated1([# Follower | _], Follower, yes) :- !.
repeated1([Follower | _], Follower, yes) :- !.
repeated1(_, _, no) :- !.
```

```
map(no, _, _, Follower, [Follower | Children], Children)
  :- !.
map(yes, Root, # Root, _, Children, Children) :- !.
    % If there is a ko ban, the Root is marked and
    % the Follower is not a Child.

other_way(left, right) :- !.
other_way(right, left) :- !.

/************************************************************
* branches(Direction, Node, History, Graph, Children,       *
* Branches). There is a Branch for each Child. In           *
* addition, there is one for a play in the environment      *
* (e) if the mast for Node has not been established.        *
* On an established mast an environmental play is           *
* redundant.                                                *
************************************************************/

branches(_, Node, History, _, [], []) :-
  mast(History, Node), !.
branches(Direction, Node, History, Graph, [], [Branch])
  :- !,       % No mast established. Play in environment.
  graph_tree(Direction, Node, e, History, Graph, Branch).
branches(Direction, Node, History, Graph, [Child |
    Children], [Branch | Branches]) :-
  graph_tree(Direction, Child, Child, History, Graph,
    Branch),
  branches(Direction, Node, History, Graph, Children,
    Branches).

/************************************************************
* mast(History, Node) determines if the current Node is     *
* on an established mast. A mast is established if          *
*     1) Node is the Root, or                               *
*     2) there has been one environmental play (e)          *
*         between Node and an earlier occurrence, or        *
*     3) there have been two environmental plays between    *
*         Node and an earlier occurrence of # Node (ko-     *
*         banned).                                          *
************************************************************/

mast([Root], Root) :- !.
mast([e | History], Node) :- !, mast1(History, Node).
mast([_ | History], Node) :- mast(History, Node).

mast1([Node | _], Node) :- !.
mast1([e | History], Node) :- !, mast2(History, Node).
mast1([_ | History], Node) :- mast1(History, Node).

mast2([Node | _], Node) :- !.
mast2([# Node | _], Node) :- !.
mast2([_ | History], Node) :- mast2(History, Node).
```

```
tree(_, Root, [], Root) :- !.
tree(_, Root, [e], Root) :- !.
tree(left, Root, Branches, Root <:: Branches) :- !.
tree(right, Root, Branches, Root ::> Branches) :- !.
```

References

1. E. R. Berlekamp: The economist's view of combinatorial games. In: R. J. Nowakowski (ed.): Games of No Chance. Cambridge University Press, NYC (1996) ISBN 0-521-57411-0
2. E. R. Berlekamp, J. H. Conway, R. K. Guy: Winning Ways for your mathematical plays. Academic Press, NYC (1982) ISBN 0-12-091101-9
3. H. Fearnley: Some positions for a "bestiary" of go (baduk, weiqi). Web page: http://www.eng.ox.ac.uk/people/Harry.Fearnley/go/bestiary/molasses_ko.html (1997)

Computer Go: A Research Agenda

Martin Müller

ETL Complex Games Lab
Tsukuba, Japan
mueller@etl.go.jp

Abstract. The field of Computer Go has seen impressive progress over the last decade. However, its future prospects are unclear. This paper suggests that the obstacles to progress posed by the current structure of the community are at least as serious as the purely technical challenges. To overcome these obstacles, I develop three possible scenarios, which are based on approaches used in computer chess, for building the next generation of Go programs.

1 A Go Programmer's Dream

In January 1998, I challenged the readers of the *computer-go* mailing list [23] to discuss future directions for Computer Go:

> Assume you have unlimited manpower at your hand (all are 7-Dan in both Go and programming) and access to the fanciest state of the art computers. Your task is to make the strongest possible Go program within say three years. What would you do?

Many of the answers I received severely criticized all currently used approaches, and advocated the development of revolutionary new techniques. I do not think such wholesale criticism is justified. In this paper I take a close look at the current state of Computer Go, and propose a more systematic use of already available, proven techniques. I claim that this will already lead to substantial progress in the state of the art. Writing programs for Go has turned out to be much more complex than for other games. The way Go programs are developed must adapt accordingly: it is necessary to scale up to larger team efforts.

The paper is organized as follows: Section 2 analyzes the state of Computer Go, identifies some of the strengths and weaknesses of the current generation of programs, and outlines a plan which draws on existing technology but still promises substantial progress within a few years. In section 3, I introduce three development models that have been used in chess, and discuss how to adapt them to Computer Go. Finally, section 4 introduces promising topics for long-term research.

2   The State of Computer Go

To the casual observer, Computer Go may seem to be in fine shape. However, a number of problems threaten the future prosperity of the field. Many of these problems are rooted in the current structure of the Computer Go community.

2.1   The Computer Go Community

Recent years have seen many developments in Computer Go. Good progress has been made in the tournament scene and the internationalization of the field. However, in several ways the field has remained immature: programs are constructed on an ad hoc basis, results are held back for commercial reasons, and the lack of support for new researchers willing to enter the field is a severe problem.

The Tournament Scene Several annual tournaments have been established. Two world championships, the Ing foundation's International Computer Go Congress and the newer FOST cup, continue to attract the elite of Go programs from all over the world. The annual European and the North American Go congresses host smaller, local computer championships. The internet-based Computer Go Ladder [21] allows Go programmers from all over the world to compete with a wide variety of opponents.

An International Activity While human Go players are still concentrated mainly in Asia, Computer Go has become a truly international activity, with serious programs being developed in at least a dozen countries. It is not unusual to see the first five places in a tournament taken by competitors from many different countries. Several strong new commercial programs have been developed, and the total number of programs to participate in tournaments easily exceeds a hundred.

There seems to be renewed interest from the general AI and the computer games community. After the world-championship performance of programs in games such as chess, checkers or Othello, many eyes have turned to Go as the 'final frontier' of computer-game research.

Lack of Support for New Researchers Only a few individuals or institutions have sufficient resources to subscribe to a full-scale Go programming effort. Indeed most new Go programmers have to start almost from scratch. Because of the overhead in getting started, it is very hard for a smaller project, such as a masters thesis, to make a significant contribution.

Given the complexity of the task, the supporting infrastructure for writing Go programs should offer more than the analogous infrastructure did for other games such as chess. However, the Go infrastructure is far inferior. The playing level of publicly available source code [18] is far below that of state of the art programs. Quality publications are scarce and hard to track down. A few of the top programmers have an interest in publishing their methods. Whereas contributions on computer chess or general game-tree search regularly appear in mainstream AI journals, technical publications on Computer Go remain confined to the proceedings of specialized conferences. The most interesting developments can only be learned by direct communication with the programmers, unfortunately, they are never published.

2.2 State of Go Programs

Computer Go constitutes a formidable technical challenge. Existing programs suggest the following difficulties:

- A competitive program needs 5-10 person-years of development.
- A typical program consists of 50-100 modules.
- It is the weakest of all these components which determine the overall performance.
- The best programs *usually* play good, master level moves, but their performance level over a *full* game is much lower because of the remaining blunders.
- A number of standard techniques have emerged. However, there is no single program which incorporates most of the currently existing successful Computer Go techniques.

Let me discuss the strengths and weaknesses of the programs in some detail.

Special Strengths of Current Programs Several of the leading programs do one specific task better than the others. For example, HANDTALK excels in overall integration, playing good shape, and in knowledge about group attack and defense. *Go 4++*, on the other hand, is most efficient in taking territory, plays the fewest unnecessary 'wasted' moves per game, and has an extensive special purpose *joseki* book for opening move sequences. Further, *Go Intellect* is a mature program with strong tactical fighting and overall Go knowledge, while *GoTools* [25] is a high dan-level Life and Death solver specialized for completely surrounded areas.

Incompleteness of Existing Programs Unfortunately, there is no one program that incorporates most of the currently existing successful Computer Go techniques. A next-generation program would need to recreate and integrate most of these individual capabilities.

It is easy to see why single-person projects are inadequate: the sheer number of necessary components. Even assuming only one month for each module, a reasonably complete program will take four to five years to build, even without considering testing and system integration.

Disappointing Sustained Performance The difference in first-play versus sustained long-run performance of programs against human players is drastic: programs typically do well in their first game against an opponent inexperienced in playing computers. For example, world champion HANDTALK has won Ing's 11 stone challenge matches against high dan-level human players, and has beaten a 1-dan player without handicap in an exhibition game at the FOST cup. However, if allowed a few practice games, humans soon spot and exploit a program's weaknesses. The same program that once beat the 1-dan regularly loses against a well-prepared 5-kyu player, even when receiving huge handicaps of up to 20 stones.

2.3 What To Blame? The Model or Its Implementations?

What is the reason behind the irregular performance of programs? How can Go programs look so good on one day and so pathetic on the next? One theory is that there is a fundamental problem in the underlying models. In this view, current Go programs are not able to capture the true spirit of Go: they may play good-looking moves, but do so without any 'real understanding' of the game, which inevitably shows sooner or later. The alternative view is that the current model is basically sound and sufficient, but programs suffer from incomplete or buggy implementations.

2.4 A Model of Go Program Components

Fotland's 'Computer Go design issues' lists about sixty components of current Go programs [8], and can be considered as defining a 'standard model'. State of the art programs contain many of the modules described in Fotland's list. I will use the following simplified classification of development tasks for a Go program:

- Mathematical foundations and Go theory
- Knowledge representation and data structures
- Search methods
- Global move decision
- Software engineering and testing
- Automatic tuning and machine learning

I will briefly discuss the issues for each group of tasks, describe the current state of their implementation, and point out promising areas for short-term research and development. A program implementing this 'standard model' as completely and technically accurate as possible would serve as an interesting milestone and allow a more meaningful analysis of its strengths and weaknesses than is currently possible.

Mathematical Foundations and Go Theory Theoretical techniques applicable to Computer Go range from abstract mathematics for group safety and endgame calculation [3, 4] to Go-specific knowledge such as the *semeai formula*. [16] gives a detailed discussion.

Standard game tree-searching methods are well established for goal-oriented tactical search in Go. In addition, new search methods such as proof-number search [1] have been successfully applied in at least one commercial program. The many potential benefits offered by theory have only partially been applied in current programs.

Knowledge Representation and Data Structures Most programs use a hierarchical model for board representation. Low-level concepts are blocks of adjacent stones and connections or 'links' between stones. Chains, groups and territories are higher-level concepts built from the primitives. Pattern matching is used to find candidate moves. Knowledge representation has been the focus of the majority of Computer Go research to date, and has reached a sophisticated level.

I expect that the quality of knowledge incorporated in programs will gradually be refined. The quantity of knowledge is rising dramatically due to large scale pattern learning methods, which are becoming increasingly popular [5, 13, 12]. However, it is unclear how computer-generated pattern databases can reach a quality comparable to human-generated ones. For comparison, it would be fascinating to develop a large corpus of human Go knowledge, to try to identify and encode the pattern knowledge of Go experts.

Search Methods Three types of search are commonly used in Computer Go: single-goal, multiple-goal, and full-board search.

Specialized searches that focus on achieving a tactical goal are some of the most important components of current Go programs. A major advantage of goal-directed search over full-board search is that evaluation consists only of a simple test, which is much faster than full territory evaluation. One use of goal-directed search is to propose locally interesting moves to a selective global move decision process. I expect the use of goal-directed minimax searches to expand widely in volume and scope.

Single-goal search uses standard game tree searching techniques for finding the tactical status of blocks, chains, groups, territories, or connections. Knowing this status improves the board representation and is a precondition for creating a meaningful scoring function.

Examples for the targets of single-goal search are given in the table below. Current programs implement many but not all of these goal-oriented searches. A complete implementation of all basic single-goal searches seems to be a straightforward development task.

Target	Reference
Single block capture	[11]
Life and death	[25, 14]
Connect or cut	[7]
Eye status	-
Local score	(Goliath)
Safety of territory	[20]
Semeai	[7]

Multiple-goal search tries to achieve a combination of basic goals, such as capturing at least one of a set of blocks. The implementation of such tasks is rudimentary in most programs. Simple double threats such as *double atari* are usually built in as special cases. However, many standard Go strategies can be understood as more general double threats [17]. The following table lists a few common themes. In [9] a more sophisticated architecture for multipurpose strategic planning is described.

Target(s)	Goal(s)
Multiple blocks	save all blocks
Territory boundary	capture *or* break through
2 groups	splitting and leaning attacks
Group	attack *and* make territory
Group	live locally *or* break out
Group	make eyes *or* win semeai

Searching each possible goal combination leads to combinatorial explosion in the number of searches. Heuristics can be used to select promising goal combinations for search. I expect a lot of progress on this problem over the next few years.

Full board search seeks to maximize the overall evaluation. Because of the complex evaluation and high branching factor of Go, full-board search has to be highly selective and shallow. I expect the use of full board evaluation to increase steadily along with improving hardware, but without playing the same dominating role it has played for other games.

Global Move Decision There is a great variety of approaches to the problem of global move decision in Go. No single paradigm, comparable to the full-board minimax search used in most other games, has emerged. Most programs use a combination of the following methods:

- Static evaluation to select a small number of promising moves
- Selective search to decide between candidate moves
- Shortcuts to play some 'urgent' moves immediately

- Recognition and following of temporary goals
- Choice of aggressive or defensive play based on a score estimate

I expect experimentation to continue, without any clear preference or 'standard model' emerging. Methods based on combinatorial game theory (section 4.1) have the potential to replace more traditional decision procedures.

Software Engineering and Testing A competitive Go program is a major software development project. Software quality can be improved by using standard development and testing techniques [15]. A wide variety of game-specific testing methods are available, including test suites, auto-play and internet-based play against human opponents.

It is hard to judge objectively, but I suspect there is much room for improvement in this area. Most leading programs have been in continuous development for ten or more years. Many of these programs may be reaching a level of internal complexity where it is difficult to make much progress. Originally designed for machines a thousand times smaller and slower, programs have grown layer upon layer of additions, patches and adjustments. Some programs have been rewritten from scratch in the meantime, but this is a daunting and extremely time-consuming task [24].

Automatic Tuning and Machine Learning Machine learning techniques are currently used in only a few programs [5, 6]. However, parameter tuning and book learning techniques seem to be in more frequent use. I predict that the applications of machine learning techniques in Computer Go will increase, for example for fine-tuning the performance of complex programs with many components.

3 A Research Plan for Computer Go

What is the real limit on the performance of Go programs imposed by current models? Should research focus on developing new models or on improving the implementation of current ones? To answer these important questions, I propose the following three lines of research and development:

- Detailed analysis of current programs' errors
- A *Dreihirn* experiment
- Large scale Go programming projects

The first two methods are designed to better understand the problems of current programs. The third proposal addresses testing the limit of current technology.

3.1 Detailed Error Analysis

Detailed error analysis of current programs can draw upon a wealth of available game records [18]. Many classifications of mistakes are possible. For our purposes, it may be sufficient to assign errors to one of two broad groups: lack of basic understanding and lack of efficiency. Lack of basic understanding can be defined as the failure to identify the current focus of a game. Examples are attacking or defending the wrong group, ignoring threats or double threats, or making wrong life and death judgments.

Efficiency errors are less drastic individually but have a large cumulative effect. Mistakes belonging to this category are: making overconcentrated shapes, taking *gote* when a *sente* move is available, or achieving the correct main goal without considering secondary effects. An example of the latter kind of problem is saving a group by connecting it on a neutral point, instead of living more profitably by surrounding enough territory to make two eyes. Another example of efficient play is *kikashi*: playing profitable forcing moves before going back to make a neccessary but unattractive defense.

Research along these lines will aim to develop automatic methods for performing such analyses, by using statistical techniques and developing suitable test position collections.

3.2 A Dreihirn Match for Go

In the *Dreihirn* chess games [2], a team of two chess computers supervised by a human 'boss' has achieved strong results against chess grandmasters. The team played markedly better than each individual program, even though the 'boss' was a relatively much weaker player. The human supervisor was able to select a promising overall direction of play and avoid some typical computer missteps.

I propose a comparable experiment in Go, with a team of several Go programs supervised by a *strong* human player. At each move, the human selects one of the moves proposed by the programs. This team is tested against a variety of opponents, including other programs and humans of different strengths. Such a test can serve to establish an upper limit of current program performance, and show whether the uneven play is due more to individual bugs in the implementations or due to more fundamental limitations of all current programs. If a series of games is played, the test would also show if human opponents can adapt as quickly to such a system as they adapt to each individual program's weaknesses.

3.3 Outline of an Architecture for Large Scale Go Projects

From the beginning, most Computer Go projects have consisted of a single programmer with occasional assistance from either scientists or Go experts. In recent years, a few commercial programs have been developed on a slightly bigger scale, with small teams of programmers and managers working on the Go engine and user interface.

I believe that the scale of these projects is not large enough, and that projects an order of magnitude larger are necessary to produce a qualitative jump in performance. Section 2 has identified a long list of tasks required to implement a complete Go program based on the current 'standard model'. However, implementing a successful large scale Go project requires a series of preliminary steps.

- Secure an existing state of the art program to build on, including an easy to use basic Go toolkit.
- Modify the program to increase its usability in a multi-programmer environment.
- Describe the model underlying the program in detail.
- Document and structure the source code extensively.
- Define an effective communication method between team members.
- Implement a well-defined process for subtask assignment, code integration and testing.

3.4 Three Proposals for Large Scale Go Projects

In chess, three approaches have been taken in recent years that may serve as an inspiration for Go:

- Large company funded teams (DEEP BLUE)
- Public domain source code (GNU CHESS, CRAFTY)
- University projects (many)

Plan 1: Large Scale Commercial Project The DEEP BLUE chess project represents a large-scale effort, one order of magnitude larger than typical competitive chess programs. Its success rests on two pillars: on the technical side, it is a complete, mature system, the result of sound engineering firmly based on a large amount of previous research. On the organizational and financial side, the DEEP BLUE project was backed by a large company with an interesting new marketing strategy. Computer chess was chosen as an advertising vehicle because it represents an attractive topic that is tied to deep myths about human and machine intelligence.

Would a similar alliance of research and big business make sense in Go? Who would be a potential sponsor, and what would be their interest? In my view it would be a world-class company with a strong interest in the Asian market, and an ambition to create or reinforce their image as an intellectual leader. The company would profit mainly from the publicity generated by exhibition games, not from sales of Go software. Given the high regard for Go as an intellectual sport, it seems possible to attract a level of attention comparable to that of the chess matches, at least in East Asia. In Go, what is an achievable goal that will fascinate the masses? World championship level play still seems far in the future. Yet a program playing at a sustained 1-dan level, which can beat professional players on 9 stones handicap, will be perceived as an intellectual achievement at least equal to that of the chess machines. Is it possible in the near future? Let us try!

Plan 2: Public Domain Go Project Source code for more than a dozen chess programs is readily available on the internet [22]. The two best-known of these programs, GNU CHESS and CRAFTY, have active user groups which are testing, discussing or directly improving the program.

In Go, several public domain projects have been attempted over the years. So far, none of these has resulted in a tournament level program. Recently, there seems to be renewed interest in such a project, which has generated a large amount of messages on the mailing list.

The characteristics of a public domain Go program are quite different from a funded project and include the following items.

- Greater fluctuation of team members.
- Less individual commitment, lower work intensity.
- Low development cost.
- Difficult moderation and integration tasks.

The project goal could be to develop a noncommercial, research-oriented tool. The program structure should allow small or medium-scale experiments, for example in machine learning, to profit from a state of the art Go engine. A less ambitious approach would aim at developing only a library of commonly used functions.

Plan 3: University Research Project Many of the strongest chess programs are developed at universities. The situation in Go is comparable: about half of the current top 20 Go programs have started as student projects. An advantage of student projects is that relatively little funds are required, and students can combine their parts of the overall programming work with their research.

The main challenge of this approach is to assemble a large group of talented students and keep their efforts coordinated over a number of years. Given the current distribution of Go players, a large-scale university Go project would probably be feasible only in an Asian country.

4 Some Issues for Long-Term Research in Computer Go

Compared to the complex reasoning processes of human Go experts, the models incorporated in current Go programs are severely limited. A goal of long-term research could be to close this gap, either by building more sophisticated models or by deriving human-like reasoning capabilities from simple models.

Another fascinating topic is modeling the high level full-board plans of human players, or advanced Go concepts such as *aji*, *korikatachi* or *sabaki*. Therefore, one direction for research is evaluation from first principles, using only search and learning, without relying on human-engineered heuristics.

Long-term machine learning topics are automatic derivation of sophisticated Go concepts from first principles, or the learning of patterns along with suitable contexts for their application.

Yet another research topic, addressed in the next subsection, is the application of combinatorial game theory to Computer Go.

4.1 Combinatorial Game Theory for Computer Go

As a framework for Computer Go, combinatorial game theory has several advantages compared to the standard minimax game-playing model. However, the finer points of this theory are as good as unknown ouside the small combinatorial games community. Several of the tools provided by this theory are well suited for analyzing Go and should be used more in Go programs.

For example, the method called *thermography* is able to model fundamental Go concepts such as *sente* and *gote* very naturally [4]. Thermography computes the temperature of each local situation, which is a measure of move urgency. Comparing the thermographs before and after a move yields an optimal temperature range for each move. These ranges may differ dramatically for both players, for example in the case of *one-sided sente* moves. Using such an analysis, programs will be able to follow the standard Go strategy of keeping *sente* moves in reserve as potential *ko* threats. The theory is able to determine precisely how long the *ambient* temperature of a game remains high enough to prevent an opponent's *reverse sente* move.

Another important concept from combinatorial game theory is *reversibility*, which allows a player to make many moves based on a local consistency argument, without any full-board analysis. The computational advantages of this idea are immediate. Thermography introduces the stronger notion of *thermographic reversibility* by which a further reduction of search can be achieved [4].

Recent research by Kao addresses handling incomplete local game trees and selective search strategies within a combinatorial game framework [10]. An important research question is to generalize the precise concepts of combinatorial game thoery to work in a heuristic setting, in analogy to the heuristic game tree search based on minimax used in other games.

4.2 Handling of Ko Fights

Ko fights are considered the most complex phase of the game, and are handled poorly by current programs. Progress in theory and in practical algorithms for thermography [4, 19] provides effective and sound methods for comparing the relative values of *ko* and non-*ko* moves. This framework also allows the evaluation of possible *ko* threats.

5 Summary

Computer Go has enjoyed a boom in recent years, but its progress is hampered by problems in the structure of the Computer Go community. An analysis of the current state of Computer Go indicates promising directions for research, both short-term and long-term. To overcome the lack of critical human resources, Computer Go would benefit from the same kind of larger scale projects that have succeeded in chess.

References

[1] L.V. Allis. *Searching for Solutions in Games and Artificial Intelligence.* PhD thesis, University of Limburg, Maastricht, 1994.
[2] I. Althöfer. A symbiosis of man and machine beats grandmaster Timoshchenko. *ICCA Journal*, 21(1):40–47, 1997.
[3] D.B. Benson. Life in the game of Go. *Information Sciences*, 10:17–29, 1976. Reprinted in Computer Games, Levy, D.N.L. (Editor), Vol. II, pp. 203-213, Springer Verlag, New York 1988.
[4] E. Berlekamp. The economist's view of combinatorial games. In S. Levy, editor, *Games of No Chance: Combinatorial Games at MSRI*. Cambridge University Press, 1996.
[5] T. Cazenave. *Systeme d'Apprentissage Par Auto-Observation. Application au jeu de Go.* PhD thesis, University of Paris, 1997. www-laforia.ibp.fr/~cazenave/-papers.html.
[6] M. Enzenberger. The integration of a priori knowledge into a Go playing neural network. cgl.ucsf.edu/go/Programs/NeuroGo.html, 1996.
[7] D. Fotland. Knowledge representation in The Many Faces of Go. Report posted on internet newsgroup rec.games.go, 1993. Available by ftp from igs.nuri.net.
[8] D. Fotland. Computer Go design issues. www.usgo.org/computer/text/-designissues.text, 1996.
[9] S. Hu and P. Lehner. Multipurpose strategic planning in the game of Go. *IEEE Transactions on Pattern Analysis and Machine Intelligence*, 19(9):1048–1051, 1997.
[10] K.Y. Kao. *Sums of Hot and Tepid Combinatorial Games.* PhD thesis, University of North Carolina at Charlotte, 1997.
[11] A. Kierulf. *Smart Game Board: a Workbench for Game-Playing Programs, with Go and Othello as Case Studies.* PhD thesis, ETH Zürich, 1990.
[12] T. Kojima. *Automatic Acquisition of Go Knowledge from Game Records: Deductive and Evolutionary Approaches.* PhD thesis, University of Tokyo, 1998.
[13] T. Kojima, K. Ueda, and S. Nagano. An evolutionary algorithm extended by ecological analogy and its application to the game of go. In *Proceedings of the Fifteenth International Joint Conference on Artificial Intelligence(IJCAI-97)*, pages 684–689, 1997. www.brl.ntt.co.jp/people/kojima/research.
[14] J. Kraszek. Heuristics in the life and death algorithm of a Go-playing program. *Computer Go*, 9:13–24, 1988.
[15] S. McConnell. *Rapid Development.* Microsoft Press, 1996.
[16] M. Müller. Game theories and Computer Go. In *Proc. of the Go and Computer Science Workshop (GCSW'93)*, Sophia-Antipolis, 1993. INRIA.
[17] M. Müller. *Computer Go as a Sum of Local Games: An Application of Combinatorial Game Theory.* PhD thesis, ETH Zürich, 1995. Diss.Nr.11.006.
[18] M. Müller. American Go Association Computer Go Pages. www.usgo.org-/computer, 1997.
[19] M. Müller. Generalized thermography: A new approach to evaluation in Computer Go. In H. Iida, editor, *Proceedings of IJCAI-97 Workshop on Computer Games on Using Games as an Experimental Testbed for AI Research*, pages 41–49, Nagoya, 1997.
[20] M. Müller. Playing it safe: Recognizing secure territories in Computer Go by using static rules and search. In H. Matsubara, editor, *Proceedings of the Game Programming Workshop in Japan '97*, pages 80–86, Computer Shogi Association, Tokyo, Japan, 1997.

[21] E. Pettersen. The Computer Go Ladder. cgl.ucsf.edu/go/ladder.html, 1994.
[22] P. Verhelst. Chess program sources. www.xs4all.nl/~verhelst/chess/sources.html, 1997.
[23] L. Weaver. COMPUTER-GO Mailing List Archive. www.hsc.fr/computer-go, 1998.
[24] B. Wilcox. Chess is easy. Go is hard. Computer Game Developers Conference. home.sprynet.com/sprynet/vrmlpro/cgdc.html, 1997.
[25] T. Wolf. Investigating tsumego problems with RisiKo. In D.N.L. Levy and D.F. Beal, editors, *Heuristic Programming in Artificial Intelligence 2*. Ellis Horwood, 1991.

Estimating the Possible Omission Number for Groups in Go by the Number of n-th Dame*

Morihiko Tajima and Noriaki Sanechika

Electrotechnical Laboratory
1-1-4, Umezono, Tsukuba-shi, 305-8568 JAPAN
{tazima,sanetika}@etl.go.jp

Abstract. This paper describes a new method for estimating the strength of a group in the game of Go. Although position evaluation is very important in Computer Go, the task is very complex, so good evaluation methods have not yet been developed. One of the major factors in evaluation is the strength of groups. In general, this evaluation is a difficult problem, so it is desirable to have a simple method for making a rough estimate. We define PON (Possible Omission Number) as a precise measure for the strength of groups, and present a simple method for the rough estimation of PON by calculating n-th dame (liberties). We report some experiments indicating the effectiveness of the method.
Keywords: Go, Strength of groups, Possible omission number, PON, n-th dame

1 Introduction

Research in computer Go is becoming increasingly popular. Unlike games such as chess or Othello, no effective way of position evaluation has been developed. Many methods for selecting candidate moves in Go are not based on position evaluation. The reason is that the game board is large and the components of position evaluation are not clear. We believe that the development of more accurate position evaluation is needed for the development of high-level playing programs and detailed position analysis. In [5], we proposed a method which estimates the value of candidate moves by directly evaluating the position variation caused by the moves. That method also needs accurate position evaluation.

One of the difficult problems in evaluating Go positions is that one should evaluate both the global configuration of stones and local semeai (mutual attacks) in a balanced way. One of the most important components for both global and local evaluation is the strength of groups. In local semeai, the relative strength of groups fighting each other decides the winner. Arranging stones in a globally optimal configuration means overall balancing of group strengths, while also considering the strength of opponent's groups.

Group strength is determined by the size and shape of the group and the configuration of nearby friendly and opponent stones. To calculate the exact

---
* We thank Dr. Martin Müller for extensively proofreading this paper, and for useful technical feedback.

strength in general, deep search or using a pattern database as in [1] and [4] is necessary. Exact evaluation is not always needed, however. In practical play, sometimes a rough estimation is enough or is forced because of shortage of time. Simple estimation methods are useful in such cases. For many playing programs, group strength evaluation is one of the most important components. Such programs typically evaluate strength empirically by the number of dame, the number of eyes, the shape of the group, etc.

In many cases, however, the resulting value is not concrete or substantial, but arbitrarily devised and abstract. In this case, its rationale is not clear, so that its application needs experience and it is not suitable for analysis. Therefore we must find a value that is substantial and applicable to estimate the strength of groups. We consider the number of possible *tenuki* (or *omission* of making moves), which was originally introduced as *X life* in [2], to be such a value. In this paper we name the number *possible omission number (PON)* and will show that we can estimate this number from the number of n-th dame [3].

To state the conclusion briefly in advance, the estimation of group strength by this method cannot replace precise search methods for determination of life and death for tightly surrounded groups. However, it provides a low-cost and practical means of evaluating the strength of immature groups which are loosely surrounded in the opening or in the early middle game.

In the next section, we define the possible omission number and how to enumerate n-th dame which are used to estimate this number. In Section 3 we experimentally find a method for approximately computing the possible omission number. We discuss the strengths and weaknesses of our method in Section 4, and conclude this paper in Section 5.

2 Possible Omission Number and n-th Dame

In the following description, we assume that it has already been determined by some other method which stones belong to the same group. For life and death of groups, we consider capturing the whole group as death, and living with at least some part of the group as life. We define the possible omission number (written as *PON* below) for a group as follows:

Possible Omission Number (PON)
Consider a group G of color C.
- (a) **Group G is neutral** (i.e. life and death depends on the next turn). PON of G is 0.
- (b) **Group G is alive.** If G becomes neutral after n opponent moves in a row but G is still alive after $n-1$ consecutive opponent moves, PON of G is n.
- (c) **Group G is dead.** If G becomes neutral after C is allowed to make n consecutive moves but G is still dead after $n-1$ consecutive moves, PON of G is $-n$.

Intuitively, PON means how often the player or the opponent can ignore a local play before the group status changes.

Although PON cannot be regarded as the very measure of the strength of groups, it is clear that PON is closely related to the strength of groups. Popma and Allis [2] made use of the number in the analysis of multi-step ko positions. The usage is for rather closely surrounded groups often seen in the end game. However, computing the number is effective not only in such positions but also in positions often seen in the opening or middle games, where two kinds of applications of PON can be considered: First, PON can be used as a criterion for making moves strengthening the player's groups and weakening the opponent's groups. Second, PON can be used to decide which group is the most urgent when there is more than one weak group. For example, in a battle between two groups of different colors, groups having similar PON are usually urgent and groups having very different PON are usually not urgent.

It is difficult to calculate PON precisely in complicated positions, e.g. when groups are fighting closely. Exhaustive search has to be used, which is impossible when analysing under time pressure. An alternative is employing lots of known patterns. Human experts can decide their moves quickly due to their extensive pattern knowledge. But in a program, it requires a lot of resources to create a big pattern database and to search patterns. We think that a method that can estimate PON easily with pretty good accuracy is necessary.

We will show such a method in the following. As a preparation, we define group, edge, n-th dame (or liberty) of a group, and kosuri (or rub).

group a set of stones of the same color which can be regarded as at least loosely connected.

edge the outermost intersections on a Go board.

n-th dame and kosuri Span antennas step by step from the stones of the group to empty points along the lines of the board. Empty points which can be reached in n steps in a manner defined below are called *n-th dame*. During these steps, empty points are labeled by their *kosuri* number, which is defined as follows: If a point is adjacent to (an) opponent's stone(s), count one kosuri for each opponent stone. Propagate the kosuri numbers along the lines on the Go board. Only points for which the kosuri number is at most one can be reached.

There are two additional rules for kosuri: For points immediately under opponent's stones on the second or the third lines, add 2 to the kosuri number. For points on the edge, add 1 to the kosuri number.

The kosuri number has the following meaning: life and death of a group depends on how easily it can escape when it is surrounded by its opponent. If a path for escape touches an opponent's stone, escape is hard. It is still possible to escape if the path touches once, but there is little possibility to escape if the path touches twice. The reason why points immediately under opponent stones on the second or the third line are regarded as having touched twice is that it is very hard to escape in such situations.

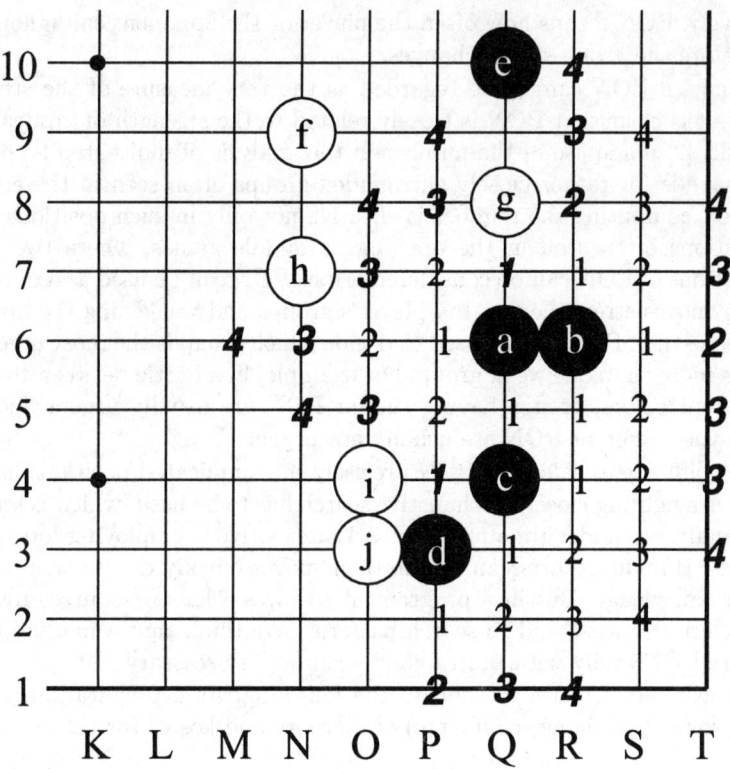

Fig. 1. An example of n-th dame

Fig. 1 shows an example of a black group. Dame up to $n = 4$ are denoted by the numbers. Black stones a, b, c, and d form one group, and the black stone e constitutes a different group. Consider the path leading straight upward from stone b. The first dame R7 is a point without kosuri. The second dame R8 is a point with one kosuri from the white stone g next to it. Points with one kosuri are denoted by an Italic-bold font. Since the kosuri number is accumulated, the third dame R9 and the fourth dame R10 are also points with one kosuri. Point Q9 to the left of R9, becomes a point with kosuri number 2 and cannot be reached because it touches the white stone g. Consider another path going downward from stone d. The second dame P1 is a point with one kosuri because it is on the edge of the board. O2 at the lower left of d, is a second dame, but it cannot be reached because it has kosuri number 2. It is immediately under the opponent's stone j, which is on the third line.

The reason for our definitions is the following: In general, groups with a lot of dame can make eyes easily. On the contrary, groups with small number of dame are hard to make alive. Among the n-th dame, dame closer to a group are more effective for making eyes.

We have defined three kinds of dame, i.e. edge points, points with one kosuri, and other points. The meaning of these kinds for PON is as follows: As described before, points with one kosuri are worse than those without kosuri because escape is more difficult. Points on the edge are preferable because making eyes there is easier.

In the following section, we will experimentally find an estimate for PON from the number of dame.

3 Designing an Evaluation Function

We made some experiments to find a method for the simple estimation of PON. We will explain the test instances used and show the results of the experiments.

3.1 Instances Applied in This Paper

We identified 23 groups, which are not completely dead or alive, in the seven opening and midgame positions of a game in [6]. These instances are shown in the Appendix. The numbers of dame, of degree from 1 to 4 of each kind were counted, and the real PON for each instance was estimated by a human expert of amateur six dan rank. The real PON of the 23 instances vary from -2 to 5 (see Fig. 2).

3.2 Experiments

Assume that we can describe the PON, N, by a function of the numbers of n-th dames. In this paper, we assume a simple approximation function

$$N = f(\sum_{i=1}^{4} w_i \sum_{k=e,t,u} w_k d_{ik})$$

where w_i, w_k, d_{ik} are the weights for the degree i, the weight for the kind k (w_e is the weight for edge points, w_t is the weight for the points with one kosuri except edge points, and w_u is the weight for other points), and the number of dame of degree i and kind k, respectively. f is an appropriate function.

Since we limit the degree of dame to 4, this approximation function has seven parameters, i.e. $w_1, w_2, w_3, w_4, w_e, w_t$, and w_u. We observe for each instance how

$$g(w_1, w_2, ..., w_u) = \sum_{i=1}^{4} w_i \sum_{k=e,t,u} w_k d_{ik}$$

changes as parameters change. It is the purpose of these experiments to observe how the value of $g(w_1, ..., w_u)$ (shown as S hereafter) corresponds to the real PON, and what the optimum values of the weights are.

Several kinds of measures could be used for evaluating the approximation function. The most important condition is that S correlates with PON. Therefore the rate of reversed combination among all pairs of instances, as defined below, is a good measure. We define *reverse order* and *the right order rate*.

reverse order and right order rate Suppose S and real PON are given for each group in a set of instances. Choose two groups g_1 and g_2 and let their S and PON values be S_1, S_2 and P_1, P_2, respectively. If $S_1 > S_2$ and $P_1 < P_2$, or $S_1 < S_2$ and $P_1 > P_2$, S of these two groups is in *reverse order*. Considering all combinations of two groups, the fraction of pairs for which the value S is not in reverse order is called *the right order rate* of the function.

For example, if we have a set of instances of 23 groups, we get 253 pairs of groups. If 10 pairs are in reverse order, the right order rate is about 0.96.

The Effect of the Weights Concerning the Degree of Dame We tested how the number of pairs in reverse order varies as we change the degree weights while keeping the kind weights constant as $w_e = w_t = w_u$. Table 1 to 3 show the number of pairs in reverse order when $w_1 = 1$ and w_2, w_3, and w_4 change. The following properties can be observed:

- The number of pairs in reverse order is 21 in the best case. This right order rate is about 92% seems insufficient for practical use.
- The smaller the degree, the greater the effect, i.e. the effect of w_2 is greater than that of w_3 and that of w_3 is greater than that of w_4.
- w_4 does have an effect, but it is small.
- If one of the weights of some degree is too small, the weights of other degrees can compensate it to some extent. For example, the optimum value of w_2 becomes greater than 1 when $w_3 = w_4 = 0$. The value becomes greater than the optimum value (approximately 0.5) in the case where neither w_3 nor w_4 is 0.
- There is no sharply defined optimum. Under the condition of $w_1 > w_2 > w_3 > w_4$, the value ranges $w_2 = 0.4$ to 0.6, $w_3 = 0.2$ to 0.3, $w_4 = 0.2$ to 0.3 are almost optimal.

In further analysis we found that the optimum values of weights are inversely proportional to the degrees.

The Effect of the Weights Concerning the Kinds of Dame This experiment was designed to determine good weights for the kinds of dame. From the results of the preceding experiment, we set the values of $w_1, \ldots, w_4$ as follows.

$$w1 = 1 \qquad w2 = 0.5 \qquad w3 = 0.35 \qquad w4 = 0.25$$

Table 4 shows the number of pairs in reverse order when $w_u = 1$ and w_t and w_e change. The following properties can be observed:

Table 1. The number of pairs in reverse order when the weights concerning the degree change

$w_4 = 0$ $w_2 \backslash w_3$	0	0.1	0.2	0.3	0.4	0.5	0.6	0.7	0.8	0.9	1
0	43	43	39	35	30	28	25	26	25	26	24
0.1	45	42	36	33	29	29	28	26	25	25	29
0.2	42	38	35	31	30	26	26	25	25	29	28
0.3	39	35	32	32	28	27	26	23	27	28	28
0.4	37	32	32	31	28	26	23	27	28	28	30
0.5	31	33	30	29	27	23	23	26	28	30	30
0.6	32	33	30	29	26	22	24	28	28	30	30
0.7	32	28	30	28	24	24	28	28	30	30	32
0.8	29	29	28	26	25	24	28	30	30	32	32
0.9	28	29	27	28	25	28	30	32	32	32	34
1	24	28	26	28	25	28	32	32	32	34	34
1.1	27	26	28	27	31	32	32	33	34	34	35
1.2	27	28	28	31	33	32	33	33	34	34	35
1.3	27	30	30	33	33	33	33	34	35	35	35

$w_4 = 0.1$ $w_2 \backslash w_3$	0	0.1	0.2	0.3	0.4	0.5	0.6	0.7	0.8	0.9	1
0	41	38	33	25	28	27	26	25	26	25	28
0.1	41	35	30	26	29	27	25	24	24	26	28
0.2	38	33	27	30	26	26	25	25	27	28	28
0.3	33	29	30	28	27	26	23	27	28	28	30
0.4	31	30	30	27	27	23	25	26	28	30	30
0.5	32	30	27	26	23	23	26	28	30	30	30
0.6	30	29	27	25	22	24	26	28	30	30	30
0.7	28	27	26	23	24	26	28	30	30	30	33
0.8	27	28	24	24	24	28	30	30	30	32	34
0.9	28	26	24	24	26	30	30	32	32	34	33
1	27	25	26	24	28	30	32	32	34	34	34
1.1	25	27	25	26	32	32	32	34	34	35	34
1.2	27	27	25	32	32	33	33	34	34	35	34
1.3	27	29	31	32	33	33	34	35	35	35	34

- The effect of w_e is great, the effect of w_t is not so great.
- Weights can not compensate for the wrong setting of the other weight. This phenomenon is different from what is seen in the case of the degree weights.
- The optimum weights are $w_e = 1.6$ and $w_t = 0.2$, and the number of pairs in reverse order is 5 for the optimum weights. The right order rate is about 98%, which seems sufficient for practical use.

As described before, the closer to the group a dame is, the more effective it is for making eyes. Points with kosuri have a disadvantage. Dame on the edge are more effective because it is easier to make eyes there. The weights we found reflect these properties.

Table 2. The number of pairs in reverse order when the weights concerning the degree change (Continued)

$w_4 = 0.2$											
$w_2 \backslash w_3$	0	0.1	0.2	0.3	0.4	0.5	0.6	0.7	0.8	0.9	1
0	38	32	24	26	25	27	24	25	25	26	28
0.1	35	28	24	27	26	25	23	25	26	28	28
0.2	32	25	27	25	24	23	24	26	25	27	28
0.3	28	27	25	25	22	22	24	25	27	29	28
0.4	30	28	24	24	22	25	26	28	30	30	29
0.5	29	27	25	22	22	26	26	30	30	30	29
0.6	28	26	23	21	23	26	28	30	30	30	30
0.7	26	25	21	23	25	26	30	30	30	30	32
0.8	28	23	21	23	25	30	30	30	30	33	33
0.9	25	23	23	25	29	30	30	32	34	33	33
1	25	24	23	25	29	30	32	34	34	33	34
1.1	27	24	25	29	31	32	34	34	34	34	34
1.2	27	24	27	31	31	33	34	34	35	34	34
1.3	26	28	31	32	32	34	34	35	35	34	34

$w_4 = 0.25$											
$w_2 \backslash w_3$	0	0.1	0.2	0.3	0.4	0.5	0.6	0.7	0.8	0.9	1
0	33	31	24	28	23	24	24	25	25	27	27
0.1	32	25	27	24	25	23	25	24	26	27	29
0.2	27	27	26	25	24	22	24	26	27	28	28
0.3	27	27	23	24	22	23	26	25	29	28	28
0.4	27	26	26	23	22	25	26	28	30	29	29
0.5	26	28	23	22	23	25	26	30	30	29	30
0.6	26	25	22	21	25	25	30	30	30	30	30
0.7	26	24	21	23	25	29	30	30	30	30	33
0.8	24	22	23	25	25	29	30	30	31	33	33
0.9	23	24	23	25	29	29	30	33	34	33	33
1	23	24	25	27	29	29	32	34	34	33	34
1.1	24	24	27	29	31	32	34	34	33	34	34
1.2	24	26	29	31	32	33	34	34	34	34	34
1.3	24	30	31	32	33	33	34	35	34	34	36

Table 3. The number of pairs in reverse order when the weights concerning the degree change (Continued)

$w_4 = 0.3$											
$w_2 \setminus w_3$	0	0.1	0.2	0.3	0.4	0.5	0.6	0.7	0.8	0.9	1
0	30	26	25	24	25	23	24	25	27	28	29
0.1	27	24	27	23	24	22	24	26	26	29	29
0.2	23	28	24	24	22	24	23	25	27	28	29
0.3	26	26	24	22	21	23	24	25	29	28	28
0.4	25	25	23	21	23	24	24	29	29	28	29
0.5	25	25	22	22	25	25	29	30	29	29	30
0.6	25	23	21	23	25	25	29	30	29	30	30
0.7	24	22	21	25	25	29	29	30	30	30	33
0.8	22	22	23	25	27	29	29	30	30	33	33
0.9	21	24	25	27	29	29	29	34	33	33	34
1	23	24	25	29	29	29	33	34	33	34	34
1.1	23	26	27	29	31	33	33	34	34	34	34
1.2	23	28	29	31	32	33	33	35	34	34	36
1.3	25	30	31	32	33	33	34	35	34	34	36

Table 4. The number of pairs in reverse order when the weights concerning the kinds of dame change

$w_e \setminus w_t$	0	0.1	0.2	0.3	0.4	0.5	0.6	0.7	0.8	0.9	1
0	35	39	38	37	38	38	42	40	42	43	44
0.2	29	33	31	35	35	36	36	37	37	40	40
0.4	29	30	28	28	28	31	35	36	36	37	39
0.6	24	24	23	25	25	26	29	31	32	34	35
0.8	20	19	20	22	23	24	24	24	25	27	28
1	15	17	16	16	19	19	19	19	20	20	22
1.2	13	12	12	14	16	17	19	19	19	19	20
1.4	9	9	8	10	10	10	10	12	15	15	17
1.6	7	6	5	8	9	9	9	9	9	9	13
1.8	8	6	6	7	7	7	8	9	9	9	10
2	8	7	6	7	7	7	7	7	9	9	9
2.2	9	9	7	7	8	9	9	9	9	9	9
2.4	11	11	10	9	9	9	9	9	9	9	9
2.6	12	10	11	11	14	10	10	10	10	10	10
2.8	15	12	11	12	14	15	13	12	10	10	10
3	17	13	11	11	13	13	14	13	13	14	12

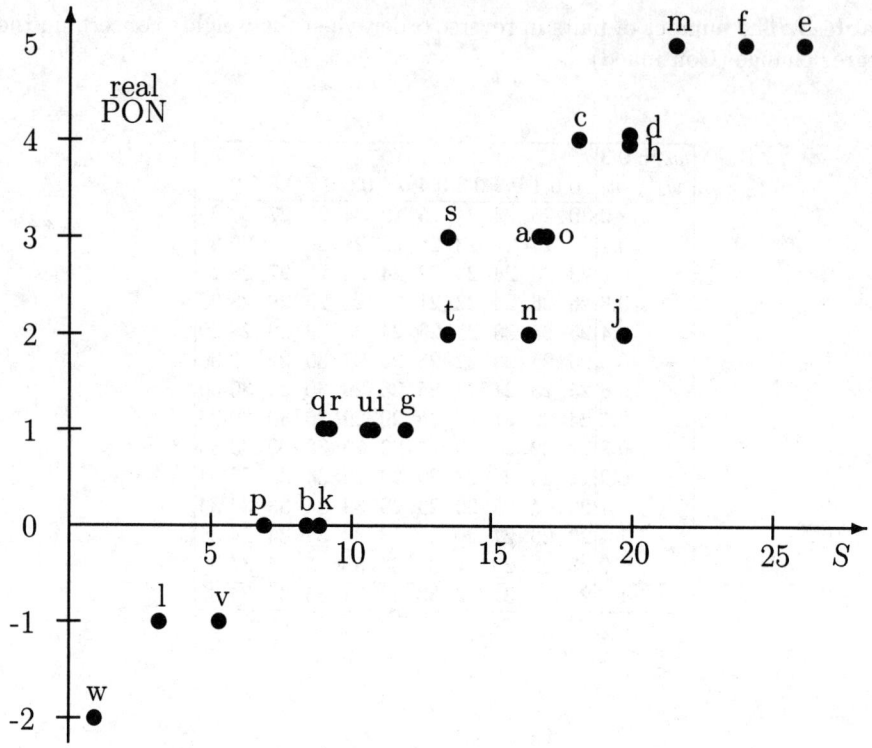

Fig. 2. Linear summation of n-th dame vs. real PON

3.3 PON Approximation Function

Fig. 2 shows the relation between the real PON and

$$S = \sum_{i=1}^{4} w_i \sum_{k=e,t,u} w_k d_{ik}$$

that is calculated by the weights which were decided in the preceding subsection. Each point corresponds to one of our 23 instances. The right order rate of 98% shows that the right order is kept in all but three instances (denoted by the points j, n, and s). One can calculate an approximate PON easily by a function $f(S)$. For example,

$$f(S) = [0.33S - 1.96]$$

where $[x]$ is the maximum integer not exceeding x, is a good approximation function. Preliminary experiments on further test instances seem to show that this function can estimate the correct PON in most cases.

4 Discussion

As stated in 3, one can estimate PON by the number of n-th dame quite well, and we think that the method is promising. However, there remain misclassified instances as shown in Fig. 2. Let us discuss the method.

4.1 Discussion of Experiments

- We computed dame of degree up to 4 and the effectiveness of dame especially of degree up to 3 was proved. Dame of degree 4 are effective to some extent. It is not clear whether further dame are effective or not, but we guess that dame of degree up to 4 are enough for a rough estimate such as the one described here.
- It was found that dame on the edge are very effective, while dame with kosuri are not.
- Three of 23 instances (points j, n, and s in Fig. 2) were not correctly estimated. The reasons are as follows:

  Instance j The real PON is smaller than expected, because group k influences group j. We treated j and k as separate groups, but the number of substantial dame of group j is smaller than computed. The dame in the lower part are not as effective as expected.

  Instance n The real PON is a little smaller than expected. This is because the shape of the group is bad. The group has spread to three directions and has a complicated form. Such groups are liable to have weak points.

  Instance 's' The real PON is one greater than expected. This is a special situation near the edge and near another friendly group. With an opponent's stone on the third line, usually a group cannot escape. But in this case, the friendly group immediately above the opponent's stone helps with the escape even after an extra *tenuki*. Since our method does not take nearby friendly groups into consideration, this kind of error sometimes occurs.

  From these exceptional instances, we see that critical cases exist that other methods are necessary to get an exact PON in such cases.

4.2 Discussions of This Method

In real games, we can classify the elements determining the strength of groups as follows:

life and death the possibility of establishing eye shape by a group itself;
escape the possibility of escaping to the outside by finding a gap in the opponent's enclosure;
counter attack the possibility of breaking through or capturing a part of the opponent's enclosure;
connection the possibility of connection to a near group of the same color.

However, only two of them, *life and death* and *escape*, are evaluated in the method proposed in this paper. This is a consequence of the simplicity of the calculation. Therefore, the following issues are topics for further research:

- Friendly nearby groups

  As seen in instance 's', the effect of such groups must be considered for more accurate estimation. For example, the difference in the number of dame between in the case of connecting two groups and in the case of being unable to connecting them has to be considered.

- Distance

  The degree of dame considered in the current method is limited to 4. A greater number of degrees might be necessary for more accurate estimation.

- Kinds of dame

  Three kinds of dame are distinguished in the current method. The effectiveness of the discrimination is clear from the experiments in Section 3. A finer discrimination might be effective. For example, dame points close to the edge are more effective than those in the center of the board from the viewpoint of making eye shape easily. And discriminating dame points by whether they are inside or outside an area enclosed by the group is possibly effective for a more precise judgement of life and death because inner dame points have a high possibility of constituting eyes.

- Recognition of groups

  In this paper, we defined the life and death of a group considering only whether one can capture the whole group or not. In some situations, however, a part of a group becomes a target of capturing. Since capturing a part of a group is easier than capturing the whole group, the PON in such a case becomes small. For example, in instance d, one can consider a different group consisting of the leftmost stone only, if the opponent's target is to capture only that stone. In this case, the PON becomes 2. Another example is that the group in instance j and the group in instance k could be combined into one group.

  This matter is related to the general problem of what set of stones should be regarded as a group. How one recognizes groups influences the calculation of PON. Recognizing groups appropriately is a precondition for the practical application of this method.

- Intrinsic limitations

  Since the method is based on a simple calculation, very accurate estimation cannot be expected. In a word the major problem of the method is its narrow scope. Generally speaking the life and death of a group may depend on all stones around the group, or even the presence and the strength of groups located far from it. Especially in the case of *semeai* (mutual attack), the life and death of a group depends critically on all opponent stones surrounding the group and all own stones surrounding the opponent's stones. These cases usually require more detailed analysis. Therefore, this method seems to work better in the case of open groups, which appear mainly in the opening and middle games, than tightly surrounded ones.

5 Conclusion

The number of times that a player or the opponent can *tenuki* without changing a group's status is a very important evaluation component in the game of Go. We defined the possible omission number, proposed a method that estimates this number by the number of n-th dame and their kinds, and made sure that one can estimate the possible omission number by our method with good accuracy. We analyzed the causes of remaining errors, and discussed the merits and possible improvements of the method. We think that the method is very promising for the evaluation of the strength of groups especially in the early stage of a game.

The following are some of the remaining issues. As for the effectiveness, one should give consideration to the fact that the application is limited to some extent. Some modification considering the configuration and the strength of other groups around is needed to have more accurate estimation. And it is better to use search or patterns together or select them when needed. It is a practical manner to use this method in usual situation and use other accurate methods in special situations. How to use this and those methods properly should be studied.

Of course the goal of the estimation of possible omission number is exact position evaluation that can help finding the best candidate moves. How to apply possible omission number to the goal is the next interesting problem.

References

1. M. Boon. A pattern matcher for Goliath, *Computer Go*, No.13, Winter, 1989-1990.
2. R. Popma and L.V. Allis. Life and death refined, *Heuristic Programming in Artificial Intelligence*, 3, 157-164, 1992.
3. Sanechika N. et. al.. The specification of Go system Gosedai, *ICOT Technical Report* TM-0618, Institute for New Generation Computer Technology, 1991. (in Japanese)
4. Sei S.. The experiment of Go program "Katsunari" using memory-based reasoning, *Proceedings of GPW'96*, 115-122, 1996. (in Japanese)
5. Tajima M. and Sanechika N.. Applicability of KPV method to Go, *Proceedings of GPW'95*, 37-46, 1995. (in Japanese)
6. Special Go game, Masao Katoh vs. Yoshiharu Habu, *Nikkei Mook Igo Taizen*, Nihon Keizai Shinbunsha, Tokyo, 6-16, ISBN4-532-18019-8, 1996. (inJapanese)

A Groups Used for PON Estimation

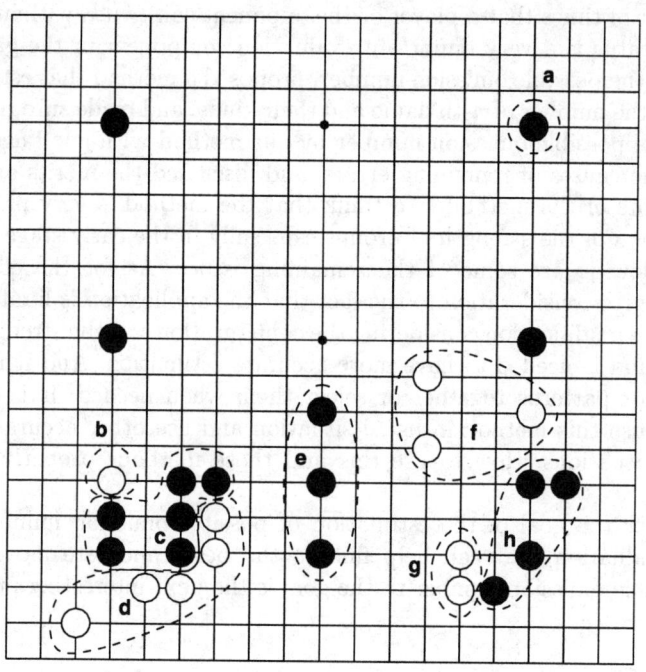

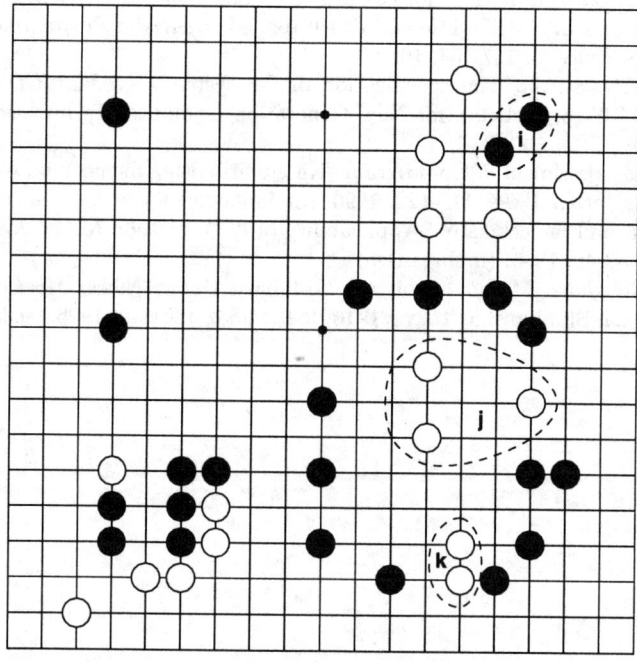

Estimating the Possible Omission Number for Groups in Go 279

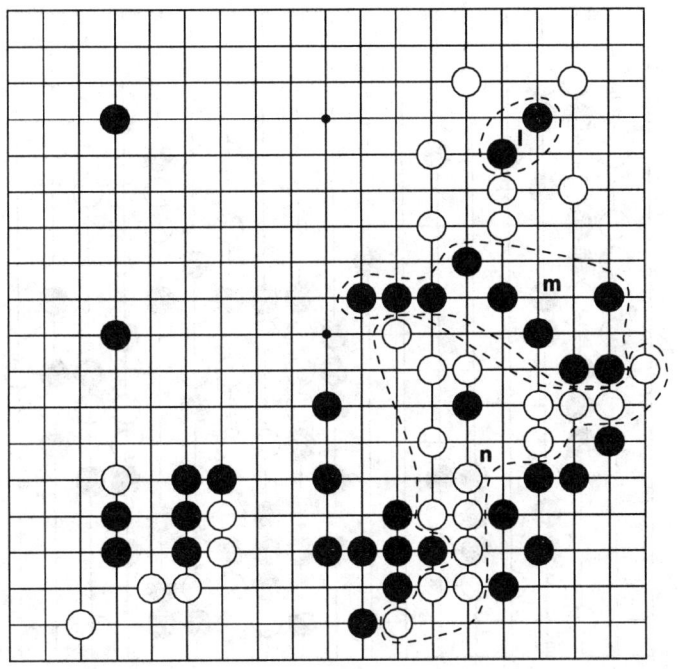

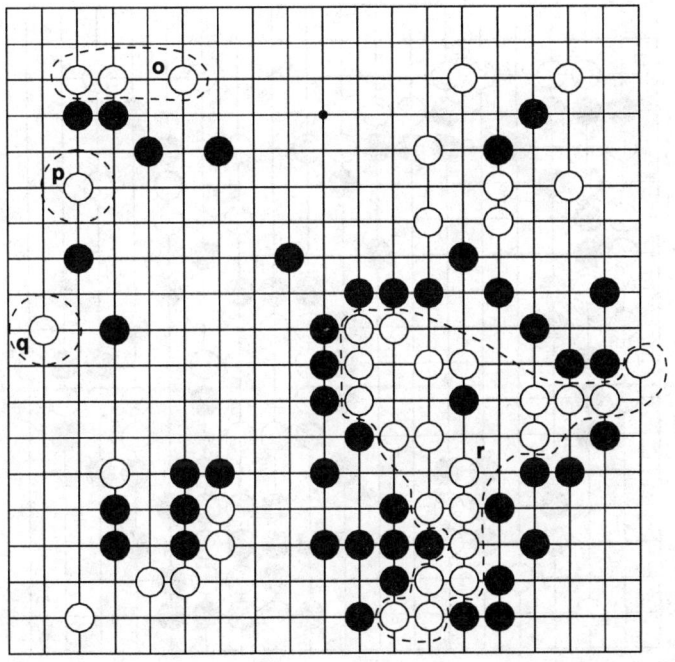

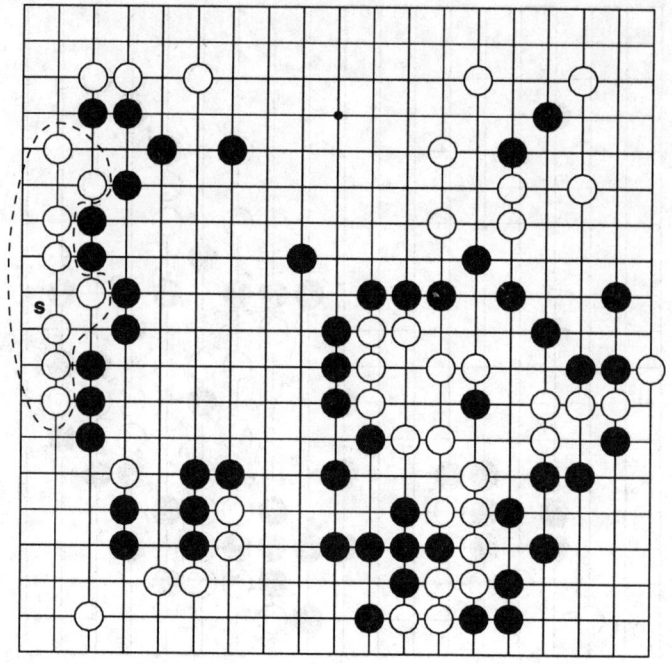

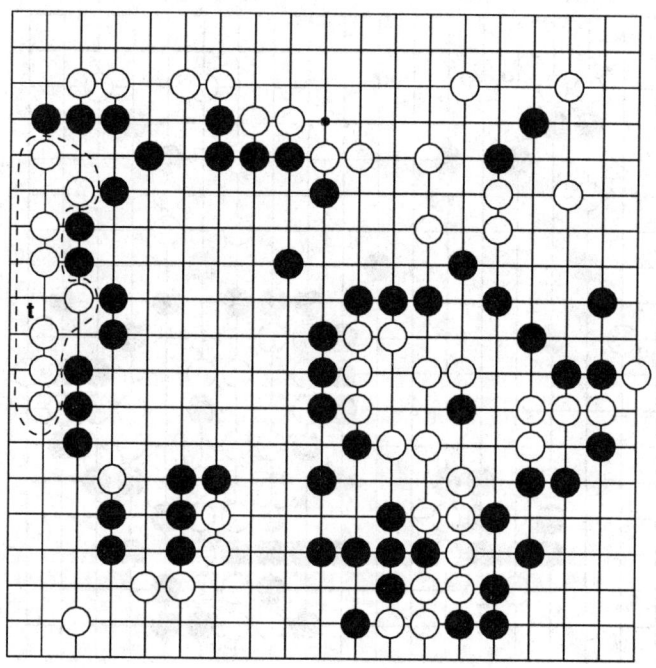

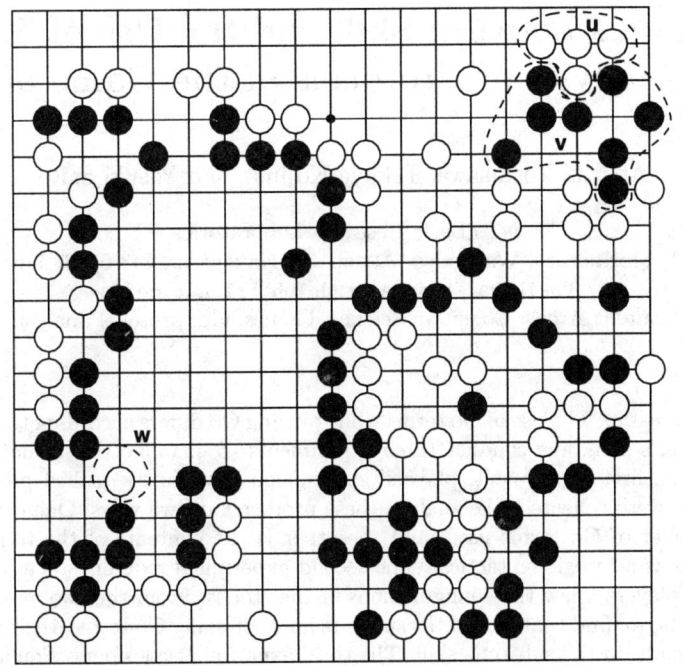

Relations between Skill and the Use of Terms
- An Analysis of Protocols of the Game of Go -

Atsushi Yoshikawa, Takuya Kojima, and Yasuki Saito

NTTBasic Research Laboratories,
3-1, Morinosato, Wakamiya, Atsugi-shi, Kanagawa, 243-0198, Japan
{yosikawa, kojima}@rudolph.brl.ntt.co.jp
Content Areas: Cognitive Science, Games, Go, protocol analysis

Abstract. The use of Go terms while playing Go differs according to the player's skill. We conduct three experiments to examine this in detail. In the first experiment, players' spontaneous utterances (called protocols) were collected. We analyze these protocols in two ways. One is the number of Go terms used, and the other is the contents of the terms, such as strategic or tactical. The second experiment examines how well the players knew the configurations of the stones. From the two experiments, we find that even if the subjects know of many Go terms, their use depends on the subject's skill. The third experiment considers "Soudan-Go," where two players form a team. They are in the same room and can freely talk to each other; their spontaneous utterances (protocols) were collected. We also analyze reports of "Houchi Soudan-Go," which is a Soudan-Go match between professional players. We find that expert players often use Go terms and they understood their partner's intentions without needing a full explanation. Intermediate level players often talked over their plan and their opponent's plan using many Go terms. From our analyses we developed a hypothesis which we call the *iceberg* model. The purpose of the model is to explain the structure of a term in the human brain from the viewpoint of the role of the term. Although this is still a hypothesis, it will become an important guide when carrying out protocol analyses and modeling the thought processes of Go players.
keyword: Cognitive science, Go, Special terms, Expert knowledge, Iceberg model

1 Introduction

Go is one of the most sophisticated two-player, complete information, board games in the world. In AI, the next grand challenge after chess is thought to be building a strong Go-playing program. Yet after more than 30 years of effort, Go-playing programs have achieved only human beginner level. There is undoubtedly more to be learned from actual human players but up until now there have been very few psychological and cognitive studies of Go-playing. Thus, we started a series of cognitive studies of Go-playing, using mainly traditional protocol analyses and an eye camera [7] [14] [8] [11]. Go-players' protocols in real matches have been gathered and analyzed. Our main purpose is to build a model of Go players' problem solving behavior. This model will help us to understand how humans cope with complex problems such as decision making in Go, and will also become a starting point in our further study of how humans acquire expertise in semantically rich and complex problem domains, such as Go. Also, the model may suggest a good approach to designing strong Go programs.

Computer Go study began in the 1960's, but Go programs are still at the beginner level. Although much effort has been made to create strong Go programs, the methods are restricted to those where programmers' introspection on Go was analyzed by themselves and the results were programmed into their Go program.

Our purpose is to make a model of Go players, specifically of strong players, by analyzing expert players. Our previous studies showed that the mechanism used by experts in solving Tsume-Go problems correctly and quickly is to use hybrid pattern knowledge [15] [16] [17]. This knowledge is constructed by concrete level knowledge such as patterns and abstract level knowledge such as the board situation and the conditions for pattern application. Also, we showed that conceptual knowledge had an important role even if subjects were beginners, by analyzing their protocols. Note that most programs use the approach of a pattern base or a search base[8] [9]. Our protocol analyses showed that all subjects consider the current situation of the board or judge the next move by using the terms specific to Go. However, the role of Go terminology is still an open question. It is also unclear as to whether word role changes with the subjects' skill. Therefore, this paper clarifies word role by examining how the subject's skill influences the use of Go terminology.

Section 2 explains the classification of Go terms proposed by Shirayanagi, which will be used later herein to classify protocols. Section 3 explains the results of protocol analyses in ordinary play. They show that the usage of Go terms depends on the player's skill. Section 4 shows the experimental results collected on Go term usage from various players, except novices. From the results of the two sections, we focus on why players use Go terms in different ways according to skill. Section 5 describes experiments on "Soudan-Go," where two players form a team and can talk to each other while playing a game. The results show that players of different level use Go terms differently. The same Go term carries different information according to the player's skill. Section 6 introduces the "iceberg" model to explain these results.

2 Shirayanagi's Classification of Go Terms

This study is based on collecting and analyzing Go terms. Go has more terms than other games such as chess or Shogi. Chess and Shogi have piece names whereas all stones identical in Go. One reason for the variety of terms created to describe the roles of the stones seems to be that the arrangement of stones is the key to success. Go has also tactical terms as does chess. First, we introduce the Go terms in the classification of Shirayanagi [13]. The simple explanations of the terms are based on [1] as modified by us.

On the board
Form and Position
Posture It is the binary relation between the newest stone and side stone.
- *Narabi*. A solid extension.
- *Kosumi*. A diagonal extension.
- *Ikken-tobi*. An on-space jump.

Relationship It is the binary relation between the newest stone and enemy's stone.
- *Nozoki*. A peep.
- *Boushi*. A capping move.
- *Tsuke*. A contact play.

Position
- *Komoku*. Any of 3-4 points in the four corners.
- *Sazan*. The 3-3 points in any of four corners.
- *Hoshi*. A star point.

Fuseki
- *San-ren-sei*. Three star point stones in a row.
- *Syusaku-ryu*. The Syusaku opening, characterized by 3-4 point moves rotating through three corners.
- *Mukai-komoku*. A linear 3-4 point opening.

Contents and Meaning
Tactical moves
- *Oiotoshi*. Capturing by creating a shortage of liberties through a series of sacrifices.
- *Uttegaeshi*. A snapback.
- *Shicho*. A ladder.

Operation and Tactics
- *Shinogi*. Saving an endangered group of stones.
- *Sabaki*. Making light, flexible shape in order to save a group.
- *Kikashi*. A forcing move requiring an answer.

Evaluation and Judgment
Configuration of stones
- *Aki-sankaku*. An empty triangle.
- *Guzumi*. A move that becomes the apex of an empty triangle.
- *Dango*. A clump of stones.

Move or stone
- *Aji-keshi*. A move which eliminates the *Aji* in a potential.
- *Honte*. A proper move.
- *Karai*. A tight move, i.e., a strongly territory-oriented move or strategy.

Go
- *Komakai-Go* A game in which there are many groups with a lot of intricate situations scattered throughout the board.
- *Taisa*. A big difference in the game.

Rule
- *Ko.* A situation of repetitive capture.
- *Seki.* A situation in which neither of two groups of opposing stones has two eyes.
- *Nakade.* Playing inside a larger eye in order to reduce it to a single eye.

Out of the board
  Psychology and Strategy
    Psychology
- *Tsurai.* Painful move.
- *Ura-wo-kaku* Outwitted opponent by doing just the opposite of what he expected.
- *Kiai* Pumped up for the game.
    Strategy
- *Amashi.* A strategy for White in a no-komi game in which he lets the opponent take good points but as compensation takes territory, aiming to outlast the opponent.
- *Oh-moyo.* An especially large frame work of territory, potential but not actual territory.

3 Protocol Analysis of Ordinary Playing Go

We collected many protocols under ordinary play. Subjects were placed in separate rooms, were asked to talk aloud when thinking and playing through a computer monitor. Table 1 shows the skill of each subject, condition of playing Go, the total number of moves and the amount of protocols transcribed in each game. The amount of protocols in a game can be very large (around 300 KB per game). In Table 1, novice means a subject who had just read a Go book for beginners. We call a kyu level player a 'beginner', an under 2-dan player 'intermediate', over 3-dan player 'advanced'.

The protocols were analyzed in two ways. One was to measure the frequency Go term usage. We used the "Chasen" Japanese morpheme analysis system, and picked the terms. The other was to classify the contents of the utterance in each rough sentence unit and then count them up. Go protocols are naturally divided into subparts by the moves made by either side. Thus this is the 'basic unit' of the analysis. Each basic unit contains several sentences. These sentences were coded according to their contents. Multiple sentences were assigned to a single code from time to time. Table 2 shows the main codes used. The most easily identifiable parts of the protocols were ⟨N⟩, ⟨CM⟩ and ⟨L⟩. We mainly used the plan, purpose, reason and evaluation parts of our protocols in the analyses reported below.

3.1 Frequency of Go Term Usage

Original Terms Formed by the Novices Very interesting results were found when examining the protocol usage of novices. Since a novice does not know Go

Table 1. Collected transcribed protocols in ordinary play of Go.

Match No.	Black Rank	White Rank	Total Moves	Result	Protocol Size	Remarks
1	4k	2k	287	White win	265KB	ordinary game
2	novice	novice	184	suspension	70KB	White wore Eye camera ordinary game
3	2d	4d	88	Black resigned	247KB	Black wore Eye camera with interviewer
4	3d	4d	182	Black resigned	313KB	with interviewer
5	1d	4d	152	Black resigned	182KB	with interviewer
6	1d	4d	247	Black won by 1 stone	371KB	with interviewer 2 stones handicap
7	3k	1d	170	Black resigned	134KB	with interviewer 2 stones handicap

terms, he/she made up original terms and used them. Some of the general Go terms were used, but not often. Indeed, very few Go terms appeared in the protocols, and those that did were elementary ones such as in Table 3.

In the 10th line of Table 3 under the Black side, he says "Ana-Futatsu (two holes)" (this means "two eyes"). This is an example of an original term. Also, he used other original terms like the following examples.

> That is a basic move. (long pause) That move is not understood, not understood. I will capture these stones. Woo. That is a stupid idea, isn't it? Which move is the best? I can not judge that. **A nursery song of the notched heart** (It is a phrase of famous Japanese song.) **notches**, I select it.

The word "notches" reflects the shape shown in Fig. 1. The novice thought that it was an easy way to make an eye (the novice called it a hole).

Note that the terms created by the novices mainly described stone posture or configuration.

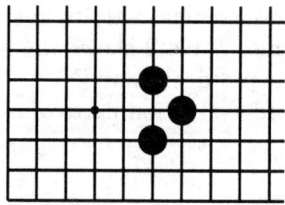

Fig. 1. The shape for "notches" created by a novice.

Table 2. Table of main codes used in our analyses.

Codes	Meanings	Additional explanation
⟨UP⟩	Understand purpose	purposes behind a move
⟨EP⟩	Explain purpose	local and global
⟨EG⟩	Global plan	global plan involving whole board
⟨N⟩	Naming	S⟨N⟩: own move, O⟨N⟩: opponent's move
⟨CM⟩	Candidate move	
⟨L⟩	Lookahead	
⟨SR⟩	Reason of selection	Why one chooses a certain candidate
⟨P-R⟩	Prediction and response	A pair of prediction and response
⟨VJ⟩	Judge winning or not	global and local
⟨VM⟩	Evaluate move (local)	S⟨VM⟩: own move, O⟨VM⟩: opponent's move

Table 3. Top 10 Go terms used by novices (ranked by frequency).

Order	Black		White	
1	Toru	(44)	Utsu	(78)
2	Tsugu	(43)	Oku	(27)
3	Utsu	(30)	Toru	(24)
4	White	(18)	Nuku	(23)
5	Kakou	(16)	Black	(14)
6	Black	(15)	White	(13)
7	Dead	(11)	Dead	(7)
8	Atari	(10)	Tsugu	(6)
9	Cut	(8)	Lose the game	(6)
10	Ana-Futatsu	(6)	Katameru	(3)

Rules Acquired by Novices The novices learned more rules heuristically while playing Go. For example, in a protocol:

> Probably, this group will not be captured if there are two holes. It is useless in case of being a corner. That group should not be captured if two groups are totally connecting. Therefore, here is a hole, and there is not a hole. There is another hole. It is useless to make a hole because this is a corner.

In this protocol, the novice noticed rules, "if there are two holes in a group, then the group is alive (this means two eyes alive)," and "in case of a corner, that rule does not work (this means a false eye)." While playing only a single game, people can learn general rules and some heuristics as these examples showed. However novices did not generate any strategic terms.

The Characteristic of Terms Appearing in Protocols The volume of protocols used by novices was very small as shown in Table 1; novices could not report what they were thinking. They always said "what could I do." The number of terms they used was very small, and many were original. Although beginners used more protocols than novices, they used only a few Go terms. Intermediate players used more protocols than beginners and they often used Go terms but only a limited variety of terms. Advanced players used a large number of protocols and many different terms were used.

The terms were classified according to Shirayanagi's classification. As a rough result, only "form and position" terms appeared in the protocols of novices. The original terms they invented were always "form and position" terms. Beginner players and intermediate players mainly used "form and position" terms, similar to novices. However, 3-dan or better players (advanced players) used "content and meaning" terms and "evaluation and judgment" terms. Especially, "operation and tactics" terms in "content and meaning" terms frequently appeared in the case of their own turn after the middle game.

3.2 Contents of Utterances

Table 4, shows the results of an analysis of the frequency of appearance of contents in the protocols from matches 2, 1 and 3. Naming ($\langle N \rangle$), which appears in most of the basic units, has been omitted from this table.

Table 4. Percentage of basic units containing the following contents(%).

Contents	novice (match 2)	beginner (match 1)	intermediate (match 3)
Purposes ($\langle UP \rangle$ $\langle EP \rangle$)	10.5	16.2	56.8
Candidate move($\langle CM \rangle$)	0.8	6.1	41.0
Lookahead($\langle L \rangle$)	4.6	12.0	18.2
Judge winning or not($\langle VJ \rangle$)	0.0	2.3	6.3
Reason of selection($\langle SR \rangle$)	0.0	0.0	2.3
Global plan($\langle EG \rangle$)	0.0	0.5	5.1

As for the novices, they used $\langle EP \rangle$ and $\langle L \rangle$ categories, and rarely used other categories. As for the beginners, they used all items except $\langle SR \rangle$. Increasing the rate of $\langle UP \rangle$ usage implies an increase in the level of "purpose".

As for the intermediate players, they used all items. Utterances of $\langle UP \rangle$, $\langle EP \rangle$ and $\langle CM \rangle$ were used much more frequently than was true for the beginners. This indicates that the intermediate players tried to understand the intention of the opponent's moves, and so they examined their next move while being aware of ⟨CM⟩. Reasons for choosing candidates were rarely explained by using Go terms[1].

The protocol usage of advanced players was similar to those of intermediate players except ⟨L⟩. ⟨N⟩, ⟨VM⟩ and ⟨P-R⟩ usage increased, although it does not appear in Table 4. Increasing ⟨P-R⟩ usage indicates that the advanced players may have typical sequential patterns.

In summary, advanced players think about their own purpose and their opponents' purpose (intention) at first, only then did they generate candidate moves and look ahead. Novices and beginners rarely judged whether they were winning or not, and they seldom explained the reason for move selection nor proposed a global plan. Purpose (both of his own and of his opponent's) seemed to be the main concern of all players.

4 Do Weak Players Know Go Terms?

The experiments in Subsection 3.1 showed that weak players (beginners and intermediate players) use a limited range of Go terms. We carried out two experiments to investigate whether the weak players knew Go terms or not.

First, we did a vocabulary test of Go terms, where subjects explained Go terms by placing stones on a board. As a result, we found that weak players have less knowledge than advanced players and that there was positive correlation between the vocabulary test and their skill.

Next, in order to see whether weak players really did not know Go terms, we conducted a recognition test, wherein subjects provided the Go term for the situation they were shown. This experiment was carried out on two subjects, whose results in the former were bad. As a result, we found that the subjects could easily recognize the situation using Go terms, even if they were not strong players.

4.1 A Vocabulary Test of Go Terms

The method for this experiment is as follows. A Go term was presented in the upper left corner of a monitor. The subjects then placed stones on the Go board displayed in the center of the monitor or explained the meaning of the term. Two advanced players, an intermediate player, and a beginner were examined. One hundred of the Go terms that are often used in Go books or commentaries were shown.

As a result, it was proved that there was a difference in the knowledge of Go terms according to subjects' skill. Table 5 shows the percentage of correct answers. Abstract terms such as "thickness" or "*katachi*" have more than one "correct answer". When a reasonable explanation was given, we judged it as "correct answer (knowing the term)". "Wrong answer" was awarded only when

---

[1] One reason why they explained the reasons may be because an interviewer sat beside the player during each game.

the answer was "don't know" or nothing could be explained at all. Even with this loose criterion, Table 5 shows that the number of terms known depends on the skill: the better one's skill is, the more terms one knows. While the difference of skill between beginner and advanced is huge, the result is not significantly different. We think that there are two reasons for this. One is that we used very famous terms in the test. Another is the loose criterion used. The protocols of the experts were more accurate than those of the beginners.

Table 5. The correct rate of a vocabulary test of Go terms.

level	correct rate
beginner	66%
intermediate	42%
advanced	92%
advanced	93%

4.2 A Recognition Test of Go Terms

This experiment was conducted as follows. Some specific board situations were presented on a monitor in front of the subject. Sometimes a static configuration of stones was presented, and sometimes a sequences of moves was displayed. Each situation had a proper name. For example, the numbered stones in Figure 2 were sequentially added to the board, and the subject was asked the name of the sequential move. The recognition test was held at least one week after the vocabulary test.

Table 6. The rate of correct answer for board situation recognition experiment

Subject	No. of Problems	No. of effective problems	No. of correct answers	Correct answer	Correct rate for original 100 words
beginner	37	34	18	53%	83%
intermediate	46	46	27	59%	79%

The procedure is explained in detail below. When the initial board situation was presented, we asked the subject what would be his choice for the next move. Sometimes the subject used the term expected when he explained his choice. After showing the sequence of moves, we asked the subject what he would call such a sequence of moves. This procedure examined whether the Go terms were developed from the board situation. The board situation recognition experiment was given to beginners and intermediate level players.

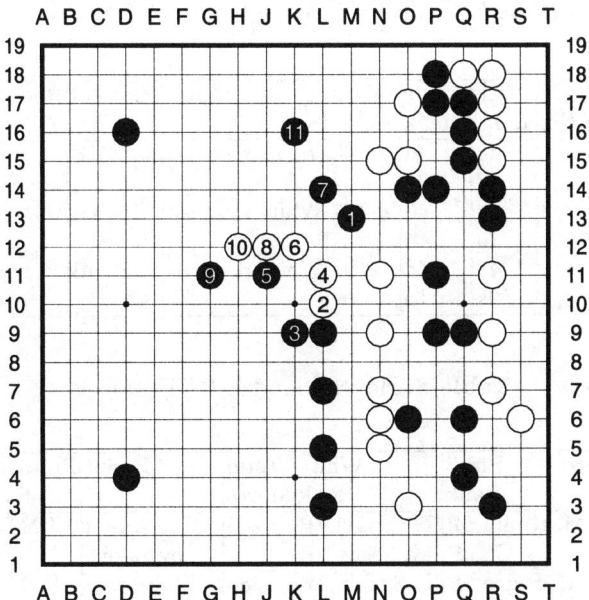

Fig. 2. Example of board situation presented to subjects (example of "Karami-zeme")

The results of the experiment are shown in Table 6. Note that a loose criterion for the correct answer was used. For example, when the subject answered only "Harazuke" instead of "Itachi-no-harazuke", we judge the answer as correct.

The problems presented in this test are those for which the subjects failed to make correct answers in the vocabulary test. [2]

"The correct rate for original 100 words" is calculated by summing the number of correct answers in this experiment and the number of correct answers in vocabulary test in Section 4.1. The result shows that the recognition rate does not change according to the skill very much.

5 Protocol Analysis of Soudan-Go

The protocol analysis of ordinary play showed that there was a difference in using terms and in the content of the subject's introspective protocols according to skill. In the above recognition test, there was no big difference in the ability to recognizing the terms from a board situation according to skill. Why is there a big difference in using Go terms in the protocol, though they have the ability to use the terms? We analyzed the "Soudan-Go" protocols in order to identify

---

[2] Some problems which were not presented in the vocabulary test were presented in this test to the beginner.

Table 7. Collected Playing Soudan-Go Protocols

Match No.	Black Rank	White Rank	Total Moves	Result	Protocol Size	Remarks
8	4d / 1d	4d / 3d	140	Black resigned	205KB	
9	4d / 1d	4d / 1d	177	White resigned	158KB	
10	4d / 2k	4d / 1d	91	White resigned	89KB	13x13

Table 8. Houchi Soudan-Go Protocols

Match No.	Black Rank	White Rank	Total Moves	Result	Protocol Size	Remarks
11	M. Kitani 6P / S. Go 5P	T. Suzuki 7P / K. Segoe 7P	245	Black won by 1stone	1.5MB	No *Komi*
12	K. Segoe 7P / S. Go 6P	T. Suzuki 7P / M. Kitani 7P	286	White won by 8.5 stones	2.2MB	3.5 poiint *Komi*
13	K. Iwamoto 6P / U. Hashimoto 6P	M. Kitani 7P / C. Maeda 6P	190	Black resigned	1.5MB	3 point *Komi*

the reason. Soudan-Go is a game between two groups, each of which consist of two players, who can talk to each other freely. The combination of the groups is shown in Table 7.

Also, we analyzed "Houchi Soudan-Go" protocols to examine the word usage of expert players. The protocols were taken from printed records as reported in the Houchi Newspaper. While there is no guarantee that the original utterances were reproduced exactly, we found that utterance category was. Therefore, we only analyzed the utterance part of the report. Table 8 shows the list of Houchi Soudan-Go.

5.1 Analysis of Soudan-Go

By comparing Table 4 to Table 9, we find that advanced players do not speak more in explaining their selection than intermediate players. One reason is that advanced players made very few candidate moves, usually only one. Another reason is that advanced player's thinking is compactly conveyed to his partner by the use of appropriate terms. One example is shown below.

[After black plays at 53-th move] **Black1**: ... if white plays *kake*, we play *keima*, (**Black2**: Yeah) and if white plays *kake* again. (**Black2**: Yes) What do we do? We have other potential power, so this black can live

Table 9. Percentage of basic units containing various contents(%)

Contents	advanced (match 9)
Purposes(⟨UP⟩ and ⟨EP⟩)	54.0
Candidate move(⟨CM⟩)	25.1
Lookahead(⟨L⟩)	3.9
Judge winning or not(⟨VJ⟩)	1.4
Reason of Selection (⟨SR⟩)	0.6
Global plan(⟨EG⟩)	3.0

easily, (**Black2:** yes) (Long pause) [white plays at 54-th move], Ah, here they come. **Black2:** Ah, they played.

In match 5, we explicitly asked players to explain their candidate moves and their reasons for their selections; they could explain their thinking using short Go terms.

The following example shows that naming, which means to identify the board situation, is very important in recognizing the board situation and selecting a strategy. "*Kake*" in following example implies that opponent should answer the *keima*. We found that terms were charged with various meanings.

> In Soudan-Go match (match 8), after the game, two parties came to the same room, and they discussed what was good and what was bad. Finally black lost the game, and they attributed the cause of their loss to white's 54-th move and 56-th move. In the protocol, black refers to these two moves as "two *kakes* in the center", while white refers to them as "two *keimas*".
> **Black1:** ...We estimated the situation too optimistically, we overlooked the effect of *kake*. **Experimenter:** Where is it? More towards the beginning? **Black1:** It was a *kake*. **Black2:** White's *kake* in the center. **Black1:** That two *kakes* in the center. We know that white wouldplay, but we did not take it seriously. **Black2:** We said "not so serious". **White1:** Ah, those two *keimas*? **Black1:** Yes, yes. **Black2:** Yes, yes. **Experimenter:** Ah, these moves. **Black1:** We underestimated their effect.

Kake and *keima* were used in that situation. Because the black players felt that white stone was superior to the black one and the white stone oppressed the black one, they called the situation *kake*. *Keima* is just a posture category term.

5.2 Analysis of Houchi Soudan-Go

Houchi Soudan-Go was held from 1936 to 1938. The sponsor was the Houchi newspaper company. The aim of the series was to present the professional players' thinking to citizens.

One example is shown below. This example is the opening of a game(match 13). The professional discussed the local strategy.

> **Kitani:** I think opponent should place a stone at a *star point* or the *komoku* point. Woo. Probably a *star point*. If they select a *star point*, most possible *star point* is at the upper right corner, then we select the diagonal *star point*, don't we? **Maeda:** But star point is too simple. Shall we select diagonal *oh-takamoku* point? **Kitani:** If we adopt your suggestion, the opponent should place a stone at *star point*. Then we have to select *kosumi* position to complete. Then that move makes this game very *busy*, I dislike the move. Of course we can continue to play after selecting the move, but *star point* may not cause our errors, I think. **Maeda:** The second move depends on the opponent's move. What is the third move of the opponent? *Takamoku* was tested before Soudan-Go, so they change the place. Probably, they select *star point* or *komoku* at the lower right corner. **Kitani:** That is the next consulting subject just after we watch the opponent's *moyo*. Anyway, we decide if the opponent select the upper right *star point*, then we select diagonal *star point*.

This example means that even when an expert player used only position terms they implied judgment and planning. For example, Kitani said "*busy*" in the middle of the quotation. "*Busy*" means the judgment of the future situation, not current. They shared the same future board situation using only two words such as *oh-takamoku* and *kosumi*. So the term *oh-takamoku* has the feature of the plan to make future image in this situation.

In the middle game, they often discussed evaluations. For example, "that is a solidly built move" and "and then opponent selects the *kosumi*, which causes unbelievable situation". That means ⟨VM⟩ is used more often in their utterance than in advanced players'. That is, professional players' utterances suggest or predict move(s) and evaluate them. This form is similar to that of advanced players.

Although, both advanced players and professionals explain their evaluations, the level of the explanation differs. While professionals use terms such as "unable to escape," or "becoming a decisive battle," advanced players say just "connecting" or terms in "form and position" category of Shirayanagi's classification, a quite different response.

6 Discussion

The recognition test showed that all subjects except novices have the ability to recognize board situations in Go terms. However, the results of the protocol analysis of ordinary Go showed that the use of the terms in the protocols deeply depends on the level of skill. There is also a difference in the utterances. Intermediate or better players can discuss purpose, evaluation, and planning. One of the differences between intermediate players and better players is in the use of utterances to explain selections. Intermediate players gave detailed explanation, while professional players used abstract reasoning terms, such as "busy" or "quick". Beginner and intermediate players evaluated their moves using sentences, while advanced players made evaluation by using Go terms. Professional players evaluated their choice by using Go terms and abstract words. Furthermore, professional players used form and position terms in order to convey their purposes and ideas to their partner.

Therefore we have to explain the role of terms. There are three constraints.

- Even form & position terms can describe the purpose.
- When a subject looks at the board, he/she is conscious of purpose and future image expressed through the Go terms.
- Unskilled players cannot use Go terms like advanced players.

We propose that the iceberg model satisfes the above constraints to explain how the differences in the usage of terms in utterances accords to the skill of the player.

6.1 Iceberg Model

The iceberg model (Fig 3) attempts to explain the observed differences in the usage of Go terms. Each Go term is an iceberg. The tip of the iceberg lies above the surface of the water and the next lies below the surface. The upper part can be easily observed, while the lower part is more difficult to directly observe.

Each Go term has a lexical meaning which has a dictionary description and is easy to define, such as a configuration of stones, i.e. This is the part above the surface of water. All subjects except novices understand the lexical meanings of Go terms as shown by the recognition test. However, we think that the terms have various meanings other than the lexical meanings, as was observed in the protocols. For example, when someone says "the opponent puts the stone at the position of *keima*, which is cutting, and then we select *deru*", "*Deru*" means not only going through the opponent's wall, but also taking advantage. The meaning, "taking advantage", emerges from the interaction of the term and the current situation of the board. In other words, we think that the term itself may suggest the emergent meaning by referring to the current situation of the board. The emergent meaning is the part below the surface of water.

Advanced players explained their reasoning at an abstract level using only Go terms. In the case of Soudan-Go, neither professional nor advanced players used concrete terms to explain their reasoning. Although the terms could be interpreted in different ways, misunderstandings did not occur amongst the partners, as was observed that they agreed the following discussion. On the other hand, intermediate players clearly explained their reasoning in detail using concrete words. Professional and advanced players used only the Go terms,

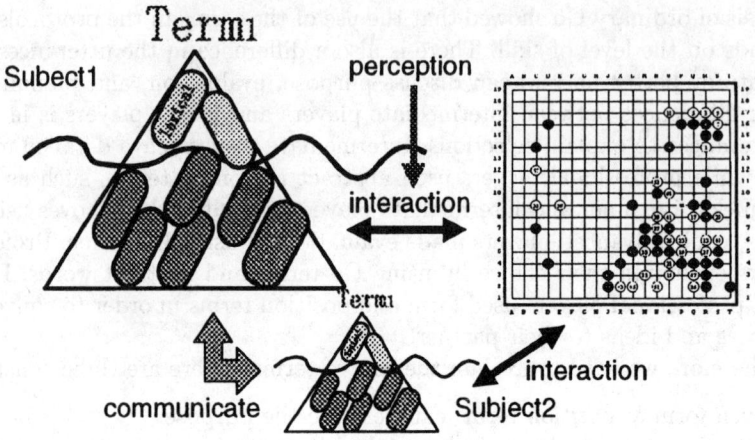

Fig. 3. Iceberg Model

which include the meaning of intermediate player's explanation. When an advanced player says a term to the other advanced player, the emergent meanings of the situation or values are also conveyed to the partner. Emergent meanings of terms differ according to the skill.

Let us consider frame representation in AI to explain emergent meaning. Image that each iceberg consists of slots. When a player understands the board situation, he fills the slots by extracting information from the board. One problem exists: whether all slot names are decided beforehand or not. Let us consider the following examples. When a player's situation is bad and he says "opponent selects *mage*, then I select *hane*, then he selects *hane*", *hane* of one's own play implies endurance or defense. When the situation is good, however, it implies offense. In short, the meaning of the terms used depends on one's situation. However, we suppose that there are not so many meanings in a term and that a class of slots may be decided beforehand for a player's level, even though the class of slots differs enormously between different level players. In other words, slot names are not defined beforehand, but the class of slots is defined. This flexibility is the source of the emergent meaning. In the original concept of frame representation, all the slots should be defined beforehand. This is the difference between the traditional frame representation and our model.

For the submerged part of the iceberg slots are filled only when the subject looks at the board. Those slots consist of on-the-fly knowledge. So, when a subject expressed the term to his partner, the partner has to construct the submerged part by looking at the board. Only when they have same slot names in this part, they can understand each other.

The size of the iceberg depends on the player's skill. Advanced players have larger icebergs and the submerged part become larger than those of intermediate players. This is the result of increasing the number of slots.

6.2 Comparison of Iceberg Model with Template Theory

A model similar to the iceberg model was proposed by Gobet and Simon[5]. It was called "template theory." The model was introduced to explain how memory chunks[2] evolve into templates. That is the big difference from our iceberg model. However it is interesting that the structure is very similar. The template also consists of frames. The main difference in the structure of the iceberg model and template theory is the slot. The slot of the iceberg model is only approximately decided, while the slot of the template theory is predefined. Of course, flexibility is possible in the template theory by choosing from among various kind of slots. The difference reflects the differences between Go and chess, which are described in the following two examples.

Our previous study showed that expert players hold the board image as a mixture level description[18]. The mixture level description was reported to be a uni-space structure. That is, stones involved with an area of focus are clearly remembered, while other areas are unified at an abstract level. More precisely, when advanced players had enough time to observe and recall the board situation, they could reconstruct the board by recalling the relationship between stones. When recall was performed under time pressure, they only recalled stones that were associated with important areas, and other areas were represented using features, such as "black area" or "white is stable." And which stone can be recalled does not depend on the distance from the focused area. Even if the board was small, such as a 9 by 9 board, the effect was the same. Another of our studies showed that human players recalled the board situation in the sequence order even when they only observed a static pattern[12]. In recall, they often verbalized form and position category words in Shirayanagi's classification. Therefore, advanced players could easily find sequential meaning. These two studies indicate that human memory about static Go boards is based on uni-space of the meaning including the concrete shape of stones.

It is well known that advanced Chess and Shogi players can easily reconstruct a board by hearing the game record. The same is not rue in Go even with a 9 by 9 board. The game record of Go consists of stone color and position. The imaging task is made easier if Go terms was well as the game record are given to the player[19]. Specifically, the correct imagining rate is high when both form and position category terms and content and meaning category terms are given in addition to the game record.

The above discussion indicates that for humans to perceive the board situation of Go, they must invest each configuration of stones with a role. In case of Chess or Shogi, the role of each piece is already given by the definition of movement. So, in Go, more basic level perception is necessary than chess. Therefore in Go, human should invest the configuration of stones with multiple roles, described by multiple terms. Thus we can not put the static relation of configuration of stones and basic function beforehand. Player should assign a function to the configuration of stones dynamically according to the board situation.

Intermediate players have a structure of knowledge quite similar to that of the template theory. Advanced and professional players, even if they use position terms, assign to the term information about the player's plan, strategy, judgment and so on, all of which are strongly dependent on the situation. Plan and strategy sometimes are contained together in the same term. Accordingly, they do not exist as independent slots.

7 Conclusion

We analyzed the Go protocols expressed during ordinary play and Soudan-Go. We carried out experiments to see how many Go terms were known to a wide range of subjects. As a result, all subjects, except the begineers, knew the Go terms, but term usage differed with the subject's skill. The iceberg model was proposed in order to explain the results.

Acknowledgment

Support for this research is gratefully acknowledged from Dr. Ken-ichiro Ishii, the executive manager of the Information Science Research Laboratory in NTT Basic Research Laboratories. Members of the Dialog Understanding Research Group gave us invaluable comments regarding this paper.

References

1. R. Bozulich : The Go Player's ALMANAC, The Ishi Press, (1992).
2. W. G. Chase and H. A. Simon : Perception in Chess , Cognitive Psychology, Vol. 4, pp. 55-81 (1973).
3. A. D. de Groot : Thought and Choice in Chess, The Hague: Mouton (1965).
4. K. A. Ericsson, and H. A. Simon : Protocol Analysis – Verbal Reports as Data (Revised Edition), The MIT Press, (1993).
5. F. Gobet and H. A. Simon : Templates in Chess Memory: A Mechanism for Recalling Several Boards , Cognitive Psychology, Vol. 31, pp. 1-40 (1996).
6. Y. Saito and A. Yoshikawa : How to make Stronger Go Programs (in Japanese), in *IPSJ SIGAI*, No. 91-7, pp. 55-64, (1993).
7. Y. Saito and A. Yoshikawa: Cognitive Studies for the Game of Go (in Japanese), in Proc. of the Game Programming Workshop in Japan, Hakone, pp. 44-55, (1994).
8. Y. Saito and A. Yoshikawa: Do Go Players think in Words?, Proc. of the 2nd Game Programming Workshop, pp. 118-127, (1995).
9. Y. Saito and A. Yoshikawa: An Analysis of Strong Go-players' Protocols, Proc. of the 3rd Game Programming Workshop, pp. 66-75 (1996).
10. Y. Saito: Psychological and Cognitive Researches on Games (in Japanese), Proc. of the 3rd Game Programming Workshop, pp. 44-55 (1996) .

11. Y. Saito: Cognitive Scientific Study of Go (in Japanese), Doctoral dissertation, University of Tokyo , (1996).
12. N. Shingaki and A. Yoshikawa : Spatial chunk and sequential chunk of Go (in Japanese), Proceedings of Japanese Society of Cognitive Science 11th annual conference, pp. 102-103 (1994).
13. K. Shirayanagi: Basic Study of Knowledge Processing for Go, *NTT Internal Technical Report*, 12967,(1986).
14. A. Yoshikawa and Y. Saito: Cognition of Board Situation in Go (in Japanese), *IPSJ SIGAI*, No. 91-6, pp. 41-53, (1993).
15. A. Yoshikawa and Y. Saito : Perception in Tsume-Go under Three-second. Time Pressure (in Japanese), Proc. of the 2nd Game Programming Workshop, pp. 105-112, (1995).
16. A. Yoshikawa and Y. Saito : Can not solve Tsume-Go problems without looking ahead? (in Japanese), Proc. of the 3rd Game Programming Workshop, pp. 76-83 (1996).
17. A. Yoshikawa and Y. Saito : Hybrid Pattern Knowledge – Go players' knowledge representation for solving Tsume-Go problems –, in 1st International Conference on Cognitive Science in Korea, pp. 134-139 (1997).
18. A. Yoshikawa : The forefront of the computer Go studies (in Japanese), Proceedings of Japanese Society of Industrial and Applied Mathematics annual conference 1998, pp. 248-249 (1998).
19. A. Yoshikawa : in preparation.

A Survey of Tsume-Shogi Programs Using Variable-Depth Search

Reijer Grimbergen

Electrotechnical Laboratory,
1-1-4 Umezono, Tsukuba-shi, Ibaraki-ken, Japan 305-8568
grimberg@etl.go.jp

Abstract. Recently, a number of programs have been developed that successfully apply variable-depth search to find solutions for mating problems in Japanese chess, called *tsume shogi*. Publications on this research domain have been written mainly in Japanese. To present the findings of this research to a wider audience, we compare six different tsume programs. To find the solutions of difficult tsume-shogi problems with solution sequences longer than 20 plies, we will see that variable-depth search and hashing to deal with a combination of transposition, domination and simulation leads to strong tsume-shogi programs that outperform human experts, both in speed and in the number of problems for which the solution can be found. The best program has been able to solve *Microcosmos*, a tsume-shogi problem with a solution sequence of 1525 plies.

Keywords: Variable-depth search, best-first search, conspiracy-number search, game playing, tsume shogi

1 Introduction

Most work on game-tree search has focused on algorithms that make the same decisions as full-width minimax search to a fixed depth [15]. Examples are alpha-beta pruning and SSS^* [30]. Human players do not use full-width fixed depth search, but a combination of shallow search and deep search [4]. A number of algorithms have been proposed that perform variable-depth search. For example, *conspiracy numbers* [20, 27], *singular extensions* [3], *proof-number search* [1, 2] and *best-first minimax* [15] are all algorithms for searching game trees without explicit bounds on the search depth.

One of the domains where search to variable depths has been very successful but where this success has been almost unnoticed by the international AI community is *tsume shogi*. Tsume shogi are mating problems in Japanese chess. Since the early 90s, several tsume-shogi programs have been developed that can quickly find the solution of problems with solution sequences of more than 50 plies. Both on the development of strong tsume-shogi programs and the characteristics of the programs, there are a number of publications in Japanese ([6, 7, 8, 9, 10, 17, 24, 25, 29]). There have been a few English publications on tsume shogi, but only Seo [28] and Kawano [13] give a description of a program for finding the solutions to difficult tsume-shogi problems [1]. Some features of a tsume-shogi program can be found in [22], but the description is very brief.

In this paper, we would like to present the results of tsume-shogi research to a wider audience. In Section 2 a short explanation of the rules of tsume shogi are given, along with its history and the relevance for a shogi-playing system. In Section 3 the computational features of a program to find solutions for tsume-shogi problems are given. In Sections 4 to 9, a number of tsume-shogi programs are described. Also, their results on different test sets of tsume-shogi problems are summarized. We end with some conclusions and thoughts on future research on a general shogi-playing system using methods discussed in this paper.

2 Tsume Shogi

2.1 Rules of Tsume Shogi

Tsume shogi are mating problems in Japanese chess. As far as the rules of Japanese chess are concerned, to understand the contents of this paper it is sufficient to know that shogi is similar to chess. The aim of the game is the same as in chess, namely the mating of the opponent's king. The shogi board is slightly bigger than the chess board, 9×9 instead of 8×8. Some pieces in shogi are the same as in chess, like the rook and the bishop, but some pieces are different. There is no queen in shogi, but instead there are golden generals, silver generals and lances. Promotion is also a little different. Most pieces can promote and can do so on any of the top three ranks of the board. The most important difference between shogi and chess is that in shogi captured pieces can be re-used. A piece captured becomes a *piece in hand* for the side that captured it. When it is a player's turn, he can either play a move with a piece on the board or put one of the pieces previously captured back on a vacant square on the board (this is called *dropping* a piece). It then becomes his own piece. Most drop moves are allowed, even dropping with check or mate is legal. Finally, it should be noted that in shogi the player to move first in the starting position is black and the other player is white (in chess this is the other way around). For a more detailed comparison of chess and shogi, see [19].

The rules of tsume shogi are simple. The goal of the attacking side (black) is to mate the king by consecutive checks; the goal of the defending side (white) is to reach a position where the attacking side has no checks. Therefore, the attacking side has to give check at every move and the defending side has to defend against these checks and prevent mate as long as possible.

2.2 History

Tsume shogi has a long history. The first tsume-shogi problems date back to the 17th century [12] and the collection of tsume-shogi problems that is still

---

[1] Seo's work is described in a Master's thesis, so not easily available.

considered to be the most brilliant ever, has been published in 1755(!) and were composed by Ito Kanju [11]. Also, there are many books with collections of tsume-shogi problems and all shogi magazines have tsume-shogi problem corners. There is even a monthly magazine called *Tsume Shogi Paradise* dedicated to tsume shogi. Of course, in tsume-shogi problems that are published in shogi magazines the artistic element is very important, just as in chess problems. Recently, there have been attempts at automatically composing tsume-shogi problems with artistically appealing features [5].

2.3 Relevance for Shogi Game Playing

Tsume shogi is not the same as perfect endgame play. It is possible that a game can end in fewer moves than by consecutive checks. However, tsume shogi is very important in the shogi endgame. In the endgame, the number of pieces in play is the same as on the first move. Furthermore, dropping pieces can have a major impact on the strength of attack or defense. Therefore, mate is the prime objective and resignation because of material deficit is rare. Usually a player resigns when he can no longer avoid mate, either because the opponent has started a tsume sequence (continuous checks leading to mate) or if there is no defense against such a tsume threat. The shogi endgame is a mating race, so finding mate and realizing that the opponent is threatening mate is vital for the endgame strength of a shogi player and also for a shogi-playing program.

2.4 The First Tsume-Shogi Program

The first tsume-shogi program was built by Ochi in 1968 (described in [10]). Ochi's program has been reported to find the solution of tsume problems with a solution sequence of 9 to 11 plies as quickly and accurately as human players. Even though the program ran on one of the fastest computers of its time, this is quite an incredible result for a program that is 30 years old. We have been unable to find the original paper with a description of the program and the supporting data for this claim. In any case, for 25 years there was no breakthrough in the development of tsume-shogi programs that made it possible to find the solution of problems with a solution of 15 plies or more. This changed with the introduction of algorithms for searching to variable depths in the early 1990s.

3 Computational Features of Tsume Shogi

A correct tsume-shogi problem should have only one solution. The search tree for a tsume-shogi problem is an AND/OR tree, or a minimax tree where the evaluation of every node can have only two values, TRUE or FALSE. At each OR-node it is sufficient to find one check for which all the defenses lead to mate. If there is one check that leads to mate from the root, the problem is solved and the search can be stopped.

A tsume-shogi program to search an AND/OR tree has to deal with several problems:
- search deep for long mating sequences
- avoid redundant search
- recognize that positions have the same mating sequence
- decide which moves to search first

We will now describe each of these problems in detail.

3.1 Problem 1: Long Solution Sequences

Tsume-shogi problems have a solution length ranging from 3 plies to hundreds of plies[2]. Seo [28] has experimentally found that the average branching factor of tsume shogi is only about 5. This is much smaller than the average branching factor of normal shogi play, which is about 80 [18]. However, even with this small branching factor, it is difficult to find the solution of tsume-shogi problems with solutions of more than 15 plies by brute force methods. Currently the tsume-shogi problem with the longest solution sequence is a problem called *Microcosmos*, which has a solution of 1525 plies. Finding long solutions of tsume-shogi problems is the first challenge for tsume-shogi programs.

3.2 Problem 2: Avoiding Redundant Search

To avoid searching the same position from different parts of the search tree, usually a standard transposition table is used. In shogi, not only transposition, but also *domination* can lead to redundant search. Domination is a concept that is used in every tsume-shogi program, but it has not been properly defined in the literature. We define it as follows:

Definition 1 *A position P is dominating position Q if the board positions of P and Q are the same, and the pieces in black's (white's) hand of P are a proper superset of the pieces in black's (white's) hand of Q and it is black's (white's) turn to play in both P and Q.*

In tsume shogi, if P dominates Q regarding the attacker's pieces and a mating sequence has been found in Q then this mating sequence will also work in P. One way to check this, is to store extra information about the pieces in hand in the hash table.

3.3 Problem 3: Finding Simulating Positions

In chess, the number of checks and defenses against these checks is very limited. In tsume shogi the possibility of dropping pieces greatly increases the number of possible moves and makes the problem much harder. For example, in Diagram 1, the defending white king on 1a is checked by the black rook on 9a[3]. The white

---
[2] A position where there is mate in one move is trivial and not considered a tsume-shogi problem.
[3] The possibility of dropping pieces makes the total number of pieces in any shogi position the same (40). In Diagram 1 only the relevant pieces are shown.

pawn on 1b blocks the king's escape to 1b. A lance in shogi is a piece that moves like a rook but only in the forward direction. The black lance on 2i therefore blocks the escape of the king to 2b and also covers 2a. In a similar chess position, this position would be mate.

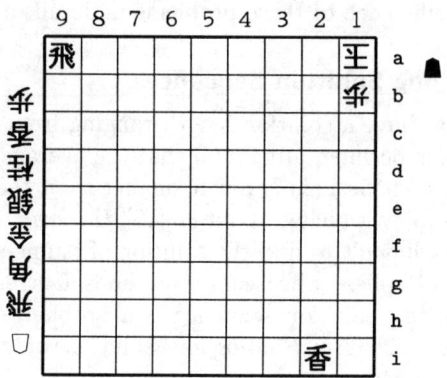

Diagram 1. An example of interpositions

However, in this shogi position the defense has all the pieces on the left side of the board in hand. The defending side can therefore drop any of these seven pieces between the checking rook and the king. Since there are seven vacant squares between rook and king and seven different pieces to drop, the number of interposing defenses is 49.

A rule of tsume shogi is that drops which do not change the solution sequence but only increase its length are not considered proper defenses. These useless interpositions are not counted as moves in the length of the solution sequence. Since all the interposing moves in Diagram 1 are useless (black can take any interposing piece immediately with the rook), this position is mate according to the rules of tsume shogi. For a tsume-shogi program to recognize when drops are useless for check or defense with as little search as possible is a non-trivial task.

Another complication of tsume shogi is that in most cases promotion of pieces is optional. For tsume shogi, this usually leads to the same move sequence, since the moves of the promoted piece are in general only slightly different from the moves of the unpromoted piece. However, even in the case where the moves of a promoted piece are a superset of the moves of the unpromoted piece, there are some special cases where promotion of a piece does not lead to mate, while the non-promotion of a piece does. As a result, a tsume-shogi program has to search the moves where a piece promotes and also the moves where the piece does not promote. Since in shogi a piece can promote on any of the top three ranks, this considerably increases the number of moves to be searched. A tsume-shogi program should avoid redundant search for optional promotion moves as much as possible.

To deal with the problems of useless interpositions and optional promotion, the concept of *simulation* has been introduced. The general definition of simulation is [13]:

Definition 2 *A position P simulates position Q if all move sequences that can be played from position P can also be played from position Q.*

For example, the moves of a promoted rook are a superset of the moves of an unpromoted rook. If P is a position with an unpromoted attacking rook and Q is the same position with a promoted attacking rook, then P simulates Q. For tsume shogi, it would seem logical that if there is a mate in position P, there is also a mate in Q. However, in the previous section it was already mentioned that there are some special cases where a position with a promoted rook does not lead to mate, while a position with an unpromoted rook does.

Also, moves may have the same meaning even though they are not exactly the same. For example, a mate in position P is found and next position Q is searched where the only difference between the two positions is that the starting square of the rook is shifted one square to the left or right. This is a case where P might simulate Q. In these cases, it is natural to try the mating sequence in P first. However, since the starting square of the rook move differs, these positions do not simulate each other according to the strict definition given above.

Therefore, usually a more general, but heuristic concept of simulation is used in tsume-shogi programs. Instead of searching for simulating positions where the move sequences *must* be the same, some shogi dependent knowledge is used to look for positions where the move sequences are *likely* to be the same. If the current position P might simulate a position Q based on these heuristics and a mate from P has been found, then the mating sequence found in P is tried first in Q. In tsume-shogi programs the following heuristics are used for simulation:

– promotion vs. non-promotion
– different starting squares for the same long range piece (rook, bishop and lance)

3.4 Problem 4: Most Promising Moves First

The final challenge for a tsume-shogi program is to guide the search for a solution in the right direction. Expert human tsume-shogi solvers are very good at selecting promising candidate moves from a position. Usually the first or the second move considered in a position is the move leading to mate. To avoid wasting time by searching moves that are not likely to lead to mate, move ordering is very important.

Now the problems a tsume-shogi program has to deal with have been discussed, we will give a description of six tsume-shogi programs:

- Noshita's T1, T2 and T3 tsume-shogi programs
- Ito's tsume-shogi program
- Kawano's tsume-shogi program
- Seo's tsume-shogi program

These are the strong tsume-shogi programs for which a detailed description of the methods used have been published. There are other strong tsume-shogi solvers, but these are part of commercial shogi software and there have been no publications on them. Especially famous in this category is a program called Morita Shogi. Results of this program [21] show that it might be just as good as the programs described here. It would be very interesting to know if this program uses different methods for finding solutions in tsume-shogi problems.

4 T1: Iterative Alpha-Beta with Selective Deepening

4.1 Method

In 1991, Noshita's T1 program [24] became the first program to use selective deepening in combination with iterative alpha-beta search to be successful finding solutions of tsume-shogi problems with solutions longer than 11 plies. T1 uses alpha-beta iterative deepening with only limited selective deepening. The selective deepening is based on the heuristic of measuring the freedom of the king. After each move, for each of the eight squares adjacent to the king it is calculated if the king can move to this square or not. If the king is very limited in its movement, the position is assumed to be close to mate and the search is extended for a maximum of 4 plies to try and find a mate.

T1 puts a lot of emphasis on move ordering. There are no less than 60 criteria for move ordering. Examples are: ordering promotions higher than non-promotions, preferring king moves away from the attacking pieces and ordering drops higher the closer they are dropped to the king. Noshita also used parameter learning to tune these move ordering criteria.

T1 has some heuristics to deal with transposition and domination, but has no hash tables. To avoid useless interposing drops, T1 uses the definition of the shogi programmer Kakinoki [16]:

Definition 3 *(i) An interposed piece is useless if capturing the piece results in a mating position. (ii) Suppose the king is mated in N moves in a certain position P without any interposing drop between checking piece and king. Then an interposed piece in P is useless if it is immediately captured, and there is a mating sequence of length N after the capture and the captured piece is not used in the mating sequence.*

In T1 this is applied as follows. If a piece is dropped between the checking piece and the king on square S and can be captured immediately, this might be a useless interposing drop. If capturing the piece results in mate, the drop was useless according to (i) and all other pieces that can be dropped on S are useless interposing drops and search can be stopped. If taking the dropped piece does not give a mating position, but a move sequence is found leading to mate after the capture without using the captured piece, then the dropped piece on S was a useless interposing drop according to (ii) and search can be stopped.

4.2 Results

T1 was implemented on a 16 MHz PC98-RA21 (a Japanese PC) and on a SUN IPC workstation. Here we will summarize the results of three tests performed with the program [24]. One was a test set of 50 tsume-shogi problems made by the professional shogi player Nakata Shodo. The length of the solution sequence was 7 to 15 plies. The performance of T1 was compared to that of human experts. T1 found the solutions of the entire problem set in about 27 minutes, while the best human expert could only solve 46 problems within a one hour time limit.

There are some unanswered questions with this test. One is that the original test set had 54 problems. It is unclear why only 50 of those problems were selected. Also, it is unclear how many human subjects there were and what the exact test conditions were.

The second test given to T1 was a set of 32 difficult tsume-shogi problems from a shogi magazine. The solution length of the problems was 7 to 9 plies. T1 found the solution of all the problems with an average solution time per problem of 55 seconds. According to the scoring table attached to the tsume problems, this is the highest possible level: "professional strength". Of course, the solutions of these problems are not long enough to really test the program. After all, the old program by Ochi mentioned before already claimed the same performance.

The final test for T1 were eight 15 ply tsume-shogi problems and eight 17 ply tsume-shogi problems. The 15 ply tsume problems were solved in 130 seconds on average, while the 17 ply tsume problems were solved in about 10 minutes on average.

Noshita's own conclusion is that T1 works well for tsume-shogi problems with a solution sequence up to 17 plies and is able to find the solution of some problems with solutions that are longer than 30 plies. However, for most of the problems with a solution longer than 17 plies, T1 is too slow to find the solution within a reasonable time limit.

5 T2: T1 Plus Hash Tables

5.1 Method

T2 is also made by Noshita and an improved version of T1, using similar ideas, which are described in [9, 25]. One improvement in T2 over T1 is that the heuristics for king safety are better. However, the big difference between the two programs is the introduction of hash tables for transposition, domination and simulation. The search for useless interposing pieces is now done differently. If a possible useless interposition is detected, the piece is taken and given back to the defending side. If there is a mate in the resulting position, then the piece was indeed a useless interposition and all moves with other pieces dropped in defense on the same square need not be searched. Here hashing is very useful if the position with the capturing piece closer to the king is searched first and stored in the hash table. For tsume shogi the introduction of hash tables can make a big difference. Noshita gave the position of Diagram 1 before the check with the rook on 9a (black rook on 9i instead of 9a) to both T1 and T2. T1 had to search 1,500,000 positions to search through all the interposing drops and find a mate, while T2 only needed to search 25,000 positions.

5.2 Results

T2 runs on a SUN SPARC IPX with 28 MB of memory. For T2 a different test set of 3 to 9 ply tsume problems was used. There were 112 problems in the test set. T2 found the solutions to the whole problem set in 1 minute. According to the author of these tsume-shogi problems, solving all problems in less than 70 minutes could be considered to be a "professional" performance. Again, the solution lengths of these test problems are too short to really test the program.

T2 has also been given 25 problems with solutions of 19 to 25 plies. T2 could solve all these problems, but needs more than 15 minutes per problem for the 23 and 25 ply tsume problems.

Finally, T2 has been given the problems in the classic book *Zoku-tsumuya-tsumazaruya* ("Mate or no mate?") [12]. In this collection the tsume problem with the shortest solution has a solution sequence of 11 plies and the longest problem has a solution of 873 plies. Even though the collection officially has 200 tsume problems, the actual number of problems is 195. A few problems are not tsume problems and a few problems have no solution because a defense against which no mate can be found was overlooked by the composer.

T2 found the solution of 70 of these hard problems. It is unclear what the maximum time per problem was, but the graphs in [9] seem to suggest more than two and a half hours.

6 Ito's Tsume Program: Best-First Search

6.1 Method

Ito's program was the first to use best-first search [9, 6]. It assigns the following numbers to the nodes of the search tree:

- Mate: 0
- No possible checks: ∞
- Leaf node: KingFreedom
- AND-node: $\sum KingFreedom_{children(n)}$
- OR-node: $\min KingFreedom_{children(n)}$

The next node to expand is the node where the freedom of the king is minimal. If multiple nodes have the same minimal value, one node is chosen randomly. There are no depth limits to the search. Transposition, domination and simulation is dealt with in the same way as in T1 and T2.

6.2 Results

Ito's program runs on a Sparc Station 10. It was given the same tests as T2. It took 2.5 minutes to find the solutions of the 112 easy problems. Of the 25 problems of 19 to 25 plies solutions depth, Ito's program could solve 19. It was much quicker than T2 in solving the 23 and 25 ply tsume problems, using only 100 seconds on average.

Ito's program was much better than T2 in solving the hard problems of Zoku-tsumuya-tsumazaruya. It could solve 120 problems.

The collection Zoku-tsumuya-tsumazaruya has a major drawback: no less than 25 problems have been shown to be incorrect, namely having more than one solution. In all but one case there was a shorter mate than the one intended by the composer of the problem. In one case there was a defense which had a solution that was 9 plies longer than the intended solution. The collection is therefore easier than the number of moves of the intended solutions suggest.

A better test set is *Tsume Zuko*, the masterpieces of the tsume composer Ito Kanju (1719-1761). These 100 tsume problems were published in 1755 and are still considered to be among the best problems ever made. In the set of 100 problems with a solution length of 9 plies to 611 plies (average length 42 plies) there are only two problems for which there is a shorter solution than the one intended. For these two problems there are repaired versions available with only small changes in the position and the solution as intended by Ito Kanju. Ito's program could find the solutions of 63 of these 100 problems [28].

Ito's program could not find the solutions of short tsume problems as well as T2, but it was a major step forward in deep search. However, like most best-first approaches, it suffered from memory problems when searching very deep, since too big a portion of the search tree had to be kept in memory. For example, Ito's program was not able to find the solution of the 611 ply *Kotobuki*[4] problem with its normal memory management. Only when the memory management was replaced by a scheme that freed large portions of the search tree that were unlikely to lead to mate, Ito's program could solve Kotobuki. However, it took 70 hours to find the solution, indicating that too often parts of the search tree had to be regenerated.

7 T3: AO* for Tsume Shogi

7.1 Method

T3 is the third tsume-shogi program by Noshita. The methods used in the program are described in [10][5]. T3 is based on AO* [23] and therefore very different from T1 and T2. There is a priority queue of leaf nodes which are stored with

---

[4] Kotobuki means "Long life" and is the final problem in the Tsume Zuko problem set.

[5] I have not found a separate publication on T3. Details about the program have been taken from Japanese papers describing different programs and co-authored by several tsume-shogi programmers.

their expectancy of mate. This value is again based on the freedom of the king, but in T3 also the mating expectancy of several ancestor nodes of the leaf nodes is used in the mate expectancy value of the leaf nodes [26]. The node that is at the top of the queue is the node to be expanded next.

The second part of the T3 algorithm is the serialization of AND-nodes. Since all AND-nodes need to be TRUE, only one of the children needs to be expanded. All other nodes are only expanded if the active node returns a TRUE value. OR-nodes are expanded in the normal way. An enhancement of this scheme used in T3 is to use a limited alpha-beta search to look for a quick mate after an AND-node is being expanded. This keeps the mating search in a local promising area as long as possible.

The third important concept in T3 is a different way of dealing with transposition, domination and simulation. T3 has a set of linked lists that connect nodes for which the mating value is related. Only the node on which all other nodes in the linked list depend is considered for expansion. All other nodes are *frozen* until this one node returns either a mating or a non-mating value. Then the other nodes in the list are revived and the mating sequence of the solved node is tried in each of the other nodes in the linked list. As a result, the search tree is no longer a proper tree, but an AND/OR graph.

The data structure (numerous sets of linked lists) necessary for this best-first search scheme is large. T3 keeps as much of it as possible in memory. However, discarding part of the search tree and the connected linked lists cannot be avoided when searching for tsume-shogi problems with very long solutions. Regenerating too many nodes will seriously slow down the program, so the speed of the program depends very much on the quality of the garbage collection. Still, connecting nodes in this way uses less memory than normal hash tables. This is because in shogi the concepts of domination and simulation make it necessary to store extra information in the hash table about pieces in hand.

7.2 Results

T3 ran on the same machine as T2, a SUN SPARC IPX with 28 MB of memory. The only published test results of T3 are on the Tsume Zuko problems. T3 could find the solution of 68 of these. Kotobuki is solved in 65 hours.

8 Kawano's Tsume Program: Priority and Simulation

8.1 Method

Kawano's tsume program [13] is based on best-first search. Each node is given a *priority*. At the leaf nodes, the priority is the same as the move ordering according to the criteria given for Noshita's programs. However, this move ordering is only done for moves by the defending side. This local ordering is generalized into a global ordering of the tree by giving the attacker's move the priority of its parent node and making the move with the highest local priority equal to the priority of the parent node. This gives an ordering of the nodes in the game tree and the leaf node to be expanded next.

Kawano deals with the problem of possible simulation by defining a choice function for the moves of the attacking side. This choice function defines when two moves are the same even though they might be textually different. Simulation is now defined as follows [13]:

Definition 4 *Position P simulates position Q if the move sequences* defined by the choice function *that can be played from P are the same as the move sequences* defined by the choice function *that can be played from Q.*

The choice function is only defined for the moves of the attacking side, so the definition of this function does not affect the outcome of the search. If two moves are defined by the choice function to be the same, but are actually different (i.e. the mating sequence of P does not work in Q), the different defense moves at the next move will show that the two positions are not simulating each other. The choice function is therefore a shogi heuristic used to guide the search in a more promising direction. The choice function defines moves with promoted pieces the same as moves with unpromoted pieces and moves with pieces from different starting squares as the same (see [13] for more details).

8.2 Results

Kawano's tsume program runs on a Dec Alpha 400 MHz with 1GB of memory. Like for T3, there are only reported results for the Tsume Zuko problems. Kawano's program can find the solution of 88 of these problems [14]. The maximum time per problem was 2 hours. In Table 1 it can be seen that 81 of these problems are solved within a 100 seconds. The solution of the Kotobuki problem is found in 6 minutes and 46 seconds.

Table 1. Results of Kawano's program on Tsume Zuko

CPU time (s)	Solved
0-1	39
1-10	30
10-100	12
100-1000	6
1000-7200	1
Total	88

9 Seo's Tsume Program: Conspiracy Numbers

9.1 Method

Seo's tsume program [28, 29] uses an algorithms called C^*. This name is not explained but probably chosen because the algorithm is based on AO* but uses conspiracy numbers to guide the search. In conspiracy-number search a heuristic value is assigned to each node in the search tree. The conspiracy number of a node N is the minimal number of leaf nodes in the subtree of N that need to change their value in order to change the minimax value of N. Nodes in the tree are then expanded in such a way that they will narrow the range of possible root values [20]. For the AND/OR trees of tsume shogi, there are only three values for every node: TRUE (mate), FALSE (no mate) and UNKNOWN. Therefore, in tsume shogi we have the following rules of assignment of conspiracy numbers (CN) to a node n:

- Mate node: $CN(n) = 0$
- Node with no possible checks: $CN(n) = \infty$
- Leaf node: $CN(n) = 1$
- AND-node: $CN(n) = \sum CN_{children(n)}$
- OR-node: $CN(n) = \min CN_{children(n)}$

To solve the problem of memory, Seo uses a depth-first iterative deepening approach. First, find a solution for a conspiracy number threshold of 1 for every node. Then, if no solution is found, set the threshold to 2 and so on until a solution is found for conspiracy number threshold n. In general, this iterative approach would result in too many regenerated nodes. To keep this to a minimum, Seo uses big hash tables to store as many positions as possible with their conspiracy number. If a position is searched again at iteration p, the conspiracy number of the node is initialized to the conspiracy number in the hash table, which is a value smaller or equal to $p - 1$ (Seo calls this *dynamic evaluation*). Also, to avoid regenerating nodes as much as possible, Seo's tsume program gives priority to the nodes that have not been expanded at previous iterations. Seo has been able to keep the average number of regenerated nodes at about 20%.

Seo also uses move ordering and looks for transposition, domination and simulation. Interposing drops closer to the king higher are ordered higher than interposing drops further from the king like in most of the other programs. Simulation is used by Seo to try the same mating sequences in positions that only differ in the interposed pieces and in cases where promotion is optional. He calls this a *killer heuristic*, similar to the concept used in chess programs.

9.2 Results

The first version of Seo's program ran on a Sun Sparc Station 20 workstation. Seo's results are very impressive. His program can find the solution of 190 of 195 problems in the Zoku-tsumuya-tsumazaruya test set and 99 out of the 100 problems in Tsume Zuko, also using a maximum of two hours per problem. Detailed results on solution speed can be found in Table 2. The Kotobuki problem was solved in 1 hour and 12 minutes. Seo's program is slower than Kawano's program, but it is more powerful in that it can find the solutions of more tsume-shogi problems with long solutions.

The only problem in the Tsume Zuko test set that Seo's program could not find a solution for was a very complicated 41 ply tsume problem with an unusually high number of long side variations leading to mate. As a result, the solution subtree has a very high conspiracy number throughout the search so there are only few node expansions in that vital part of the tree.

Table 2. Results of Seo's program

CPU time (s)	ZokuTT	Tsume Zuko
0-1	21	2
1-10	44	23
10-100	68	39
100-1000	39	29
1000-7200	18	6
Total	190	99

Diagram 2. Microcosmos

The holy grail for tsume-shogi programs is the Microcosmos problem (Diagram 2). This problem was composed by Hashimoto in 1986 and has a solution length of 1525 plies. Since the publication of his master's thesis in 1995, Seo has improved his program and ported it to a 166 MHz Pentium with 256 MB memory. This new version was able to solve Microcosmos in about 30 hours in April 1997.

10 Conclusions and Further Research

In this paper we have discussed the following features of a good tsume-shogi program:

- an algorithm that can search deeply to variable depths;
- hash tables to not only deal with transposition, but also with domination and simulation;
- move ordering based on freedom of the king.

Table 3. Results of all programs on the two hard test sets

Program	Author	ZokuTT	Tsume Zuko	Year
T1	Noshita	-	-	1991
T2	Noshita	70	-	1992
Ito	Ito	135	63	1992
T3	Noshita	-	68	1992
Kawano	Kawano	-	88	1994
Seo	Seo	190	99	1995

In Table 3 the results on the two major test sets for the programs discussed in this paper are summarized. It is not easy to compare the performance of the programs, since they are running on very different platforms. However, two things are clear from this table and from the data summarized in the previous sections. One is that most of the current tsume-shogi programs perform better than human experts. The two tsume-shogi collections are considered to be very hard and the general opinion among shogi players is that no human player will be able to solve more than 80% of these problems. The recognition of the performance of the tsume-shogi programs is further supported by the fact that shogi magazines these days use tsume-shogi programs to aid in the analysis of difficult endgames played by top professional players.

The second conclusion is that Seo's tsume program is clearly the best of the tsume programs discussed in this paper. Since different hardware is used, comparing Ito, T3 and Kawano's tsume program is almost impossible. Especially Kawano's tsume program is running on one of the fastest and biggest machines currently available. It would be interesting to test T3 again on such a machine. However, even with Kawano's extra computing power, Seo's program still finds more solutions of long tsume-shogi problems. Seo's program is able to find the solution of almost any tsume problem and it will be hard to make a program to improve it. Still, even though Microcosmos has been solved, there remain a number of other long tsume-shogi problems for which Seo's program cannot find the solution in a limited time.

One of the possible improvements of Seo's program might be to use proof-number search instead of conspiracy-number search. Proof-number search is an improved version of conspiracy-number search designed especially to solve AND/OR trees [2]. We have started to develop a tsume-shogi program for our shogi-playing system SPEAR based on proof-number search. Although work on this tsume-shogi program has not been finished yet, preliminary results show that for a significant number of test problems, smaller search trees are built than in Seo's program. Improving the tsume-shogi program in SPEAR is a future work.

It is interesting that complexity of a tsume-shogi problem cannot be defined by the length of the solution sequence. For the 17 problems in the range from 9 plies to 19 plies, Seo's program on average needed 150 seconds per problem, while for the 18 problems from 31 plies to 39 plies, Seo's program took only 82 seconds per problem. Furthermore, there was a relatively short 23 ply tsume-shogi problem in the Tsume Zuko test set that could not be solved by Kawano and for which Seo took almost 1.5 hours. Variable-depth search is clearly having problems with other features of a tsume problem. For human players also, the difficulty of a tsume problem is not necessarily related to the length of the solution. It would be interesting to see if there is a correlation between the difficulty of tsume-shogi problems for variable-depth search and human experts.

Tsume shogi no longer seems to be a hard problem. In this paper we have discussed methods for building a strong tsume program. However, there is still a good number of problems for which the strongest programs cannot find a solution. Furthermore, the time limit of two hours for the problems in the test set is long, even though it has become the standard time limit for most tests with the programs discussed. Algorithmic improvements to get the same results with a stricter time limit are another challenge left for tsume-shogi programs. Finally, there is one problem remaining with most variable-depth tsume-shogi programs. It is possible that a promising check at node N is searched deeply and that a mate is found in p moves ($p > 1$). This will end the search at this OR-node, even if there is a mate in one at a sibling node of node N. Most programs have some heuristics to avoid this problem as much as possible, but except for the alpha-beta search in T2, all programs discussed in this paper from time to time give solution sequences that are too long. The simple solution to this problem would be to regenerate the search tree after a mate is found and search all child nodes of OR-nodes that were not expanded for shorter mating sequences. It is unclear how much search overhead this will cause.

The important question is of course how these successful results can be used outside the domain of tsume shogi. After all, tsume shogi is only important in the final stages of a shogi game. Can the techniques discussed in this paper help in building a strong shogi-playing program? For the time being, this remains an open question. As said, the branching factor of normal shogi (80) is much larger than the branching factor of tsume shogi (5). The size of the search tree for a 40 ply tsume-shogi problem is therefore about the same as a 15 ply search in a normal shogi position. Such a deep search is not out of reach for tsume-shogi programs, so the methods discussed in this paper might be interesting for a normal shogi-playing program as well. Conspiracy-number search has been shown to be applicable to normal minimax game trees. One big problem is of course to set the correct search target which is trivial in tsume shogi. This problem is illustrated by Seo, who has made a shogi program based on his work in tsume shogi. His program thus far cannot compete with the strongest programs, despite its obvious strength in tsume shogi. We believe that variable-depth search algorithms are worth further investigating for shogi and we intend to develop a shogi program based on these ideas in the future.

Acknowledgements

I would like to thank Hitoshi Matsubara for helping me understanding the Japanese papers on tsume shogi and Kohei Noshita for his patient explanations of his tsume shogi programs.

References

[1] L.V. Allis. *Searching for Solutions in Games and Artificial Intelligence*. PhD thesis, The Netherlands: University of Limburg, 1994. ISBN 90-9007488-0.

[2] L.V. Allis, M. van der Meulen, and H.J. van den Herik. Proof-number search. *Artificial Intelligence*, 66:91–124, 1994.

[3] T. Anantharaman, M.S. Campbell, and F. Hsu. Singular extensions: Adding selectivity to brute-force searching. *Artificial Intelligence*, 43:99–109, 1990.

[4] A.D. de Groot. *Thought and Choice in Chess*. The Hague, The Netherlands: Mouton & Co, 1965.

[5] M. Hirose, H. Matsubara, and T. Ito. The composition of tsume-shogi problems. In *Advances in Computer Chess 8*, pages 299–319, Maastricht, Holland, 1996. ISBN 9062162347.

[6] K. Ito. Best-first search to solve tsume shogi problems. In H. Matsubara, editor, *Computer Shogi Progress*, pages 71–89. Tokyo: Kyoritsu Shuppan Co, 1996. ISBN 4-320-02799-X. (In Japanese).

[7] K. Ito, Y. Kawano, and K. Noshita. On the algorithms for solving tsume-shogi with extremely long solution-steps. *Journal of the Information Processing Society of Japan*, 36(12):2793–2799, 1995. (In Japanese).

[8] K. Ito, Y. Kawano, M. Seo, and K. Noshita. Tsume shogi. In H. Matsubara and I. Takeuchi, editors, *BIT special issue: Game Programming*, pages 130–138. Kyoritsu Shuppan Co., Tokyo, Japan, 1997. ISBN 00110-2-57035. (In Japanese).

[9] K. Ito and K. Noshita. Two fast programs for solving tsume-shogi and their evaluation. *Journal of the Information Processing Society of Japan*, 35(8):1531–1539, 1994. (In Japanese).

[10] T. Ito, Y. Kawano, M. Seo, and K. Noshita. Recent progress in solving tsume-shogi by computers. *Journal of the Japanese Society of Artificial Intelligence*, 10(6):853–859, 1995. (In Japanese).

[11] Y. Kadowaki. *Tsumuya-tsumazaruya, Shogi-Muso, Shogi-Zuko*. Tokyo: Heibon-sha, 1975. ISBN 4-582-80282-0. (in Japanese).

[12] Y. Kadowaki. *Zoku-tsumuya-tsumazaruya*. Tokyo: Heibon-sha, 1978. ISBN 4-582-80335-0. (in Japanese).
[13] Y. Kawano. Using similar positions to search game trees. In R.J. Nowakowski, editor, *Games of no chance (Combinatorial games at MSRI, Berkeley 1994)*, pages 193–202. Cambridge: University Press, 1996.
[14] Y. Kawano. Personal communication, 1998.
[15] R.E. Korf and D.M. Chickering. Best-first minimax search. *Artificial Intelligence*, 84:299–337, 1996.
[16] Y. Kotani, T. Yoshikawa, Y. Kakinoki, and K. Morita. *Computer Shogi*. Tokyo: Saiensu-sha, 1990. (in Japanese.).
[17] H. Matsubara. *Shogi to computer (Shogi and computers)*. Kyoritsu Shuppan Co., Tokyo, Japan, 1994. ISBN 4-320-02681-0. (In Japanese).
[18] H. Matsubara and K. Handa. Some properties of shogi as a game. *Proceedings of Artificial Intelligence*, 96(3):21–30, 1994. (In Japanese).
[19] H. Matsubara, H. Iida, and R. Grimbergen. Natural developments in game research. *ICCA Journal*, 19(2):103–112, June 1996.
[20] D.A. McAllester. Conspiracy numbers for min-max search. *Artificial Intelligence*, 35:287–310, 1988.
[21] K. Morita. Personal communication, 1996.
[22] Y. Nakayama, T. Akazawa, and K. Noshita. A parallel algorithm for solving hard tsume shogi problems. *ICCA Journal*, 19(2):94–99, June 1996.
[23] N. Nilsson. *Problem Solving Methods in Artificial Intelligence*. New York: McGraw-Hill, 1971.
[24] K. Noshita. How to make a quick and accurate tsume-shogi solver. *Shingaku-Kenkyukai, COMP91*, 56:29–37, 1991. (In Japanese).
[25] K. Noshita. The tsume shogi solver T2. In H. Matsubara, editor, *Computer Shogi Progress*, pages 50–70. Tokyo: Kyoritsu Shuppan Co, 1996. ISBN 4-320-02799-X. (In Japanese).
[26] K. Noshita. Personal communication, 1998.
[27] J. Schaeffer. Conspiracy numbers. *Artificial Intelligence*, 43:67–84, 1990.
[28] M. Seo. The C* algorithm for and/or tree search and its application to a tsume-shogi program. Master's thesis, Faculty of Science, University of Tokyo, 1995.
[29] M. Seo. A tsume shogi solver using conspiracy numbers. In H. Matsubara, editor, *Computer Shogi Progress 2*, pages 1–21. Tokyo: Kyoritsu Shuppan Co, 1998. ISBN 4-320-02799-X. (In Japanese).
[30] G. Stockman. A minimax algorithm better than alpha-beta? *Artificial Intelligence*, 12:179–196, 1979.

Retrograde Analysis of the KGK Endgame in Shogi: Its Implications for Ancient Heian Shogi

Hiroyuki Iida[1], Jin Yoshimura[2], Kazuro Morita[3], and Jos W.H.M. Uiterwijk[4]

[1] Department of Computer Science
Shizuoka University
3-5-1 Juhoku
Hamamatsu, 432 Japan
iida@cs.inf.shizuoka.ac.jp

[2] Department of Systems Engeneering
Shizuoka University 3-5-1 Juhoku
Hamamatsu, 432 Japan
jin@sys.eng.shizuoka.ac.jp

[3] Random House Co. Japan
1-6-20 Matsugaoka, Tsurugashima, 350-02

[4] Department of Computer Science
Universiteit Maastricht
P.O. Box 616
6200 MD Maastricht, The Netherlands
uiterwijk@cs.unimaas.nl

Abstract. This paper explores evolutionary changes of Shogi (Japanese chess) using game-theoretic analyses by computer. Heian Shogi is an ancient game only briefly described in the literature. Therefore, it is impossible to know exactly how it was played. Through game-theoretic analyses of rules, we estimate the historical changes of this ancient game. Our method provides a new innovative approach to guess logically how these ancient games actually have been played. This paper focuses upon the game results of the KGK endgame on $N{\times}N$ boards, applying game-programming methods. Then it determines the size of the boards in which the side of King and Gold always wins except trivially drawn cases with the Gold being captured. Based on the analyses, we discuss the rules of Heian Shogi. We specifically provide a logical interpretation of the shift from the 8×8 board to the 9×9 board in the evolutionary history of Shogi.
Keywords: evolution of games, retrograde analysis, KGK endgame, Shogi, Heian Shogi

1 Introduction

Shogi (Japanese chess) is especial among chess games in the world, because it is the only chess variant which has a rule to reuse captured pieces. This reuse rule of Shogi makes the game extremely complex. (For an introduction to Shogi, see [1]). Because of such uniqueness in Shogi, it is very interesting to know how Shogi has been *invented* or *evolved* from ancient types of chess without rule of reusing the pieces. To know the history of the present-day version of Shogi (further called modern Shogi), it is very important to know what kind of Shogi variations have been played in the old days.

According to the famous old literature of Shogi called *Nichureki* (see Appendix A), Heian Shogi, an old style Shogi, has already been played in the Heian Era (794–1191). Heian Shogi is considered as an archive of modern Shogi, because of the basic similarity of the game structure. However, it is distinctively different from modern Shogi in a few aspects.

Interestingly, at least two types of Heian Shogi are known [3,7]. One type, shown in Fig.1 as Type (1), is very similar to modern Shogi, except it lacks a Rook and a Bishop. It also uses a 9×8 board, instead of the 9×9 board used by modern Shogi. The other type, shown in Fig.2 as Type (2), has only one Gold and uses an 8×8 board. In those days, Heian Shogi has no rule of reusing captured pieces, unlike modern Shogi shown in Fig.3.

Fig. 1. The initial position of Heian Shogi, Type (1)

It is played without Rooks and Bishops on a 9×8 board and does not reuse captured pieces.

1.1 Rules of Heian Shogi

There is no good reference that describes the rules of Heian Shogi. Nichureki (see Appendix A) is the only one that briefly describes its rule how a game is ended (hereafter called the **Nichureki Rule**) as follows: "The player wins if he takes all the pieces of the opponent, except the King."

From the Nichureki Rule, we consider four rules of Heian Shogi, as described below (R_1 to R_4).

From these rules, we evaluate how the outcomes (win, loss or draw) depend on the size of the board.

Fig. 2. The initial position of Heian Shogi, Type (2)

It is played without Rooks and Bishops, and has only one Gold for each side on an 8×8 board, and does not reuse captured pieces.

Fig. 3. The initial position of modern Shogi

It has Rooks and Bishops, and two Golds for each side on a 9×9 board, and can reuse captured pieces.

R_1 : We assume that the rules about the movement and promotion of pieces and forbidden moves are the same as in modern Shogi. However, captured pieces are not used again.

R_2 : If the player has no legal moves, he loses the game.

R_3 : The player wins if he takes all the opponent's pieces except the King.

R_4 : The game is drawn if both players have only a King (we define this condition as a **trivial draw**, to contrast it with other draws). The repetition of an identical position is also a draw.

1.2 The Shift from 8×8 Boards to 9×9 Boards

When two experienced players (e.g., Shogi grandmasters) play a game of Heian Shogi, they often reach a "King and Gold vs King" endgame, which we denote as a **KGK endgame**.

Furthermore, they may also sometimes reach a "King and Pawn vs King" endgame which will lead to a KGK endgame by the promotion of the Pawn, except the evident cases of the Pawn being captured. Therefore, the results of games ("win, lose, and draw") in Heian Shogi should be highly related to the results of the KGK endgame.

In this paper, we analyze the results of KGK endgames on $N \times N$ boards, using the retrograde-analysis method [6]. We determine the size of the boards in which the side of King and Gold always wins except evident draw cases of the Gold being captured. Based on the analyses, we discuss the rules of Heian Shogi. We specifically provide a logical interpretation of the shift from the 8×8 boards to the 9×9 boards in the evolutionary history of Shogi.

2 Retrograde Analyses and the Classification of KGK Endgames

In this section, we classify all the positions of the KGK endgame by the retrograde-analysis method [6]. For clarity, the two players are distinguished as the **attacker** for the "King and Gold" side and the **defender** for the "King alone" side.

We first describe all the positions-in-mate, the final winning conditions for the attacker. Then we expand the positions-to-mate backward by the retrograde-analysis method, and count all the attacker's winning positions. In these analyses we also take into account the size of the board, and we determine on what size the attacker always wins, except the evident drawn cases in which the Gold is captured.

2.1 Retrograde Analyses of the Attacker's Winning Positions

Let **X** be the set of all legal positions of the KGK endgame on an $N \times N$ board. Within **X**, let W_0 be the set of all the attacker's won positions, i.e., positions-in-mate. Note that, according to rule $\mathbf{R}_2$, any position where the defender is unable to move is a position in W_0.

Next, let W_1 be the set of all positions from which at least one move by the attacker will be able to lead to some position in W_0. In other words, any position in W_1 is a position where the attacker is to move and has at least one legal move that leads to a position in W_0.

Then, let W_2 be the set of all positions from which every defender's move leads to some position in W_1.

In this manner, we expand the attacker's winning positions by retrograde analysis [6], from positions-in-mate to positions-to-mate-in-n, with n some positive integer.

In this procedure, to avoid repetitions, we construct the sets W_i such that they are disjoint, i.e., such that for any i, j $(0 \leq i < j)$

$$W_i \cap W_j = \emptyset.$$

Because the set of all positions of the KGK endgame **X** is finite, retrograde analysis always converges to some integer m, such that

$$W_m \neq \emptyset \quad \text{and} \quad W_{m+1} = \emptyset. \tag{1}$$

From any position in W_m the attacker will be able to win in at most m steps, irrespective of the defender's responses. Because m is the length of the longest path for the attacker's definite win, we call such a position a **longest position-to-mate** for a given $N \times N$ board. For two board sizes ($N = 9$ and 10), examples are shown in Appendix B.

Now we can determine the set of all possible winning positions, denoted as $\bigcup_{i=0}^{m} W_i$ in which the attacker can always win, irrespective of the defender's responses.

2.2 Retrograde Analyses of Trivial Draws

It is trivial that a draw happens when the defender captures the attacker's Gold (called a **trivial draw**). Let D_0 be the set of all positions of trivial draws, i.e., all positions in which the lone King can immediately capture the Gold.

We now expand D_i from D_0 in the same manner as we did for W_i. Let D_1 be the set of all positions in which any move by the attacker leads to some position in D_0.

Then, let D_2 be the set of all positions in which the defender has at least one move that leads to a trivial drawn position in D_1, etcetera.

In these retrograde analyses of D_i, we again require that for i, j $(0 \leq i < j)$

$$D_i \cap D_j = \emptyset.$$

As in W_i, because the set of all possible positions **X** is finite, there exists a finite integer n such that

$$D_n \neq \emptyset \quad \text{and} \quad D_{n+1} = \emptyset. \tag{2}$$

Now we can determine the set of all positions, denoted as $\bigcup_{j=0}^{n} D_j$, in which the defender can always achieve a trivial draw, irrespective of the attacker's moves.

2.3 Analyses of Non-trivial Draws

Previous analyses consider the positions that lead to either the attacker's win, or a trivial draw in which the defender captures the Gold. We know by experience that there are some positions which do not belong to either case. In the following, we consider such inconclusive positions.

Some positions may belong to neither $\bigcup_{i=0}^{m} W_i$ nor $\bigcup_{j=0}^{n} D_j$. Let P be one of such inconclusive positions where the attacker is to move. In P, the attacker has no move that leads to $\bigcup_{i=0}^{m} W_i$. Meantime, $P \notin \bigcup_{j=0}^{n} D_j$. Therefore, there always exists a move $P \to P_1$ such that $P_1 \notin \bigcup_{j=0}^{n} D_j$, and the attacker should choose such a move.

Thus we get the position P_1 which satisfies:

$$P_1 \notin \bigcup_{i=0}^{m} W_i \cup \bigcup_{j=0}^{n} D_j. \qquad (3)$$

Similarly, in P_1 where the defender is to move, he has no move that leads to any position of $\bigcup_{j=0}^{n} D_j$. Therefore, there always exists a move $P_1 \to P_2$ such that $P_2 \notin \bigcup_{i=0}^{m} W_i$, and the defender should choose such a move. Thus we get the position P_2 that satisfies:

$$P_2 \notin \bigcup_{i=0}^{m} W_i \cup \bigcup_{j=0}^{n} D_j. \qquad (4)$$

By repeating this procedure, we obtain a (potentially infinite) sequence of positions:

$$P \to P_1 \to P_2 \cdots \to P_l. \qquad (5)$$

It is easy to show that this sequence always contains a loop for any P, as explained below. Note that the following condition holds for any such P_k:

$$P_k \notin \bigcup_{i=0}^{m} W_i \cup \bigcup_{j=0}^{n} D_j. \qquad (6)$$

Since any position P_k from such a sequence is a KGK position, it must be contained in the finite set $\mathbf{X}$. Therefore, some position P_l must be identical to some position P_j with $j < l$, leading to a loop.

Thus any position P_k that satisfies the relation (6) has at least one move that leads to a loop. For the attacker, such a move is to avoid a trivial draw and for the defender, to avoid the attacker's win. Therefore, the game is a draw, according to rule $\mathbf{R}_4$, because it returns to the same position repeatedly.

2.4 The Definition of Deterministic Wins

Now we consider the relationship between board size and the attacker's winning positions. First we evaluate the size of the boards in which the attacker is always able to win a KGK endgame. We introduce the concepts of *deterministic win* and *complete deterministic win*.

A KGK endgame on a given $N \times N$ board is called a **deterministic win** if the set of all KGK endgame positions (say $\mathbf{X}$) satisfies:

$$\mathbf{X} = \bigcup_{i=0}^{m} W_i \cup \bigcup_{j=0}^{n} D_j. \qquad (7)$$

Relation (7) means that the set of non-trivial draws (loop cases) is empty.

A KGK endgame on the 3×3 board is called a **complete deterministic win**, since all draw sets D_j are empty.

For every deterministic won KGK endgame the sets in the right-hand side of equation (7) are disjoint. Therefore, the number of all the elements $|\mathbf{X}|$ of the whole set $\mathbf{X}$ satisfies:

$$|\mathbf{X}| = \sum_{i=0}^{m} |W_i| + \sum_{j=0}^{n} |D_j| \qquad (8)$$

We show a relation between board size and the length of the mating sequence of a longest position-to-mate in Theorem 1.

Theorem 1.
Let P and $m(N)$ be a longest position-to-mate and the number of its steps for a given $N \times N$ board, respectively. If two board sizes (say N_1 and N_2) satisfy condition (7), then

$$N_1 < N_2 \quad \Rightarrow \quad m(N_1) < m(N_2). \qquad (9)$$

Proof
Suppose position P on an $N \times N$ board. A mating sequence for P always will lead to a mate, in which the defending king is mated on an edge or in a corner (no mate at the middle of the board is possible). Constitute the $N+1$ size board position from P, by adding one row and one file to the $N \times N$ board at the edge or corner where the mating process will take place (if mated at the upper or lower edge, the column may be added arbitrarily at the left or right side; similarly for a mate at the left or right edge). Then, from the proposed longest mating sequence on the $N \times N$ board the defending King may escape to the additional space at the $(N+1) \times (N+1)$ board, giving rise to at least some additional moves to be mated. Since no trivial draw is involved, the only other possibility would be that the defending side can enter a non-trivial draw loop, in contradiction with equation (7). As a consequence, the longest mating sequence on an $(N+1) \times (N+1)$ board without non-trivial draws is strict longer than such a sequence on an $N \times N$ board. □

Let Δ be the set of non-trivial draws. The following relation obviously is derived from Theorem 1.

$$m(N) \geq m(N+1) \quad \Rightarrow \quad \Delta \neq \emptyset \qquad (10)$$

We can determine the maximum size of the board for which KGK is a deterministic win, by looking for the first occurrence of the following condition:

$$m(N) \geq m(N+1) \qquad (11)$$

during the counting process of the attacker's winning positions in the retrograde analyses.

2.5 Counting All Positions

We have writen a computer program that optimally plays KGK endgames on an $N \times N$ board. This program examines all possible moves for each player and determines the best move in each step. We have analysed the KGK endgame on $N \times N$ boards for $N = 3, \ldots, 15$, by counting all positions in W_i and D_j defined in Sections 2.1 and 2.2, respectively.

Table 1 shows the result of counting all attacker's winning positions, trivial-draw positions and non-trivial-draw positions.

Table 1. The number of all the attacker's winning positions and draw positions on an $N \times N$ board.

N	|X|	|W|	|D|	|D'|	m(N)
3	382	294	88	0	6
4	3916	3066	850	0	16
5	19446	15708	3738	0	28
6	66928	55162	11766	0	40
7	183910	153524	30386	0	54
8	433452	364876	68576	0	72
9	913486	773334	140152	0	94
10	1767616	1502328	265288	0	122
11	3197358	18252	472254	2706852	103
12	5475820	10932	799360	4665528	39
13	8962822	10877	1297120	7654825	37
14	14121456	10876	2030620	12079960	35
15	21536086	10885	3082108	18443093	35

N represents the board size. $|X|$ represents the size of the set $\mathbf{X}$ of all legal positions of the KGK endgame on an $N \times N$ board. Similarly, $|\mathbf{W}|$, $|\mathbf{D}|$ and $|\mathbf{D}'|$ represents the number of all winning positions, the number of all trivial-draw positions, and the number of all non-trivial-draw positions, respectively. $m(N)$ represents the number of steps for each longest position-to-mate.

From Table 1 we see that the KGK endgame is a deterministic win on boards of sizes up to $N = 10$, and that the endgame includes non-trivial draws for board sizes with $N \geq 11$.

The data are graphically depicted as percentages in Fig.4. We clearly see that from $N = 11$ upwards the percentage of won positions drastically decreases from aproximately 80% to almost zero, whereas non-trivial draws behave exactly opposite. Trivial draws always constitute some 20%, irrespective of board size.

3 Discussion

From these results of the KGK endgames, we can infer some results on the evolutionary history of Heian Shogi.

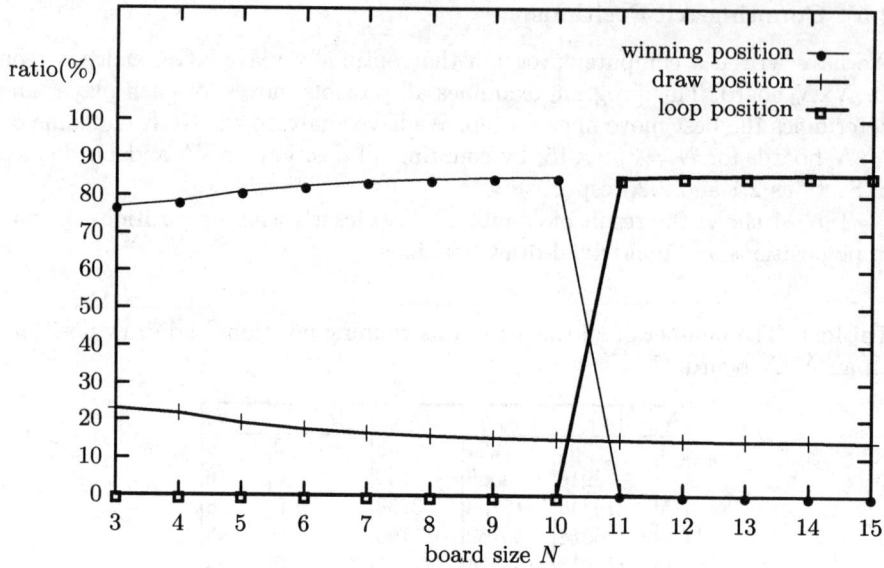

Fig. 4. The ratio of three different types of positions as a function of board size.

3.1 Symmetry of the Initial Position

In modern Shogi played on the 9×9 board the initial position is left-right symmetric around the center file (the file of Kings), except for the Rooks and Bishops that are symmetric to each other. However, at least two different sizes of the board are known in Heian Shogi: the 8×8 and 8×9 boards (Fig.1 and Fig.2).

The initial position cannot be left-right symmetric on the 8×8 board because of the even number of files. However, from the viewpoints of the conceptual beauty, perfection and excellence of games, left-right symmetry seems very important in chess-like games. The initial position in the 8×9 board is left-right symmetric around the King's file. It is therefore natural to imagine that the 8×8 board is a primitive type of Heian Shogi and the 8×9 board an advanced type, being a transition to the 9×9 board of modern Shogi.

Another interesting point in the initial positions of Shogi games is the symmetry between the lower side and the upper side. There are two types of symmetry: point symmetry to the center of the board and line symmetry between the two sides. The initial position of the 8×8 board is point symmetric, but not line symmetric, because of the placement of Kings and Golds (see Fig.2). In contrast, the 8×9 board is both point symmetric and line symmetric (see Fig.1). Modern Shogi is both point symmetric and line symmetry except for the Rooks and Bishops that are not used in Heian Shogi.

Suppose that we add one row between the two sides to the 8×8 and 8×9 boards, spacing three squares between the opposing Pawns, instead of two squares. Because Heian Shogi does not reuse the pieces, such games would follow mimic play (the second player can never lose if he moves exactly line-symmetric to the first player, since the first player will never put a piece beyond the fourth row, because it simply would be captured by the second player). Such games therefore necessarily would lead to draws [5].

Furthermore, Heian Shogi has no Rooks and Bishops. Therefore, it is logical for Heian Shogi to have only eight rows. From the technical points of view, it is natural and logical to suppose that the addition of one row to make nine rows is accompanied with the introduction of both Rooks and Bishops during the transition from Heian Shogi (eight rows) to modern Shogi (nine rows).

3.2 Adding Rooks and Bishops to Heian Shogi

Suppose that Heian Shogi would be played with Rooks and Bishops, but without the reuse rule for captured pieces. Such a game is highly likely to end up with the KGK endgame. Suppose that a player has at least a piece more than the other player. Then, as a strategy, he should try to exchange pieces, aiming at the KGK endgame (or better). This exchange rule often is a good strategy when ahead in many other chess-like games also.

As a consequence of this fact it seems logical that the introduction of Rooks and Bishops has been accompanied by the immediate or soon introduction of the reuse rule for captured pieces.

3.3 Harmony between the Perfection of Beauty and the Evolution of Complexity

In the KGK endgame of Heian Shogi, the maximum board size for the attacker's deterministic wins (the attacker wins except trivial draws by capturing the Gold) is the 10×10 board. However, if the board size is even, the left-right symmetry of the pieces in the initial position is broken (see Fig.1 and Fig.2). To maintain the left-right symmetry the board size should be odd. Therefore, the maximum board size for the attacker's deterministic wins with left-right symmetry is 9×9.

From a mathematical as well as a historical viewpoint, this explanation seems to be highly in line with the use of the 9×9 board in modern Shogi. The history of games seems to be a development of the harmonic balance between the perfection of beauty and the evolution of complexity. Here, beauty implies some form of simplicity and often opposes to complexity. Modern Shogi is a result of the nearly 1000-year perfection from Heian Shogi. We could suspect that, during such a long history of perfection, in order to fulfil the desire for strategic complications, Shogi became being played at the 9×9 board.

In this, the desire for strategic complications can be satisfied by the maximization of the board size, under the constraint that the attacker still has a deterministic win in the KGK endgame. If the condition of a deterministic win is not kept and the size of board is maximized beyond, Shogi becomes indeterministic and chaotic. Thus we could say that, from the viewpoint of the harmony between perfection and complications [2], 9×9 Shogi has become a highly matured game compared with other chess-like games with different board sizes.

3.4 Heian Dai-Shogi: A Dead-End of the Early Evolution

In order to deal with the necessity of strategic complications, it may be an obvious idea to enlarge the board size while adding other kinds of pieces and/or increasing the number of pieces, by which the average number of legal moves easily grows. In fact, Heian Dai-Shogi, described in Nichureki (Appendix A), is a large-size variation of Heian Shogi, being played at a 13×13 board. We judge it to have been a dead-end of the early evolution. This is because Heian Dai-Shogi had little advantage concerning strategic complications by increasing the number of legal moves, while showing a big disadvantage with respect to the beauty of the rules by requiring a coarse-grained rule to judge the ending like rule R_3.

3.5 The Rules of Heian Shogi

In the beginning we proposed as rules of Heian Shogi R_1, R_2, R_3 and R_4. The rules of Heian Shogi should have evolved gradually over many centuries and should have been polished to become a set of necessary and sufficient rules.

From this historical view, we consider the relationships among the proposed rules of Heian Shogi:

- R_3, a rule of Nichureki, contradicts with R_4 in the following two cases. First, the repetition of an identical position in KGK is considered a non-trivial draw according to R_4. However, R_3 treats this position as an attacker's win. Second, trivial draws belonging to $\bigcup_{j=1}^{n} D_j$ according to R_4 are treated as attacker's wins, following R_3.
- R_3 includes all the positions of R_2, since (unlike western chess) any piece in Shogi other than the King always can move when on the board. Thus if one side has no legal moves, he must have a King only. All the positions defined as attacker's win by R_2 are already defined according to R_3. Thus R_2 is superfluous when R_3 is included in the set of rules.

From these considerations we expect that the following two combinations of rules are logically consistent:

1. Rule Set 1: $\{R_1, R_3\}$
2. Rule Set 2: $\{R_1, R_2, R_4\}$

The advantage of Rule Set 1 is that the conditions for win and loss are easily defined and checked for an arbitrary board size (especially effective for $N \geq 11$). However, Rule Set 1 has a strong disadvantage in that some positions reasonably to be considered as a draw are treated as attacker's wins. Fig.5 is one such case, where the "King and Gold" side is declared winner immediately by Rule Set 1. This is in agreement with the Nichureki Rule saying that "The King only means a loss even if just one move behind." In modern Shogi such a position naturally is a drawn position, which feels intuitively correct.

On the other hand, Rule Set 2 has some logical background that is similar to the rules of modern Shogi. A minor disadvantage is that it is highly tedious and

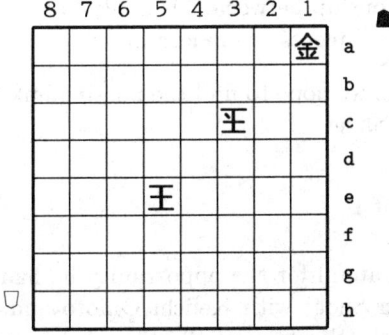

Fig. 5. An example of a position where the Gold can always be captured by the lone King, even if the attacker is to move.

difficult to judge the game results when it is played on a big board, e.g., Dai-Shogi (a type of Heian Shogi that is played on a 13×13 board). In a game played on a 13×13 board, we have to remember all the positions of a KGK endgame in order to detect non-trivial draws. Otherwise, we need a repetition rule like the 50-moves rule in western chess.

Thus Rule Set 1 seems more primitive, but due to its simplicity more suitable for Heian Shogi which is being played on a large variety of board sizes. In contrast, Rule Set 2 is more elaborated, so that it is suitable for the 9×9 board used in modern Shogi. Therefore the original rules of Heian Shogi might be similar to Rule Set 1, while Rule Set 2 is a later developed rule set being a transition (Pre-Shogi) between Heian Shogi and modern Shogi.

4 Conclusions and Future Work

In this paper we have gathered some evidence for evolutionary changes of Shogi using game-theoretic analyses by computer. Heian Shogi is an ancient game only briefly described in the few literature. Therefore, it is impossible to know the exact methods how it is played. Through game-theoretic analyses of proposed rules, we feel to have more insight into how these ancient games actually have been played.

We demonstrated that 10×10 is the largest board size on which the KGK endgame of Heian Shogi is a deterministic win. Based on the analyses of KGK endgames, we further showed that Rule Set 1, $\{\mathbf{R}_1, \mathbf{R}_3\}$, contains the rules most likely used in primitive Heian Shogi, whereas Rule Set 2, $\{\mathbf{R}_1, \mathbf{R}_2, \mathbf{R}_4\}$, is a more elaborated set and probably a transition to modern Shogi.

Future projects include:

− To investigate the structural attributes determining why for small boards all KGK positions are wins (except for trivial draws), but that on larger boards non-trivial drawn KGK positions are possible.

- To study the relationship between this analysis of KGK endgames and the reuse rule of captured pieces in modern Shogi.

Through these analyses, we hope to find the missing link between modern Shogi and the ancient Heian Shogi.

Acknowledgements

The senior author is grateful for the opportunity he had to discuss the origins and evolution of Heian Shogi with Keiichi Omoto and his collaborators (the members of the project "Strategy of Shogi and Japanese culture", of which the senior author also is a member) at the International Research Center for Japanese Studies (Nichibunken). Among them, Kwoichi Tandai suggested him the possibility of the current project. This work was supported by the Hayao Nakayama Foundation for Science & Technology and Culture.

References

1. J. Fairbairn. *Shogi for Beginners*. The Shogi Association, Ltd., London, 1984.
2. H. Iida. The Origin of Shogi Viewed from Game-Theoretic Strategies. *Human Sciences and Computer*, 34(2):7–12. Information Processing Society of Japan, 1997. (in Japanese)
3. H. Masukawa. *Shogi*. Housei University Press, 1977. (in Japanese).
4. K. Tandai. The Mystery of Heian Shogi. *Tsumeki Mate* 23:81–82. Tsumeshogi Kenkyukai, 1996. (in Japanese)
5. K. Tandai. The Mystery of Heian Shogi (2). *Tsumeki Mate* 24:70–71. Tsumeshogi Kenkyukai, 1997. (in Japanese)
6. K. Thompson. Retrograde Analysis of Certain Endgames. *ICCA Journal*, 9(3):131–139, 1986.
7. I. Umebayashi. *Shogi of the World: from the Ancient to the Modern Time*. Shogi-Tengoku-sha, 1996. (in Japanese).

A The Description of Heian Shogi in Nichureki

Nichureki was said to be edited by Tameyasu Miyoshi, a scholar of mathematics, in the late Kamakura Era (14th Century), combining Shochureki and Kaichureki that were supposed to be written in the Late Heian Era (12th Century). Nichureki describes briefly two kinds of Shogi: Heian Shogi and Heian Dai-Shogi. These two Shogi games are thought to be prototypes of modern Shogi. The following is a translation of an excerpt from Nichureki describing Heian Shogi and Heian Dai-Shogi:

From this description, Tandai [4,5] proposed the following hypothesis:

"Shogi games in the Heian Era used (were played on) the 8×8 board and Pawns located in the second rows. The reasons for the use of the 8×8 board is that, if the 9×9 board is used, the second player can never lose if he moves exactly

- the King can move in all directions;
- the Gold cannot move to the two diagonal squares behind;
- the Silver cannot move to the left, the right and straight behind;
- the Knight can move to the squares directly ahead the front corners;
- the Lance can move to any square straight ahead;
- the Pawn can only move one square ahead;
- all pieces are promoted to Gold by entering the three rows of the opponent side;
- the attacker wins as soon as the opponent becomes the King alone.

line-symmetric to the first player's moves, i.e., by a mimic strategy. Next, from the description: "All pieces are promoted to Gold by entering the three rows of the opponent side,' the third row can be considered as the boarder line or intersection; therefore the Pawns should line up in the second row." (translated from Japanese by the authors)

Nevertheless, it is impossible to specify the rules of Heian Shogi precisely from the description of Nichureki.

B Examples of Longest-Positions-to-Mate

For $N = 9$ and $N = 10$, examples of a longest position-to-mate (X_m) and the resulting checkmate position X_0 are shown below.

B.1 $N = 9$

When N = 9, the longest position-to-mate has 94 steps. One such example is shown in Fig.6. The mating sequence from X_m (Fig.6) to X_0 is shown below.

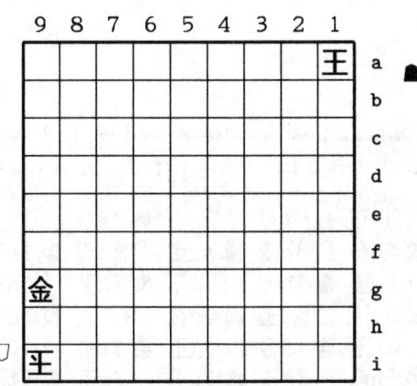

Fig. 6. An example of a longest position-to-mate on the 9×9 board (King-alone side to move).

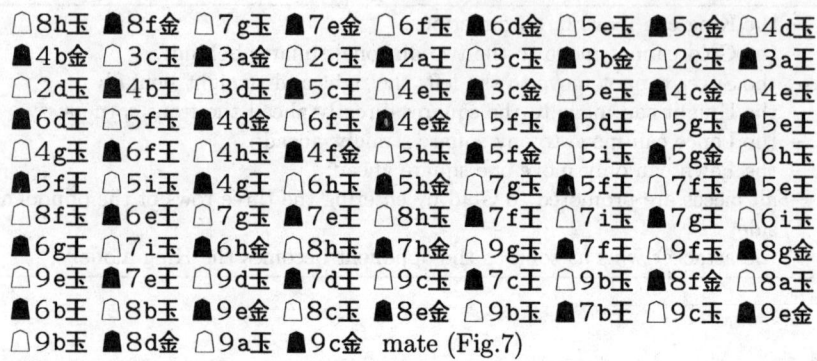

△8h玉 ▲8f金 △7g玉 ▲7e金 △6f玉 ▲6d金 △5e玉 ▲5c金 △4d玉
▲4b金 △3c玉 ▲3a金 △2c玉 ▲2a王 △3c玉 ▲3b金 △2c玉 ▲3a王
△2d玉 ▲4b玉 △3d玉 ▲5c王 △4e玉 ▲3c金 △5e玉 ▲4c金 △4e玉
▲6d王 △5f玉 ▲4d金 △6f玉 ▲4e金 △5f玉 ▲5d玉 △5g玉 ▲5e玉
△4g玉 ▲6f玉 △4h玉 ▲4f金 △5h玉 ▲5f金 △5i玉 ▲5g金 △6h玉
▲5f玉 △5i玉 ▲4g玉 △6h玉 ▲5h金 △7g玉 ▲5f玉 △7f玉 ▲5e玉
△8f玉 ▲6e玉 △7g玉 ▲7e玉 △8h玉 ▲7f玉 △7i玉 ▲7g玉 △6i玉
▲6g玉 △7i玉 ▲6h金 △8h玉 ▲7h金 △9g玉 ▲7f玉 △9f玉 ▲8g金
△9e玉 ▲7e玉 △9d玉 ▲7d玉 △9c玉 ▲7c玉 △9b玉 ▲8f金 △8a玉
▲6b玉 △8b玉 ▲9e金 △8c玉 ▲8e金 △9b玉 ▲7b玉 △9c玉 ▲9e金
△9b玉 ▲8d金 △9a玉 ▲9c金 mate (Fig.7)

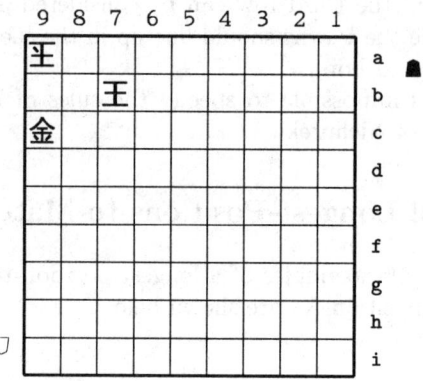

Fig. 7. The position-in-mate derived from the position in Fig.6 (King-alone side to move but unable, i.e., mate).

B.2 $N = 10$

When $N = 10$, the longest position-to-mate has 122 steps. Fig.8 is one such example. The mating sequence from X_m (Fig.8) to X_0 is shown below:

△9i玉 ▲9g金 △8h玉 ▲8f金 △7g玉 ▲7e金 △6f玉 ▲6d金 △5e玉
▲5c金 △4d玉 ▲4b金 △3c玉 ▲3a金 △2c玉 ▲2a王 △3c玉 ▲3b金
△2c玉 ▲3a王 △2d玉 ▲4b玉 △3d玉 ▲5c王 △4e玉 ▲3c金 △5e玉
▲4c金 △4e玉 ▲6d王 △5f玉 ▲4d金 △6f玉 ▲4e金 △5f玉 ▲5d王
△5g玉 ▲5e王 △6g玉 ▲4f金 △5g玉 ▲5f金 △6g玉 ▲4f王 △7f王
▲5g金 △7g玉 ▲5h金 △6f玉 ▲4e王 △7g玉 ▲5f王 △7h玉 ▲5g王
△7g玉 ▲6h金 △7f玉 ▲5f王 △7e玉 ▲5e王 △8f玉 ▲6f王 △8e玉
▲6e王 △8d玉 ▲7g金 △9e王 ▲7h金 △8f玉 ▲8h金 △8e玉 ▲8g金
△8d玉 ▲8f金 △7c玉 ▲7e金 △6c玉 ▲6d金 △5b玉 ▲5d王 △4b玉

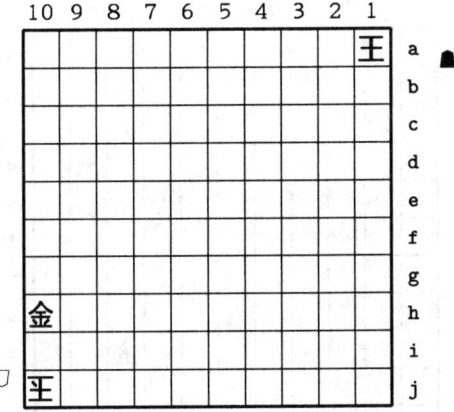

Fig. 8. An example of a longest position-to-mate on the 10×10 board (King-alone side to move).

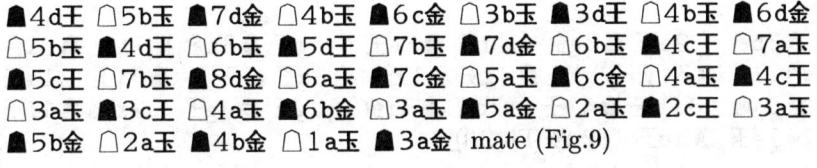

▲4d玉 △5b玉 ▲7d金 △4b玉 ▲6c金 △3b玉 ▲3d玉 △4b玉 ▲6d金
△5b玉 ▲4d玉 △6b玉 ▲5d玉 △7b玉 ▲7d金 △6b玉 ▲4c玉 △7a玉
▲5c玉 △7b玉 ▲8d金 △6a玉 ▲7c金 △5a玉 ▲6c金 △4a玉 ▲4c玉
△3a玉 ▲3c玉 △4a玉 ▲6b金 △3a玉 ▲5a金 △2a玉 ▲2c玉 △3a玉
▲5b金 △2a玉 ▲4b金 △1a玉 ▲3a金 mate (Fig.9)

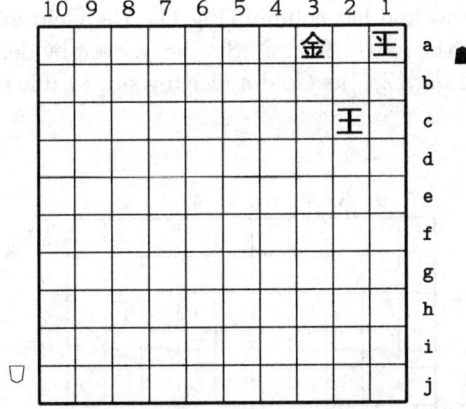

Fig. 9. The position-in-mate derived from the position in Fig.8 (King-alone side to move but unable, i.e., mate).

C An Example of Non-trivial Drawn Positions

Let us show, in Fig.10, an example of a non-trivial drawn position described in Section 2.3.

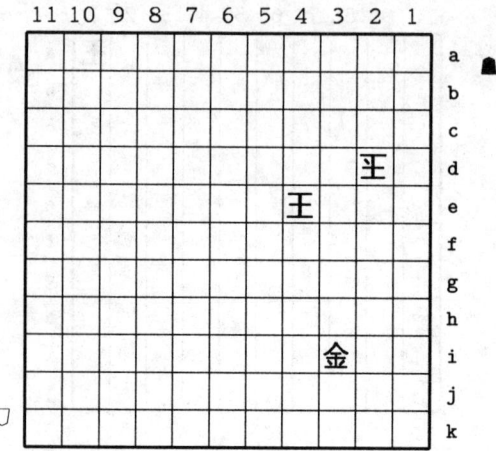

Fig. 10. An example of a non-trivial drawn position (King-and-Gold side to move).

▲3h金 △2e玉 ▲3g金 △1f玉 ▲3e玉 △1g玉 ▲3h金 △1h玉 ▲3i金
△1i玉 ▲3f玉 △1j玉 ▲3g玉 △1k玉 ▲3h玉 △2j玉 ▲4i玉 △1i玉
▲3j金 △1h玉 ▲4h玉 △2h玉 ▲3i金 △2g玉 ▲4g玉 △2f玉 ▲4f玉
△2e玉 ▲4e玉 △2d玉(Fig.10)

We should note that at the position obtained from the position in Fig.10 by deleting the upper row and left column (Fig.11), the Gold-and-King side is able to win starting with the move ▲2g金. Similarly, even by deleting the lower row and left column (see Fig.12), the Gold-and-King side is able to win starting with the move ▲3h金.

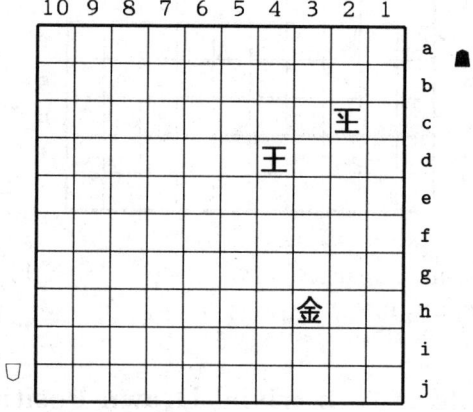

Fig. 11. Position obtained from the position in Fig.10, by deleting the upper row and the left column (King-and-Gold side to move).

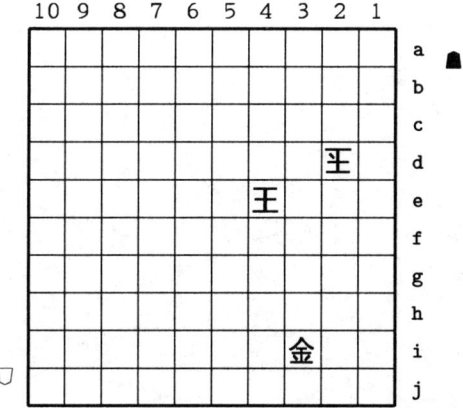

Fig. 12. Position obtained from the position in Fig.10, by deleting the lower row and the left column (King-and-Gold side to move).

Author Index

L. Victor Allis 25
David Basin 50
Donald F. Beal 113
Yngvi Björnsson 15
Daniel Borrajo 183
Dennis M. Breuker 25
Arie de Bruin 195
Alan Bundy 93
Michael Buro 126
Marcel Crâşmaru 222
Aviezri S. Fraenkel 205, 212
Ian Frank 50
Xinbo Gao 74
J. Ignacio Giráldez 183
Reijer Grimbergen 300
H. Jaap van den Herik 25, 74
Hiroyuki Iida 74, 318
Andreas Junghanns 1
Takuya Kojima 146, 282
John Levine 93

Tony Marsland 15
Kazuro Morita 318
Martin Müller 252
Wim Pijls 195
Ofer Rahat 212
Julian Richardson 93
Yasuki Saito 282
Noriaki Sanechika 265
Nobusuke Sasaki 167
Yasuji Sawada 167
Jonathan Schaeffer 1
Martin C. Smith 113
William L. Spight 232
Morihiko Tajima 265
Jos W.H.M. Uiterwijk 25, 74, 318
Steven Willmott 93
Atsushi Yoshikawa 146, 282
Jin Yoshimura 167, 318
Dmitri Zusman 205

Springer and the environment

At Springer we firmly believe that an international science publisher has a special obligation to the environment, and our corporate policies consistently reflect this conviction.

We also expect our business partners – paper mills, printers, packaging manufacturers, etc. – to commit themselves to using materials and production processes that do not harm the environment. The paper in this book is made from low- or no-chlorine pulp and is acid free, in conformance with international standards for paper permanency.

Lecture Notes in Computer Science

For information about Vols. 1–1490
please contact your bookseller or Springer-Verlag

Vol. 1491: W. Reisig, G. Rozenberg (Eds.), Lectures on Petri Nets I: Basic Models. XII, 683 pages. 1998.

Vol. 1492: W. Reisig, G. Rozenberg (Eds.), Lectures on Petri Nets II: Applications. XII, 479 pages. 1998.

Vol. 1493: J.P. Bowen, A. Fett, M.G. Hinchey (Eds.), ZUM '98: The Z Formal Specification Notation. Proceedings, 1998. XV, 417 pages. 1998.

Vol. 1494: G. Rozenberg, F. Vaandrager (Eds.), Lectures on Embedded Systems. Proceedings, 1996. VIII, 423 pages. 1998.

Vol. 1495: T. Andreasen, H. Christiansen, H.L. Larsen (Eds.), Flexible Query Answering Systems. IX, 393 pages. 1998. (Subseries LNAI).

Vol. 1496: W.M. Wells, A. Colchester, S. Delp (Eds.), Medical Image Computing and Computer-Assisted Intervention – MICCAI'98. Proceedings, 1998. XXII, 1256 pages. 1998.

Vol. 1497: V. Alexandrov, J. Dongarra (Eds.), Recent Advances in Parallel Virtual Machine and Message Passing Interface. Proceedings, 1998. XII, 412 pages. 1998.

Vol. 1498: A.E. Eiben, T. Bäck, M. Schoenauer, H.-P. Schwefel (Eds.), Parallel Problem Solving from Nature – PPSN V. Proceedings, 1998. XXIII, 1041 pages. 1998.

Vol. 1499: S. Kutten (Ed.), Distributed Computing. Proceedings, 1998. XII, 419 pages. 1998.

Vol. 1500: J.-C. Derniame, B.A. Kaba, D. Wastell (Eds.), Software Process: Principles, Methodology, and Technology. XIII, 307 pages. 1999.

Vol. 1501: M.M. Richter, C.H. Smith, R. Wiehagen, T. Zeugmann (Eds.), Algorithmic Learning Theory. Proceedings, 1998. XI, 439 pages. 1998. (Subseries LNAI).

Vol. 1502: G. Antoniou, J. Slaney (Eds.), Advanced Topics in Artificial Intelligence. Proceedings, 1998. XI, 333 pages. 1998. (Subseries LNAI).

Vol. 1503: G. Levi (Ed.), Static Analysis. Proceedings, 1998. IX, 383 pages. 1998.

Vol. 1504: O. Herzog, A. Günter (Eds.), KI-98: Advances in Artificial Intelligence. Proceedings, 1998. XI, 355 pages. 1998. (Subseries LNAI).

Vol. 1505: D. Caromel, R.R. Oldehoeft, M. Tholburn (Eds.), Computing in Object-Oriented Parallel Environments. Proceedings, 1998. XI, 243 pages. 1998.

Vol. 1506: R. Koch, L. Van Gool (Eds.), 3D Structure from Multiple Images of Large-Scale Environments. Proceedings, 1998. VIII, 347 pages. 1998.

Vol. 1507: T.W. Ling, S. Ram, M.L. Lee (Eds.), Conceptual Modeling – ER '98. Proceedings, 1998. XVI, 482 pages. 1998.

Vol. 1508: S. Jajodia, M.T. Özsu, A. Dogac (Eds.), Advances in Multimedia Information Systems. Proceedings, 1998. VIII, 207 pages. 1998.

Vol. 1510: J.M. Zytkow, M. Quafafou (Eds.), Principles of Data Mining and Knowledge Discovery. Proceedings, 1998. XI, 482 pages. 1998. (Subseries LNAI).

Vol. 1511: D. O'Hallaron (Ed.), Languages, Compilers, and Run-Time Systems for Scalable Computers. Proceedings, 1998. IX, 412 pages. 1998.

Vol. 1512: E. Giménez, C. Paulin-Mohring (Eds.), Types for Proofs and Programs. Proceedings, 1996. VIII, 373 pages. 1998.

Vol. 1513: C. Nikolaou, C. Stephanidis (Eds.), Research and Advanced Technology for Digital Libraries. Proceedings, 1998. XV, 912 pages. 1998.

Vol. 1514: K. Ohta, D. Pei (Eds.), Advances in Cryptology – ASIACRYPT'98. Proceedings, 1998. XII, 436 pages. 1998.

Vol. 1515: F. Moreira de Oliveira (Ed.), Advances in Artificial Intelligence. Proceedings, 1998. X, 259 pages. 1998. (Subseries LNAI).

Vol. 1516: W. Ehrenberger (Ed.), Computer Safety, Reliability and Security. Proceedings, 1998. XVI, 392 pages. 1998.

Vol. 1517: J. Hromkovič, O. Sýkora (Eds.), Graph-Theoretic Concepts in Computer Science. Proceedings, 1998. X, 385 pages. 1998.

Vol. 1518: M. Luby, J. Rolim, M. Serna (Eds.), Randomization and Approximation Techniques in Computer Science. Proceedings, 1998. IX, 385 pages. 1998.

1519: T. Ishida (Ed.), Community Computing and Support Systems. VIII, 393 pages. 1998.

Vol. 1520: M. Maher, J.-F. Puget (Eds.), Principles and Practice of Constraint Programming - CP98. Proceedings, 1998. XI, 482 pages. 1998.

Vol. 1521: B. Rovan (Ed.), SOFSEM'98: Theory and Practice of Informatics. Proceedings, 1998. XI, 453 pages. 1998.

Vol. 1522: G. Gopalakrishnan, P. Windley (Eds.), Formal Methods in Computer-Aided Design. Proceedings, 1998. IX, 529 pages. 1998.

Vol. 1524: G.B. Orr, K.-R. Müller (Eds.), Neural Networks: Tricks of the Trade. VI, 432 pages. 1998.

Vol. 1525: D. Aucsmith (Ed.), Information Hiding. Proceedings, 1998. IX, 369 pages. 1998.

Vol. 1526: M. Broy, B. Rumpe (Eds.), Requirements Targeting Software and Systems Engineering. Proceedings, 1997. VIII, 357 pages. 1998.

Vol. 1527: P. Baumgartner, Theory Reasoning in Connection Calculi. IX, 283. 1999. (Subseries LNAI).

Vol. 1528: B. Preneel, V. Rijmen (Eds.), State of the Art in Applied Cryptography. Revised Lectures, 1997. VIII, 395 pages. 1998.

Vol. 1529: D. Farwell, L. Gerber, E. Hovy (Eds.), Machine Translation and the Information Soup. Proceedings, 1998. XIX, 532 pages. 1998. (Subseries LNAI).

Vol. 1530: V. Arvind, R. Ramanujam (Eds.), Foundations of Software Technology and Theoretical Computer Science. XII, 369 pages. 1998.

Vol. 1531: H.-Y. Lee, H. Motoda (Eds.), PRICAI'98: Topics in Artificial Intelligence. XIX, 646 pages. 1998. (Subseries LNAI).

Vol. 1096: T. Schael, Workflow Management Systems for Process Organisations. Second Edition. XII, 229 pages. 1998.

Vol. 1532: S. Arikawa, H. Motoda (Eds.), Discovery Science. Proceedings, 1998. XI, 456 pages. 1998. (Subseries LNAI).

Vol. 1533: K.-Y. Chwa, O.H. Ibarra (Eds.), Algorithms and Computation. Proceedings, 1998. XIII, 478 pages. 1998.

Vol. 1534: J.S. Sichman, R. Conte, N. Gilbert (Eds.), Multi-Agent Systems and Agent-Based Simulation. Proceedings, 1998. VIII, 237 pages. 1998. (Subseries LNAI).

Vol. 1535: S. Ossowski, Co-ordination in Artificial Agent Societies. XV; 221 pages. 1999. (Subseries LNAI).

Vol. 1536: W.-P. de Roever, H. Langmaack, A. Pnueli (Eds.), Compositionality: The Significant Difference. Proceedings, 1997. VIII, 647 pages. 1998.

Vol. 1537: N. Magnenat-Thalmann, D. Thalmann (Eds.), Modelling and Motion Capture Techniques for Virtual Environments. Proceedings, 1998. IX, 273 pages. 1998. (Subseries LNAI).

Vol. 1538: J. Hsiang, A. Ohori (Eds.), Advances in Computing Science – ASIAN'98. Proceedings, 1998. X, 305 pages. 1998.

Vol. 1539: O. Rüthing, Interacting Code Motion Transformations: Their Impact and Their Complexity. XXI,225 pages. 1998.

Vol. 1540: C. Beeri, P. Buneman (Eds.), Database Theory – ICDT'99. Proceedings, 1999. XI, 489 pages. 1999.

Vol. 1541: B. Kågström, J. Dongarra, E. Elmroth, J. Waśniewski (Eds.), Applied Parallel Computing. Proceedings, 1998. XIV, 586 pages. 1998.

Vol. 1542: H.I. Christensen (Ed.), Computer Vision Systems. Proceedings, 1999. XI, 554 pages. 1999.

Vol. 1543: S. Demeyer, J. Bosch (Eds.), Object-Oriented Technology ECOOP'98 Workshop Reader. 1998. XXII, 573 pages. 1998.

Vol. 1544: C. Zhang, D. Lukose (Eds.), Multi-Agent Systems. Proceedings, 1998. VII, 195 pages. 1998. (Subseries LNAI).

Vol. 1545: A. Birk, J. Demiris (Eds.), Learning Robots. Proceedings, 1996. IX, 188 pages. 1998. (Subseries LNAI).

Vol. 1546: B. Möller, J.V. Tucker (Eds.), Prospects for Hardware Foundations. Survey Chapters, 1998. X, 468 pages. 1998.

Vol. 1547: S.H. Whitesides (Ed.), Graph Drawing. Proceedings 1998. XII, 468 pages. 1998.

Vol. 1548: A.M. Haeberer (Ed.), Algebraic Methodology and Software Technology. Proceedings, 1999. XI, 531 pages. 1999.

Vol. 1550: B. Christianson, B. Crispo, W.S. Harbison, M. Roe (Eds.), Security Protocols. Proceedings, 1998. VIII, 241 pages. 1999.

Vol. 1551: G. Gupta (Ed.), Practical Aspects of Declarative Languages. Proceedings, 1999. VIII, 367 pgages. 1999.

Vol. 1552: Y. Kambayashi, D.L. Lee, E.-P. Lim, M.K. Mohania, Y. Masunaga (Eds.), Advances in Database Technologies. Proceedings, 1998. XIX, 592 pages. 1999.

Vol. 1553: S.F. Andler, J. Hansson (Eds.), Active, Real-Time, and Temporal Database Systems. Proceedings, 1997. VIII, 245 pages. 1998.

Vol. 1555: J.P. Mueller, A. Rao, M.P. Singh (Eds.), Intelligent Agents V. Proceedings, 1998. XXIV, 455 pages. 1999. (Subseries LNAI).

Vol. 1557: P. Zinterhof, M. Vajteršic, A. Uhl (Eds.), Parallel Computation. Proceedings, 1999. XV, 604 pages. 1999.

Vol. 1558: H. J.v.d. Herik, H. Iida (Eds.), Computers and Games. Proceedings, 1998. XVIII, 337 pages. 1999.

Vol. 1559: P. Flener (Ed.), Logic-Based Program Synthesis and Transformation. Proceedings, 1998. X, 331 pages. 1999.

Vol. 1560: K. Imai, Y. Zheng (Eds.), Public Key Cryptography. Proceedings, 1999. IX, 327 pages. 1999.

Vol. 1561: I. Damgård (Ed.), Lectures on Data Security.VII, 250 pages. 1999.

Vol. 1563: Ch. Meinel, S. Tison (Eds.), STACS 99. Proceedings, 1999. XIV, 582 pages. 1999.

Vol. 1567: P. Antsaklis, W. Kohn, M. Lemmon, A. Nerode, S. Sastry (Eds.), Hybrid Systems V. X, 445 pages. 1999.

Vol. 1569: F.W. Vaandrager, J.H. van Schuppen (Eds.), Hybrid Systems: Computation and Control. Proceedings, 1999. X, 271 pages. 1999.

Vol. 1570: F. Puppe (Ed.), XPS-99: Knowledge-Based Systems. VIII, 227 pages. 1999. (Subseries LNAI).

Vol. 1572: P. Fischer, H.U. Simon (Eds.), Computational Learning Theory. Proceedings, 1999. X, 301 pages. 1999. (Subseries LNAI).

Vol. 1575: S. Jähnichen (Ed.), Compiler Construction. Proceedings, 1999. X, 301 pages. 1999.

Vol. 1576: S.D. Swierstra (Ed.), Programming Languages and Systems. Proceedings, 1999. X, 307 pages. 1999.

Vol. 1577: J.-P. Finance (Ed.), Fundamental Approaches to Software Engineering. Proceedings, 1999. X, 245 pages. 1999.

Vol. 1578: W. Thomas (Ed.), Foundations of Software Science and Computation Structures. Proceedings, 1999. X, 323 pages. 1999.

Vol. 1579: W.R. Cleaveland (Ed.), Tools and Algorithms for the Construction and Analysis of Systems. Proceedings, 1999. XI, 445 pages. 1999.

Vol. 1580: A. Včkovski, K.E. Brassel, H.-J. Schek (Eds.), Interoperating Geographic Information Systems. Proceedings, 1999. XI, 329 pages. 1999.